Tsunamis

Tsunamis
Detection, Monitoring, and Early-Warning Technologies

Dr. Antony Joseph
Senior Scientist
Marine Instrumentation Division
National Institute of Oceanography
(Council of Scientific & Industrial Research)
Dona Paula
Goa - 403 004, India

Amsterdam • Boston • Heidelberg • London • New York • Oxford
Paris • San Diego • San Francisco • Singapore • Sydney • Tokyo

Academic Press is an Imprint of Elsevier

Academic Press is an imprint of Elsevier
30 Corporate Drive, Suite 400, Burlington, MA 01803, USA
525 B Street, Suite 1900, San Diego, CA 92101-4495, USA
The Boulevard, Langford Lane, Kidlington, Oxford, OX51GB, UK

First edition 2011

Library of Congress Cataloging-in-Publication Data
Joseph, Antony.
  Tsunamis : detection, monitoring, and early-warning technologies / Dr. Antony Joseph.
    p. cm.
  ISBN 978-0-12-385053-9
  1. Tsunamis. 2. Tsunamis—Forecasting. 3. Tsunamis—Safety measures. 1. Title.
  GC221.2.J67 2011
  551.46'37–dc22

                                        2010038155

British Library Cataloguing-in-Publication Data
A catalogue record for this book is available from the British Library

ISBN: 978-0-12-385053-9

For information on all Academic Press publications
visit our website at books.elsevier.com

Working together to grow
libraries in developing countries

www.elsevier.com | www.bookaid.org | www.sabre.org

ELSEVIER   BOOK AID International   Sabre Foundation

Contents

After the December 2004 Indian Ocean tsunami that penetrated into the entire world oceans and devastated the Indian Ocean coastal/island regions, spewing terror and destruction in many countries including Thailand, Indonesia, Sri Lanka, and India, the subject of tsunamis and the technologies used for their detection and early warning received tremendous attention the world over, even in areas where tsunamis are relatively rare. Although considerable developments have taken place during the last few decades in the technology of measuring sea levels with more precision for studies of global climate change and early warning of marine hazards, such as storm surges and tsunamis, a vast collection of research findings and the technological innovations on various aspects of tsunamis are largely buried in research journals, patent documents, or conference/symposia proceedings, as well as, to a certain extent, IOC-UNESCO technical manuals. Considering this void, the great need for a comprehensive book on tsunamis was recognized. It was important that the book provides information on the broad scientific aspects of tsunamis and detailed coverage of the latest developments (state-of-the-art) in the technological aspects of their detection, monitoring, and real-time reporting, together with a glimpse of some upcoming technologies.

Since any marine hazard is linked to sea-level anomalies, the technologies used for tsunami warning and inundation detection, including telemetry systems, can be used for storm surges as well. For the latter, meteorological monitoring systems also need to be included in the warning systems. In recent years, the technologies available for "multihazards" have become recognized as important aspects of early warning systems. Taking this into account, this book addresses technological issues related to both tsunamis and storm surges.

This book consists of 21 chapters. Before addressing the technological aspects of detection, monitoring, and early warning of tsunamis (which is the major thrust of this book), readers are first introduced to the more general and easy-to-grasp topics on the science of tsunamis. These include a general introduction on the historical aspects of tsunamis, analyses of various mechanisms of tsunami generation, tsunami databases, geophysical tsunami hydrodynamics, the impact of tsunamis on coastal and island habitats, the protective role of coastal ecosystems, and earthquake detection and monitoring for early warning of seismogenic tsunamis. Subsequent to a reasonably broad coverage of these topics, which are targeted primarily to the nontechnical reader, the more specialized subjects of tsunami monitoring and early warning technologies are covered in the remaining chapters. In these technology sections, topics such as instrumented methods for tsunami early warning, the role of IOC-UNESCO in administering such a global network, earthquake monitoring for early tsunami warning, open ocean tsunami detection using seafloor pressure measurement and orbiting satellite altimeters, and the upcoming technologies of optical devices in satellites/low-flying aircrafts and orbiting microwave radars/radiometers are addressed. Due importance is given to the less technologically complex aspect of land-based measurements of inundation to confirm tsunamigenesis to save the lives of those residing close to the source of a tsunami. Other important aspects such as the technology of end-to-end communication and the methodologies used for effectively conveying the warning messages to the public are given adequate attention. A detailed description is given of the tsunami early warning systems in the global network maintained by IOC-UNESCO member nations.

In the second half of the book, the technology aspects of tsunamis are discussed in more detail. The topics covered include the challenges of detecting tsunamis using remotely located measuring stations; sea-level measurements from coasts, islands, and deep-sea regions; the telemetry of sea-level data; monitoring and real-time reporting technologies; and optimum technologies that are suitable under specific situations. Due attention has been paid to the method used for extraction of tsunami signals from sea-level records. A comparative evaluation and assessment of sensors available for sea-level measurement and tsunami detection have been included to permit a choice of sensors suitable for a given application in a given location. This book is designed to lead the reader on a journey through the relatively simpler topics that are of interest to most people and then progress to more difficult topics in the domain of technology. Throughout this book, a great deal of attention has been paid to conveying the discussion of technology to the nonspecialist reader in an easy-to-understand language.

I believe that the information contained in this book is of relevance globally, so I hope it will be of immense value to sea-level and tsunami scientists and technologists, emergency/disaster managers, academicians, students, and nonspecialists who have an interest in the subject of sea level, tsunamis, and storm surges.

Antony Joseph

Acknowledgments

The inspiration for writing this book came primarily from the realization that there is a severe dearth of literature on the technological aspects of tsunami detection, monitoring, and real-time reporting. The topic of sea-level observations is also of great interest in terms of sea-level rise and climate change, which is a hot issue at present. I am proud to state that during the preparation of the manuscript for this book, I have been abundantly supported and encouraged by several internationally acclaimed experts in tsunami and sea-level studies. Dr. A. B. Rabinovich (P.P. Shirshov Institute of Oceanology, Russian Academy of Sciences (RAS), Moscow) provided me with several colorful illustrations on geophysical and meteorological tsunamis and a voluminous set of literature on these topics. Dr. O. Kamigaichi (Japan Meteorological Agency), Dr. T. N. Ivelskaya (Sakhalin Tsunami Warning Center, Russia), Professor C. G. von Hillebrandt-Andrade (NOAA Caribbean Tsunami Center, University of Puerto Rico at Mayaguez), Dr. A. K. Andreev (Federal Environmental Emergency Response Centre, Russia), and Dr. T. S. Kumar (INCOIS, India) provided considerable amounts of materials on the Tsunami Warning Centers in their respective countries and a glimpse of their experiences. Dr. Y. Tsuji (Earthquake Research Institute, University of Tokyo, Japan) and Dr. T. Tachibana (Soil Engineering Corporation, Okayama, Japan) provided several illustrations concerning a hitherto unreported consequence of tsunamigenic earthquakes in the neighborhood of gas hydrate reserves located below the seabed. The Ports and Harbours Bureau of Ministry of Land, Infrastructure, Transport and Tourism of Japan provided illustrations on the Japanese ocean wave information network. Dr. P. L. Woodworth (Proudman Oceanographic Laboratory, United Kingdom) supported me by reviewing the manuscript on three different occasions at its different stages of development and by providing many useful suggestions, together with some literature. Professor K. Satake (Japan Marine Science and Technology Center) and Dr. E. Bernard (Pacific Marine Environmental Laboratory, United States) provided valuable suggestions during my personal interactions with them at the tsunami symposia held at Cape Town (South Africa) and Novosibrisk (Siberia, Russia), respectively, in 2009. Feedback from Dr. D. A. Walker (University of Hawaii) helped me to include sections based on theoretically sound and experimentally proven principles having the potential for development into a technology of the future for remote detection of tsunamis from the open ocean. Dr. A. G. Marchuk (Russian Academy of Sciences) provided input on the process of tsunami wave generation at the earthquake source. Mr. P. Foden (Proudman Oceanographic Laboratory, United Kingdom) and several anonymous reviewers provided

xi

many valuable suggestions. A large number of scientists and academicians in India and abroad encouraged me in completing this work at the earliest, including Dr. P. L. Woodworth (POL, UK); Professor H. B. Menon (Goa University); Drs. M. Baba and N. P. Kurian (Center for Earth Science Studies); Dr. B. Mathew (Naval Physical Oceanography Laboratory); Dr. R. Sajeev (Cochin University of Science and Technology); Professor S. N. Dalvi (Pune University); Dr. P. Tkalich (National University of Singapore); and Prof. V.K. Gusiakov (Institute of Computational Mathematics and Mathematical Geophysics, SB RAS, Russia). I am profoundly grateful to all of them and all the reviewers whose feedback helped me to improve the content of this book.

Antony Joseph

Chapter 1

Introduction

The oceans, which provide us with living and nonliving resources and regulate the weather, can be hostile at times. The August 27, 1883, tsunami (which was generated by the Krakatau volcano explosion in Indonesia) is the first known global tsunami (Symons, 1888; Choi et al., 2003; Pelinovsky et al., 2005). The May 22, 1960, Chilean tsunami is one of the most destructive trans-Pacific tsunamis and one of two known global tsunamis (Berkman and Symons, 1960; Wigen, 1960). The catastrophic and highly destructive December 26, 2004, Sumatra tsunami is the third known global tsunami, which severely damaged the coastal regions of the Indian Ocean, including those of 12 countries: Indonesia, Thailand, Myanmar, Malaysia, India, Sri Lanka, Bangladesh, Maldives, Somalia, Kenya, Tanzania, and South Africa. This tsunami killed more than 230,000 people; injured almost 283,000 people from about 60 countries; caused the deaths of nationals from 73 countries in this age of globalization and ecotourism; and left millions homeless and displaced (Stein and Okal, 2005; Lay et al., 2005; Titov et al., 2005; Rabinovich and Thomson, 2007). This is just one example of a tsunami's severe striking power. The Indian Ocean tsunami of December 26, 2004, with its sequel of death and destruction, showed us nature at its horrendous worst. The long-term psychosocial and intergenerational impact of the Indian Ocean tsunami is likely to be experienced for decades. In association with the 2004 Sumatra tsunami, the coastal areas of the Indian Ocean received the brunt of destruction and loss of life, with the most distant recorded death having occurred in Port Elizabeth (in the Republic of South Africa), about 8000 km from the earthquake epicenter.

The Indian Ocean experienced its most devastating natural disaster through the action of the 26 December 2004 tsunami. This event revealed the destructive effects of tsunamis, with a maximum runup exceeding 30 meters in Banda Aceh and 10 meters at several sites in Sri Lanka (Inoue et al., 2007). Sri Lanka was found to be most vulnerable to tsunamis generated from the Sunda Arc region, located approximately 1500 km to the east (See Figure 1.1a). The narrow continental shelf and the steep continental slope mean that Sri Lanka is extremely vulnerable to the action of tsunamis (See Figure 1.1b), because the shoaling effect occurs over a shorter distance and there is negligible amount of energy dissipated over the continental shelf region. There are also no major

Tsunamis: Detection, Monitoring, and Early-Warning Technologies. DOI: 10.1016/B978-0-12-385053-9.10001-8
1

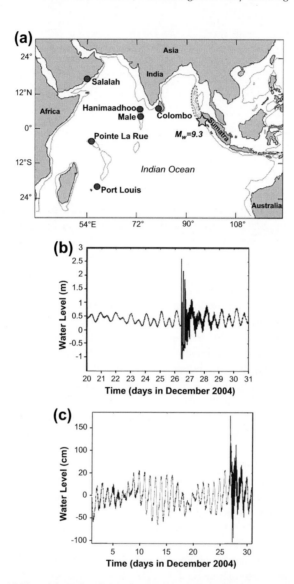

FIGURE 1.1   (a) Map of the Indian Ocean showing the location of the Mw = 9.3 December 2004 earthquake (star), the aftershock zone (dotted area) together with positions of 6 selected sea-level stations. *Source: Rabinovich et al., 2006; Reproduced with kind permission of Springer Science and Business Media;* (b) Time-series of the sea-level record from Colombo, Sri Lanka, from 20 to 31 December, 2004 and (c) from Geraldton, Western Australia, for December, 2004 showing the scale of the tsunami with respect to the tidal range and also the presence of high-frequency oscillations subsequent to the tsunami. *Source: Pattiaratchi and Wijeratne, 2009; Reproduced with kind permission of Springer Science and Business Media.*

submarine topographic features between the tsunami source region and Sri Lanka to dissipate or refract the tsunami waves. As a result, Sri Lanka was the second-most affected country for the 2004 tsunami event (Pattiaratchi and Wijeratne, 2009). Although the north-western region of Australia is located close to the Sunda Arc region, the wide continental shelf was found to have a large influence on tsunami impacts at the shoreline. It was noticed that presence of the Exmouth Plateau influences tsunami propagation through refraction, with a concentration of energy along some regions of the coastline, especially in the Geraldton region (See Figure 1.1c).

The powerful tsunami of 26 December 2004 was recorded all over the World Oceans, from Antarctica in the south to the far North Pacific and the North Atlantic Oceans, including most of the sites in Chile, Brazil, Peru, Mexico, California, British Columbia, Alaska, Hawaii, the Aleutian and the Kuril Islands. Figure 1.2 shows time-series of tsunami wave heights as recorded at selected sea-level stations in the three major ocean basins and some island locations. Exceptions are the stations located deeply inside straits and inlets, stations sheltered from open ocean waves (e.g., Seattle, Tacoma, Vancouver and Queen Charlotte City [See Figure 1.3]), and open coast stations that are strongly affected by wind waves and swell, and consequently having very low signal-to-noise ratio (Rabinovich, et al., 2006).

Historically, there have been several other tsunamis as well, which proved to be disastrous at different regions at varying levels at different occasions in the past; having killed tens of thousands of people and caused severe damage to the coastal areas of the world. As noted by Rabinovich et al., (2006), it is, perhaps, not a coincidence that the first known global tsunami event, i.e., the tsunami associated with the Krakatau Volcano eruption of August 27, 1883, and the third historically documented global tsunami event (i.e., December 2004 Sumatra tsunami) also occurred in the Indonesian Archipelago. The fact that both the 1883 and the 2004 global tsunami events originated in the same region demonstrates the potential of this region as a major source for worldwide catastrophic tsunamis. Although more than 75 percent of all tsunamis in the World Ocean have originated in the Pacific Ocean, two of the three most destructive global tsunamis have originated in the Indian Ocean. It is to be noted that the 2004 tsunami event has been followed by annual occurrence of oceanwide tsunamis (i.e., those which influence areas far from the generation region) in 2005 (earthquake magnitude 8.6 on 28 March), in 2006 (earthquake magnitude 7.7 on 17 July), and in 2007 (earthquake magnitude 8.5 and 7.9 on 12 September). The directions of tsunami wave propagation during these earthquake events were different, but oriented primarily towards the south/south-west direction (Figure 1.4). The extent of the 2005 and 2007 Sumatra tsunamis permeated as far as Sri Lanka (Pattiaratchi and Wijeratne, 2009) and even into the Arabian Sea.

The September 2007 tsunami was recorded in the Lakshadweep archipelago in the Arabian Sea, and Goa on the west coast of India (Prabhudesai et al.,

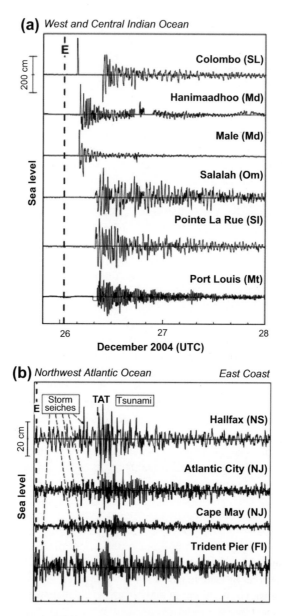

FIGURE 1.2 Tsunami records for the 2004 Sumatra tsunami for selected sites in (a) Indian Ocean; (b) Northwest Atlantic Ocean; (c) North Pacific Ocean; and (d) some Island locations. Dashed vertical line labeled "E" denotes the time of the main earthquake shock. "TAT" is the tsunami arrival time; arrows and numbers denote arrivals of the first and second wave trains. *Source: Rabinovich et al., 2006; Reproduced with kind permission of Springer Science and Business Media.*

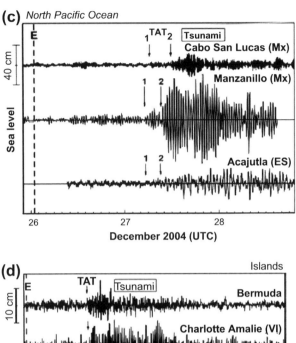

FIGURE 1.2 Continued

2008) although it was too weak to cause casualties. In recent years, the Indian Ocean continues to be haunted by earthquake-generated tsunamis, the latest being the August 10, 2009 earthquake at the Andaman Island in the Indian Ocean. The most recent in 2009 is the September 29, 2009 earthquake (magnitude 8.0) at the Samoa Islands region of the South Pacific Ocean. The Pacific Tsunami Warning and Mitigation System coordinated by UNESCO's Intergovernmental Oceanographic Commission (IOC) issued a clear warning throughout the region 16 minutes after the earthquake. People who have been able to receive the warning messages evacuated to upper grounds and their lives were saved. The response was described as good, and many people seem to

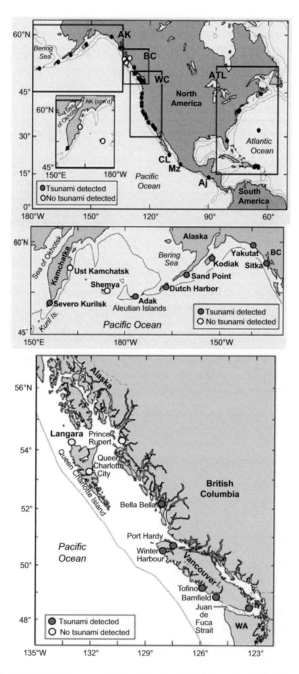

FIGURE 1.3 Maps indicating several sites in the Pacific Ocean where December 2004 Indian Ocean tsunami were detected and a few sites where this tsunami was not detected. *Source: Rabinovich et al., 2006; Reproduced with kind permission of Springer Science and Business Media.*

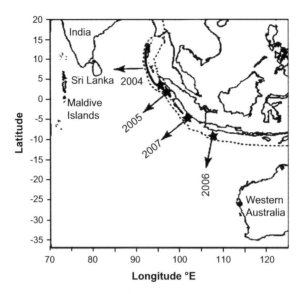

FIGURE 1.4 Map of the eastern Indian Ocean showing the location of the earthquake epicenters (indicated by stars) and direction of the primary wave propagation for the 2004, 2005, 2006, and 2007 Indian Ocean tsunamis. Note that there were two earthquakes in September 2007 within 12 hours. The dashed line indicates the plate boundary; the thick black line indicates the area of rupture for the 2004 tsunami. *Source: Pattiaratchi and Wijeratne, 2009; Reproduced with kind permission of Springer Science and Business Media.*

have learned the lessons from the 2004 Indian Ocean tsunami. However, in regions very close to the earthquake site, the short time available for warning and evacuation of coastal populations posed a major challenge to local authorities; with coastal residents in some areas not having received the requisite information in time.

Several sea-level gauges that were positioned all over the world recorded the tsunami waves caused by the Great Sumatra-Andaman Earthquake on December 26, 2004, revealing its truly global reach. Although no direct damage or destruction has been reported for this tsunami event outside the Indian Ocean, sea-level records available from different parts of the world demonstrate the ability of tsunami waves to penetrate all the world oceans (Merrifield et al., 2005; Titov et al., 2005; Woodworth et al., 2005). This tsunami is one of a small number known to have been observed around the world. It is also the first global-scale tsunami in the "instrumental era," and was measured by an unprecedented number of tide gauges (sea-level gauges) throughout the world oceans. The total number of known records of this tsunami in the Indian, Atlantic, and Pacific Oceans is about 250, from which relatively high-quality worldwide sea-level measurements are available (Titov et al., 2005; Rabinovich, 2009). More than 50% of these records are for regions where tsunamis had never before been recorded (the Atlantic coasts of Africa and South America, the Antarctica coast, and a major part of the Indian Ocean) and were made by instruments that were not designed to measure tsunami waves. The 2004 tsunami also was recorded by a number of bottom pressure gauges that had been installed on the shelf and deep in the ocean within the framework of nontsunami projects, including those examining oceanic circulation, climatic change, ocean stratification and transport, and related ocean drilling studies. These observations have yielded valuable scientific information on the physics, dynamics, and energy characteristics of tsunami waves. In addition, such multifarious tsunami measurements collected by different types of instruments with different sampling intervals and deployed at sites strongly dissimilar in their local topographic and geographical properties are invaluable for specialists working on tsunami measurements and early tsunami warnings (Rabinovich, 2009).

The 2004 Sumatra tsunami is now recognized as the most globally distributed and accurately measured tsunami in recorded history. Tsunami amplitudes along many Indian Ocean coastlines were measured in meters and in some cases the waves were large enough to destroy the sea-level recording equipment (Merrifield et al., 2005). For example, the sea-level gauge at Colombo (Sri Lanka) was damaged by the first tsunami wave and did not operate for 3 hours and 30 minutes, coinciding with the arrival of the highest waves (Pattiaratchi and Wijeratne, 2009), and eyewitness reports had to be relied upon (Rabinovich and Thomson, 2007). Likewise, the sea-level gauge at Nagapattinam, located on the most strongly affected sector of the Indian coast, was heavily damaged by the tsunami waves so that the 2004 tsunami record could not be retrieved (Nagarajan et al., 2006).

Available records from the African continent suggest that the Sumatra tsunami showed its presence in African countries as well, although its intensity deteriorated significantly on the western coastline of this continent (Joseph et al., 2006; Rabinovich and Thomson, 2007). Amplitudes along the Atlantic coastlines, especially those of North and South America, were often tens of centimeters (Rabinovich, 2005). The Atlantic Ocean historically has had no tsunami warning systems (TWSs), and no standard instruments designed for tsunami measurements existed. Unlike the Pacific Ocean, the Atlantic Ocean is not bordered by major subduction zones, which are the main sources of large tsunamis (Lockridge et al., 2002; Gusiakov, 2006). So infrequent were tsunamis in the Atlantic that in the period immediately following the 2004 Sumatra tsunami, few experts had expected that this tsunami would be recorded outside the Indian Ocean. Nevertheless, a few days after the earth-quake, tsunami waves were identified in several sea-level gauge records on the coast of North America. On the South American coast in the Atlantic, detec-tion of the 2004 tsunami includes sites in Brazil (Candella, 2005; Melo and Rocha, 2005; Franca and de Mesquita, 2007) and Argentina (Dragani et al., 2006), which are regions where tsunamis had never been recorded previously.

The 2004 tsunami was also detected in several records from the South and Central Atlantic, in particular at Signy (South Orkney Islands), Port Stanley (Falkland Islands), and St. Helena, among a few other locations in the Atlantic. The maximum tsunami wave heights observed on the coasts of Uruguay and southeastern Brazil were more than 1 meter high (Candella et al., 2008). Their findings reveal that outside the Indian Ocean, the highest waves were recorded in the South Atlantic and not in the Pacific, as has been previously suggested (Murty et al., 2005a) based on the limited data available at the time. In the Atlantic Ocean, the tsunami signals at most locations were usually large enough to be clearly identified within the normal spectrum of sea-level variability due to the ocean tide and nontidal processes such as storm surges. However, at a few other Atlantic locations, amplitudes were only several centimeters or less, thereby making it difficult to decide if a genuine tsunami signal was present among the generally large wind-waves. The records of many UK sea-level gauges contained small tsunami signals, varying with location and usually without one clear wave-period (period generally between 20 and 45 minutes) (Woodworth et al., 2005). These records, together with those from other countries, demonstrate that this particular tsunami was a truly global event, in varying levels of energy. According to Kowalik and colleagues (2007), the total inflow of 2004 tsunami energy into the Pacific Ocean was approximately 75% of the total energy inflow to the Atlantic Ocean. The event was truly unprecedented in its global reach (Titov et al., 2005a; Geist et al., 2006).

Subsequent to the catastrophic Indian Ocean tsunami of December 2004, there has been increasing interest among both the scientific and nonscientific communities the world over in acquiring general knowledge about tsunamis, along with technical knowledge in sensors and the methods used for detection, monitoring, and real/near-real time telemetry reporting of tsunamis. Given the ever increasing pace of technology development in the area of sea-level measurement and telemetry, it is hard to keep up with the new technologies that have become available since the Indian Ocean tsunami of 2004. Since that time, the oceanographic community has been determined to build tsunami sampling capability into conventional sea-level measurement stations. This requires an understanding of how existing and new measurement systems operate and how best to acquire that data in a timely fashion.

Dr. Eddie Bernard (2009), director of the National Oceanic and Atmospheric Administration's Pacific Marine Environmental Laboratory (NOAA/PMEL), stated the following:

"What emerges from the 2004 Indian Ocean tsunami and society's response is a call for research that will mitigate the effects of the next tsunami on society. The scale of the 2004 tsunami's impact (227,000 deaths, $10B damage), and the world's compassionate response ($13.5B), requires that tsunami research focus on applications that benefit society. Tsunami science will be expected to develop standards that ensure mitigation products are based on state-of-the-science. Standards based on scientifically endorsed procedures assure the highest-quality application of this science. Community educational activities will be expected to focus on preparing society for the next tsunami. An excellent starting point for the challenges ahead is education, at all levels, including practitioners, the public, and a new generation of tsunami scientists".

In an endeavour to educate the new generation of scientists, various "science" aspects of tsunamis have been discussed in Volume 15 of *The Sea: Tsunamis*. This volume largely captures the technical elements of tsunami "state-of-the-science" today. Additional information on the "science" aspects of tsunamis, including paleotsunamis, can be found in the Springer book *Physics of Tsunamis* by Boris Levin and Mikhail Nosov.

Unfortunately, there is a severe dearth of literature on the "technological" aspects of tsunami detection, monitoring, and real-time reporting. The topic of sea-level observations is also of great interest in terms of sea-level rise and climate change, which is a hot issue at present. With a view to catering to the knowledge and information needs of those who want to understand the technology behind sea-level measurements and in particular tsunami measurement and reporting techniques (an important but neglected niche), this book attempts to bring all the available technical information together under one umbrella and to provide a coherent description of up-to-date information on the present state-of-the-art equipment for sea-level observation and communication of their data to TWSs. To place the main topic of discussion (i.e., technology aspects) in perspective, and to provide a smooth takeoff, some interesting historical and scientific aspects of tsunamis have also been introduced in brief for the benefit of the nonspecialist readers.

Tsunami Generation and Historical Aspects

Both tsunamis and tsunami-like waves are generated as a result of various causes, such as an undersea earthquake (also known as a *seaquake*) rupture process or, more frequently, the secondary triggered phenomena, such as landslides (submarine or surface) and/or other geodynamic phenomena, such as rockslides, large-scale gas emissions from the seafloor (e.g., degradation of gas hydrates), volcanic eruptions, intense atmospheric disturbances, and asteroid impacts (although an asteroid impact is not of geophysical origin and hopefully occurs rarely). The most devastating tsunamis on record are listed in Table 2.1. According to the tsunami database, seaquakes, submarine landslides, volcanic eruptions, and atmospheric disturbances have been responsible for approximately 82%, 6%, 5%, and 3%, respectively, of tsunamis (Figure 2.1).

Paleo-tsunami deposits or some other evidence of tsunamis can provide useful insights about past tsunamis. For example, the tsunami event deposits that were found in a tidal flat at Miura peninsula in the south of Tokyo contained a great deal of marine planktonic diatoms (continuous samples of tidal flat sediments taken by array coring survey using a 3-meter-long Geoslicer), which indicate a recurrence of tsunamigenic earthquakes in this region (coseismic uplift and interseismic subsidence). The great Kanto earthquakes that occurred in 1703 (Genroku era) and 1923 (Taisho era) in association with

TABLE 2.1 The Most Devastating Tsunamis on Record

Casualties	Location	Year
230,000	Indian Ocean (Indonesia, India, Sri Lanka, Thailand, Somali, and others)	2004
100,000	Lisbon earthquake (Portugal, Morocco)	1755
100,000	Awa, Japan	1703
70,000	Mesina, Italy	1908
40,000	South China Sea, Taiwan	1782
36,000	Krakatau volcanic eruption (Sunda Strait)	1883
30,000	Tokaido-Nankaido, Japan	1707
27,000	Japan	1826
25,674	Chile	1868
22,070	Sanriku, Japan	1896
15,030	Kyushu Is, Japan	1792
13,486	Ryukyu Trench, Japan	1771
5,233	Tokaido-Kashima, Japan	1703
5,000	Nankaido, Japan	1605

Source: Gusiakov. V.K. ITDB/WLD: Integrated Tsunami Database for the World Ocean, Version 6.52 of October 31, 2007, CD-ROM, Tsunami Laboratory, ICMMG SD RAS, Novosibirsk, Russia.

the subduction of the Philippine sea plate along the Sagami trough took place in this area (Shimazaki et al., 2009).

In the case of tsunamis of submarine/coastal volcanic origin, near-source tsunami deposits in the tsunami inundation areas were found to consist of volcanic materials such as pumice (light, porous lava) boulders, volcanic ash, and the like. When a tsunami runs up over a beach where pumice (density ~ 0.8 g/cm^3) was deposited on the sand, the inundation limit of the tsunami is usually inferred by a line of pumice that had been carried up by the tsunami and redeposited on the ground surface. Past tsunamis in tectonically active regions such as Hokkaido-Kuril-Kamchatka can be identified and evaluated through investigation of the spatial distribution and sedimentary features of pumice-rich and other tsunami deposits in these areas. In such surveys, the tsunami heights are indicated by the deposit distribution (Nishimura and Nakamura, 2009).

The long water-waves (tsunamis) generated by impulsive geophysical events evolve substantially through three-dimensional spreading, both spatially

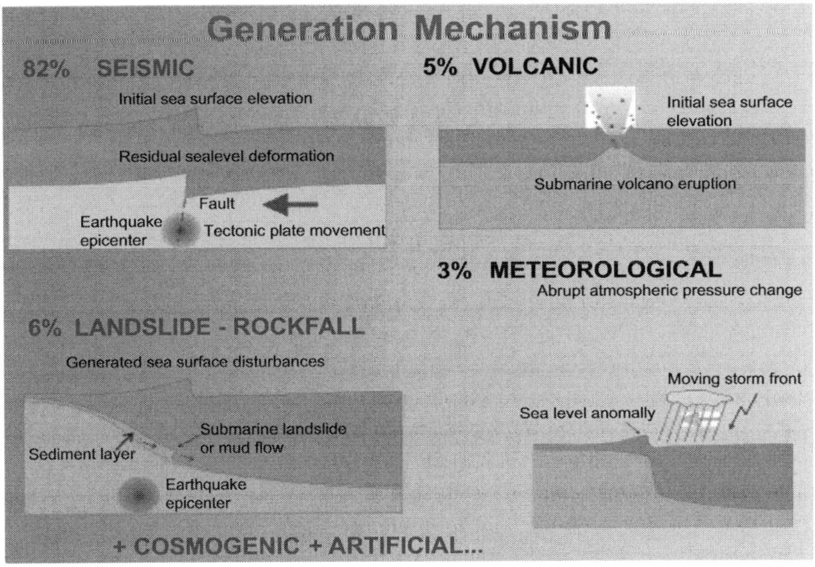

FIGURE 2.1   Schematic illustration of tsunami generation mechanisms. *Source: http://www. theplasmaverse.com/pdfs/meteorological-tsunamis-destructive-atmosphere-induced-waves-observed-in-the-worlds-oceans-seas-seiches-abiki-milghuba.pdf*

and temporarily, from their initial strip. This is referred to as their directivity and focusing. In tsunami literature both of these processes have been explained analytically (e.g., Ben-Menahem, 1961; Ben-Menahem and Rosenman, 1972; Marchuk and Titov, 1990; Marchuk, 2009). It has also been demonstrated that there exists a distinct difference between the directivity patterns of landslide- and earthquake-generated tsunamis (e.g., Ben-Menahem, 1961; Okal, 1982; Okal, 1988). Kânoğlu and colleagues (2009) explored numerically the existence of focusing, as described by Marchuk and Titov (2005), and discussed some of the geophysical implications.

2.1. TSUNAMIS GENERATED BY SEAQUAKES

Most of the historic seaquakes (i.e., earthquakes under the seafloor) have taken place at subduction zones. The presence of a deep-water trench, such as the Aleutian trench in the Alaska-Aleutian subduction zone and the Kuril-Kamchatka trench in the Kuril-Kamchatka zone, is the main characteristic of the subduction zone. The depth at the trench axis is approximately twice as large as an average depth of the Pacific Ocean. Earthquake hypocenters are mainly located under the continental bottom slope. In this case, the opposite slope of this deep-water trench works like an optical lens for tsunami waves. Due to the wave refraction above the bottom slope, the greater part of the tsunami's energy will be concentrated in the shoreward and the seaward directions (Marchuk, 2009).

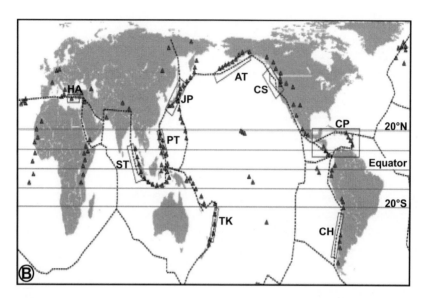

FIGURE 2.2 Summary of main plate boundaries on earth (dashed lines). Important tectonic subduction zones where earthquakes occur and tsunami are triggered include: HA — Hellenic Arc; ST — Sunda Trench; PT — Philippines Trench; JP — Japan Plate Boundary; TK — Tonga-Kermadec Trench; AT — Aleutian Trench; CS — Cascadia Subduction Zone; CH — Chilean Subduction Zone; and CP — Caribbean Plate. Triangles indicate the locations of volcanic activity. *Source: Dominey-Howes and Goff, 2009; Natural Hazards and Earth System Sciences; Copernicus Publications.*

Major interplate joints are known to be sensitive to interplate thrust earthquakes (Figure 2.2). The global tsunami waves that were generated due to a massive undersea earthquake off Sumatra in Indonesia in December 2004 were the result of two plates in the Earth's crust (the Indian plate and the Burma plate) grinding against each other. The Indian plate usually moves northeast at about 6 cm every year. The trench runs roughly parallel to the western coast of Sumatra, about 200 km offshore. At the trench, the Indian plate gets subducted—that is, the Indian plate is forced into the Earth's interior and is overridden by southeast Asia. The contact between the two plates is called a megathrust. The two plates do not glide smoothly past each other along the megathrust but instead move in "stick-slip" fashion. This means that the megathrust remains locked for centuries and then suddenly slips a few meters, generating a large earthquake. Thrust-type earthquakes that occur along subduction zones that cause vertical movement of the ocean floor and thereby perturbation of the water body overlying the rupture zone tend to be tsunamigenic. The world's largest recorded earthquakes were all megathrust events and occurred where one tectonic plate subducted beneath another. These include the magnitude 9.5 earthquake in Chile in 1960; the magnitude 9.2 Prince William Sound Alaska earthquake in 1964; the magnitude 9.1 Andrean earthquake in 1957; and the magnitude 9.0 Kamchatka earthquake in 1952. Megathrust earthquakes often generate large tsunamis that cause damage over a much wider area than what is directly affected by ground shaking near the earthquake's rupture. In the case of the December 2004 Sumatra-Andaman earthquake, as the Indian plate pulled down on the Burma plate, the two plates slid about 15 meters in a short interval. It is estimated that about 1200 to 1300 km of the Burma plate forced a massive displacement of water in the Indian Ocean. This motion generated waves that spread in all directions, moving as fast as 800 km/h.

Normal fault-type earthquakes can also generate moderate tsunamis. Strike-slip earthquakes that cause horizontal movement of the ocean floor are not tsunamigenic. However, the oblique-slip/dip-slip component in them can generate weak tsunamis. Spreading zones such as Carlsberg ridge, Ninety-East ridge, and so forth are sites of such earthquakes. On November 30, 1983, the Chago ridge east of Carlsberg ridge caused a local tsunami due to the thrust component of motion for a $M_w = 7.7$ earthquake that occurred near Diego Garcia (Rastogi, 2007).

In the past, most seaquakes took place in the Pacific region. For example, two large seaquakes occurred in the later part of the nineteenth century in southern Peru and northern Chile. Their estimated magnitudes are of the order of 9, rupturing contiguous segments of nearly 500 km each. Both of them accommodated the convergence between Nazca and South American plates, producing large seafloor and coastal elevation changes and thereby generating significant tsunamis that affected most of the coastlines of the Pacific basin. Reported local run-ups reached 20 meters. Both trans-Pacific tsunamis were recorded at a sea-level gauge in Fort Point in the Presidio area of San Francisco Bay, California. The June 2001 $M_w = 8.4$ earthquake ruptured a fraction of the 1868 event (Barrientos and Ward, 2009). Several great interplate earthquakes occurred along the Kurile trench due to subduction of the Pacific plate beneath the Kurile Islands, and large tsunamis have been generated by those earthquakes. The great Kurile interplate earthquake ($M_w = 8.3$) occurred off Simushir Island along the central Kurile subduction zone on November 15, 2006. Other great interplate earthquakes that occurred along the central Kurile region were the 1963 Kurile earthquake ($M_w = 8.3$), which occurred off Urup Island, and the 1918 Kurile earthquake ($M_w = 8.2$). The megathrust Sumatra earthquake that took place on December 26, 2004 ($M_w = 9.3$) along the Sumatra subduction zone in the Indonesian region (Figure 2.3) was the largest earthquake in 40 years. The earthquake was so powerful that it wobbled the Earth's rotation (Lay et al., 2005). A major tsunami triggered by this seismic event swept across the Indian Ocean at speeds up to 800 km/h, with succeeding waves reaching heights of up to 30 meters. Along with vast numbers of people, man-made and natural structures and habitats were destroyed or damaged, including coral reefs, beaches, seagrass beds, and other coastal vegetation. This tsunami, which destructively impacted the coastal

FIGURE 2.3 Seaquake-vulnerable Indonesian region in the Indian Ocean. *Source: http://www.theplasmaverse.com/pdfs/meteorological-tsunamis-destructive-atmosphere-induced-waves-observed-in-the-worlds-oceans-seas-seiches-abiki-milghuba.pdf*

regions of the Indian Ocean, reveals the vulnerability of Indian Ocean rim countries to powerful tsunamis.

Seaquakes can be of a tsunamigenic or nontsunamigenic nature. The strongest tsunamigenic seaquakes in the recent past are listed in Table 2.2. The majority of tsunamis that affected the African continent were generated by seaquakes located either in the Mediterranean Sea, the Atlantic Ocean, or the Indian Ocean, except the eruption of Krakatau in 1883 and a few tsunamis with unknown causes (Dunbar and Stroker, 2009). The focal parameters of the seaquakes are the locations of their epicenters, focal depths (shallow or inter-mediate depth), and their magnitudes. The lower magnitude threshold required for the generation of a large tsunami is considered to be about 6 on the Richter scale. It has been found that, in general, shallow seaquakes have increasing probability for large-magnitude tsunami generation. However, several excep-tions to this empirical rule based on the focal mechanisms of the seaquakes (which compose the initial tsunami decision matrix) have been reported. For example, a seaquake that occurred in the Pacific Ocean off the coast of California in 1999, recording 7.0 on the Richter scale, did not produce tsunami waves. In another instance, in 2008 several strong earthquakes, both shallow and intermediate-depth and exceeding 6 in magnitude, occurred near Greece, particularly along the Hellenic arc and trench system, which is the most tsunamigenic area in the European-Mediterranean region. However, none of them was reported to have produced even small tsunami-like sea-level distur-bances (Charalampakis et al., 2009).

TABLE 2.2 The Strongest Tsunamigenic Seaquakes in the Recent Past

Date	Country	Location	Magnitude
May 22, 1960	Chile	Central Chile	9.5
March 28, 1964	USA	Prince William Sound, AK	9.2
March 09, 1957	USA	Andreanof Islands, AK	9.1
November 04, 1952	Russia	Kamchatka	9
December 26, 2004	Indonesia	Off western coast of Sumatra	9
January 31, 1906	Ecuador	Off coast	8.8
February 27, 2010	Chile	Off southern coast	8.8
June 15, 1911	Japan	Ryukyu Islands	8.7
February 04, 1965	USA	Rat Islands, Aleutian Islands, AK	8.7
March 28, 2005	Indonesia	Indonesia	8.7
November 11, 1922	Chile	N. Chile	8.5
February 01, 1938	Indonesia	Banda sea	8.5
October 13, 1963	Russia	S. Kuril Islands	8.5
March 02, 1933	Japan	Sanriku	8.4
June 23, 2001	Peru	S. Peru	8.4
September 12, 2007	Indonesia	Sumatra	8.4
June 25, 1904	Russia	Kamchatka Peninsula, Russia	8.3
June 26, 1917	Samoa	Samoa Islands	8.3
August 15, 1918	Philippines	Celebes sea	8.3
April 30, 1919	Tonga	Tonga Islands	8.3
February 03, 1923	Russia	Kamchatka	8.3
April 14, 1924	Philippines	E. Mindanao Island	8.3
January 24, 1948	Philippines	Sulu sea	8.3
July 10, 1958	USA	SE. Alaska, AK	8.3
November 06, 1958	Russia	S. Kuril Islands	8.3

(Continued)

TABLE 2.2 The Strongest Tsunamigenic Seaquakes in the Recent Past—cont'd

Date	Country	Location	Magnitude
October 04, 1994	Russia	S. Kuril Islands	8.3
September 25, 2003	Japan	Hokkaido Island	8.3
November 15, 2006	Russia	S. Kuril Islands	8.3
August 17, 1906	Chile	Central Chile	8.2
September 07, 1918	Russia	S. Kuril Islands	8.2
November 10, 1938	USA	Shumagin Islands, AK	8.2
April 06, 1943	Chile	Central Chile	8.2
May 04, 1959	Russia	Kamchatka	8.2
May 16, 1968	Japan	Off east coast of Honshu Island	8.2
August 11, 1969	Russia	S. Kuril Islands	8.2
February 17, 1996	Indonesia	Irian Jaya	8.2
September 14, 1906	Papua New Guinea	Solomon Sea	8.1
April 14, 1907	Mexico	S. Mexico	8.1
June 03, 1932	Mexico	Central Mexico	8.1
December 07, 1944	Japan	Off southeast coast of Kii Peninsula	8.1
April 01, 1946	USA	Unimak Island, AK	8.1
August 04, 1946	Dominican Republic	Northeastern coast	8.1
December 20, 1946	Japan	Honshu: S coast	8.1
August 22, 1949	Canada	British Columbia	8.1
March 04, 1952	Japan	SE. Hokkaido Island	8.1
October 17, 1966	Peru	Central Peru	8.1
January 01, 1967	Solomon Islands	Solomon Islands	8.1
October 03, 1974	Peru	Central Peru	8.1
August 16, 1976	Philippines	Moro Gulf	8.1
April 21, 1977	Solomon Islands	Solomon Islands	8.1

TABLE 2.2 The Strongest Tsunamigenic Seaquakes in the Recent Past—cont'd

Date	Country	Location	Magnitude
May 23, 1989	Australia	Macquarie Island	8.1
December 23, 2004	Australia	Macquarie Island	8.1
January 13, 2007	Russia	S. Kuril Islands	8.1
April 01, 2007	Solomon Islands	Solomon Islands	8.1
December 09, 1909	USA territory	Guam, Mariana Islands	8
May 01, 1917	New Zealand	Kermadec Islands	8
December 26, 1939	Turkey	Black sea	8
November 27, 1945	Pakistan	Makran coast	8
January 14, 1976	New Zealand	Kermadec Islands	8
August 19, 1977	Indonesia	Sunda Islands	8
March 03, 1985	Chile	Central Chile	8
September 19, 1985	Mexico	Mexico	8
May 07, 1986	USA	Andreanof Islands, AK	8
July 30, 1995	Chile	N. Chile	8
October 09, 1995	Mexico	Mexico	8
November 16, 2000	Papua New Guinea	New Ireland	8
May 03, 2006	Tonga	Tonga	8
August 15, 2007	Peru	S. Peru	8
September 29, 2009	Samoa	Samoa Islands	8

AK: Alaska.
Source: NOAA National Geophysical Data Center, USA.

When describing the tsunami generation, the following conventional assumptions are usually made (Nosov and Kolesov, 2009):

- An earthquake instantly causes permanent deformations of the ocean floor.
- The displacement of the ocean floor is simultaneously accompanied by the formation at the water surface of a perturbation (initial elevation), whose shape is fully similar to the vertical permanent deformations of the bottom.

The initial elevation, thus obtained, is then applied as the initial condition to resolving the problem of tsunami propagation. The initial field of flow velocities is assumed to be zero. The imperfectness of this widespread approach is due to at least two reasons. First, even if the ocean bottom is horizontal and the bottom deformations are instantaneous, the displacement of the water surface and the vertical residual bottom deformation will not be equal to each other, in the sense that the displacement of the water surface will be smoother. Thus, the "fine spatial structure" of the bottom displacement is not manifested on the water surface. Second, in the case of a sloping (nonhorizontal) bottom, the horizontal deformation components can also significantly contribute to the displacement of the water surface. Nosov and Kolesov (2009) have proposed that a logical development of the approach in question consists in calculation of the initial elevation from the solution of the 3D problem, with allowance of all three components of the bottom deformation vector and the distribution of depths in the vicinity of the source. They believe that if the bottom deformation process turns out to be long, then the initial conditions must include, in addition to the initial elevation, the initial distribution of flow velocities as well. In cases of instantaneous deformation, the evolutionary problem can be reduced to a simpler static problem. Nosov and Kolesov (2009) have applied this method for calculating the initial elevation in the source of the Central Kuril Islands tsunamis of November 15, 2006, and January 13, 2007. Different researchers have experimented with different methods for the analysis of earthquake-generated tsunami waves.

The results of the numerical modeling and analysis by Rabinovich and colleagues (2008) highlighted the fact that earthquakes of comparable magnitude from adjacent source regions do not necessarily generate identical tsunami wave fields. Differences in the spatial extent, depth, and fault structure of closely related earthquakes can lead to distinct tsunami responses, including differences in maximum wave amplitude, dominant frequency content, leading wave polarity, and directions of energy propagation. These differences can subsequently lead to markedly different resonant responses at near- and far-field locations. Such factors must be taken into account when modeling the ocean response to earthquakes and when attempting to use tsunami observations to delineate possible source parameters.

2.2. TSUNAMIS GENERATED BY SURFACE/SUBMARINE LANDSLIDES AND ROCK AVALANCHES

Local offshore or onshore earthquakes produce landslides (submarine or surface), although landslides do not always require an earthquake to trigger them. The December 2004 Sumatra-Andaman earthquake of magnitude 9.3 generated a 30-meter-high tsunami following the occurrence of a 15-meter slip of the ocean floor along a 1300-km-long and 160- to 240-km-wide rupture. The tsunami power was enhanced by large landslides (several kilometers across) over the rupture zone of this earthquake. The force of the displaced water was such that blocks of rocks, weighing millions of tons, were dragged as much as 10 km. An oceanic trench several km wide was also formed (Rastogi, 2007).

Large "impulse type" (short duration) surface/submarine landslides (or episodes of several discrete events) have generated tsunamis. Although a submarine landslide or a subaerial landslide that flows into a large body of water can cause a tsunami, surface landslides are much more effective tsunami generators than the slower submarine landslides. It has been observed that submarine landslides generate tsunamis only when the volume of the material moved is substantial and moves at a great speed. The characteristics of these tsunamis are different from those of earthquake-generated tsunamis, which displace seabeds. Whereas the maximum energy of earthquake-generated tsunamis is focused perpendicularly to the strike of the fault and decreases in intensity along the strike (of the fault), the tsunamis generated by landslides (which move in a downslope direction) are more focused and propagate both upslope and parallel to the slide (Chadha, 2007).

Landslide-generated tsunamis represent major natural hazards for coastal communities and waterways in the fjord regions of British Columbia (Murty, 1979; Prior et al., 1982), Alaska (Miller, 1960; Kulikov et al., 1996), Norway (Longva et al., 2003), and Greenland (Dahl-Jensen et al., 2004). A high percentage of tsunamis recorded in southeast Alaska and northern British Columbia (Canada) are generated locally by landslides or rockslides. The most extreme event occurred on July 9, 1958, when a landslide at the head of Lituya Bay in southeast Alaska (triggered by a magnitude 8.3 earthquake) caused a giant wave that rose to a height of 525 meters at the shore (Miller, 1960). There are many other cases of tsunamis generated by landslides in Lituya Bay, Russel Fjord (southern Alaska), Kitimat Inlet, and the region of the Fraser River in British Columbia (Murty, 1979; Hamilton and Wigen, 1987; Lander and Lockridge, 1989; Weichert et al., 1994). The tsunami of November 3, 1994, in Skagway Harbor, Alaska, was generated by an underwater landslide formed during the collapse of a cruise ship wharf that was undergoing construction at the head of Taiya Inlet. This event was not associated with a regional seismic event or incoming oceanic tsunami (Kulikov and Rabinovich, 1996). The landslide was accompanied by a series of large-amplitude, high-frequency solitary wave oscillations estimated by eyewitnesses to be 5 to 6 meters high in the inlet and 9 to 11 meters high at the shoreline. According to eyewitnesses, immediately after the collapse of the wharf, a "wall of water" propagated toward the shore.

Taiya Inlet is a relatively long (23 km), narrow (2.32 km), and deep (385 m) inlet. Sea-level gauge measurements from the inlet (considerably attenuated because of the poor response of the gauge at short-period tsunamis) revealed oscillations with periods of about 3 minutes and maximum recorded trough-to-peak heights of 2 meters lasting for about 1 hour. Detailed analysis by Kulikov and Rabinovich (1996) revealed that the basin seiche generated by this tsunami was highly tuned to the inlet oscillations at the mouth of the basin, since the natural period of oscillation of the basin (~ 3 min) closely matched the period of the cross-channel seiche. Model studies by Rabinovich and colleagues (1999) could be used to explain many of the eyewitness accounts and to reproduce the observed dominant oscillations, including the persistent (~ 1 h) 3-minute oscillations in Skagway Harbor. Bottom friction, scattering from the rugged topography, radiation into the inlet, and dispersive effects caused the observed gradual decay of the tsunami wave motion. It is noteworthy that whereas large-scale (long-period) tsunamis can propagate for long distances while maintaining their initial form, small-scale (short-period) tsunamis attenuate quickly because of dispersion effects and give rise to small-amplitude oscillations (Gonzalez and Kulikov, 1993). As a rule, landslide tsunamis have much smaller horizontal scales than tsunamis generated by earthquakes. This is why most known landslide tsunamis have been observed in comparatively localized regions only (Kulikov and Rabinovich, 1996).

Submarine landslides have been found to be predominantly associated with low tide. The tsunami waves in Skagway Harbor occurred about 30 minutes after low tide. Similarly, the catastrophic slide in Kitimat Inlet of April 27, 1975, (Murty, 1979) occurred 53 minutes after low tide. Bjerrum (1971) described six major submarine slides in Norwegian fjords for the period 1888–1952 and showed that they all occurred around the time of low tide. A detailed discussion on how a low tide causes excess pore water pressure and might trigger slides is given by Karlsrud and Edgers (1980). Based on model studies, Rabinovich and colleagues (1999) showed that the occurrence of the landslide at Skagway Harbor is linked to critical overloading of the slope materials at a time of extreme low tide.

In some cases, landslides are triggered by earthquakes. For example, the Grand Banks landslide-generated tsunami of November 18, 1929, was triggered by a $M = 7.2$ earthquake that occurred at the southern edge of the Grand Banks, located 280 km south of Newfoundland, at an estimated depth of 20 km beneath the seafloor (Fine et al., 2005; Clague, 2001). The earthquake triggered a large submarine slope failure (200 km^3), which was transformed into a turbidity current carrying mud and sand up to 1000 km at estimated speeds of 60–100 km/h, breaking 12 telegraph cables. The tsunami generated by this failure killed 28 people, making it the most catastrophic tsunami in Canadian history. Tsunami waves also were observed along other parts of the Atlantic coast of Canada and the United States. Waves crossing the Atlantic (Figure 2.4) were recorded on the coasts of Portugal and the Azores Islands. These tsunami waves had amplitudes of 3–8 meters.

In contrast to rigid landslides, which move as single, consolidated bodies, preserving their size and form, viscous slides normally spread and flatten as they move downslope. The concentrated and focused moving-slide mass (similar to snow/ice avalanches in mountains) was apparently the main reason for the numerous cable breaks.

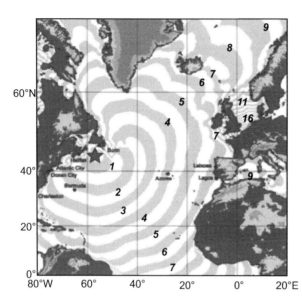

FIGURE 2.4 Source and propagation of the 1929 Grand Banks landslide-generated tsunami waves in the North Atlantic Ocean. *Source: Fine et al., 2005; Reproduced with kind permission of Elsevier.*

The African coast is also subject to local landslide-generated tsunamis that can occur near any part of the coastline, especially areas that are near the mouths of large African rivers. In fact, historical chronicles abound of several instances of unusual wave run-ups observed in the area near the Gulf of Guinea, off central West Africa. It has been suggested that in many cases of tsunamis accompanied by seaquakes in tropical countries, the tsunamis are not only induced directly by the seaquakes but are also spurred by submarine landslides (i.e., submarine slope sliding triggered by the earthquake shaking). An example of such a tsunami is one that occurred in the sea area north of Aitape City on the northern coast of Papua New Guinea in the wake of the June 15, 1998, earthquake of magnitude 7.0, during which seawater rose 15 meters above the mean sea level and more than 2000 people perished. It was pointed out that a huge submarine landslide was induced by the main shock, and the main part of the tsunami was generated by the landslide and not by the seaquake itself. In this case, the secondary tsunami was larger than the first tsunami, which was caused by the crustal motion of the sea bed (Tsuji, 2009).

There had been instances where large tsunamis were generated by rock avalanches, which is a geodynamic phenomenon. For example, the First Nations (Da'naxda'xw) village of Kwalate, Knight Inlet, British Columbia, Canada (Figure 2.5), was completely swept away by a 2- to 6-meter-high tsunami that formed when an 840-meter-high, 30×10^6 m^3 subaerial rock avalanche (Figures 2.6 and 2.7) descended into the water on the opposite side of the fjord in the mid-nineteenth century (Bornhold et al., 2007). The devastating rock avalanche and associated tsunami destroyed a large aboriginal community and forever altered the history of First Nations peoples.

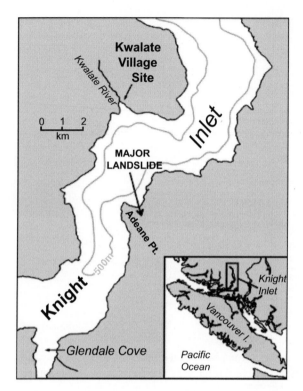

FIGURE   2.5 Rock avalanche-generated tsunami site at Knight Inlet, the former First Nations village of Kwalate. *Source: Bornhold et al., 2007; Reproduced with kind permission of Canadian Meteorological and Oceanographic Society (CMOS).*

2.3. TSUNAMIS GENERATED BY VOLCANIC ERUPTIONS

Volcanic eruption is one of the generation mechanisms for tsunamis; approximately 5% of all tsunamis are generated this way. The infamous global tsunami caused by the volcanic eruption at Krakatau Island in Indonesia on August 27, 1883 (Murty, 1977; Bryant, 2001), is a classic example (Figure 2.8). The Krakatau Island volcano exploded with devastating fury to such an extent that the overlying land and the surrounding seabed collapsed into its underground magma chamber (Rastogi, 2007). A series of large tsunami waves was generated by the explosion, some reaching a height of over 35 meters above sea level, and took a toll of 36,000 lives in western Java and southern Sumatra. On the facing coasts of Java and Sumatra, the sea flood went several km inland and caused such vast loss of life that one area was never reoccupied and is currently the Ujung Kulon nature preserve.

The 1883 Krakatau tsunami was recorded by 35 sea-level gauges in the Indian, Pacific, and Atlantic Oceans, including gauges in Le Havre (France), Kodiak Island (Alaska), and San Francisco (California) (Pelinovsky et al., 2005). Whereas the tsunami waves recorded at near-field sites originated from direct water waves that propagated from the source area, there have been

FIGURE 2.6   Rock-avalanche area north of Adeane Point, Knight Inlet; the source of the Kwalate avalanche-generated tsunami. *Source: Bornhold et al., 2007; Reproduced with kind permission of Canadian Meteorological and Oceanographic Society (CMOS).*

suggestions (Eving and Press, 1955; Garrett, 1970) that those recorded at far-field sites originated from coupling between the ocean surface and the explosion-induced atmospheric waves that circled the globe three times (Murty, 1977). As indicated in Chapter 4.2, the propagation speed (c) of tsunami waves is given by: $c \cong \sqrt{gD}$ where g is the acceleration due to gravity and D is the local water depth. This means that the tsunami waves travel faster as the water depth increases. Based on the preceding formula, the fastest tsunami waves (for example, those in Pacific Ocean, where water depth is the largest) travel at a speed of approximately 800 km/h. However, atmospheric sound waves travel much faster (at a speed of approximately 1224 km/h). From this knowledge, it was possible to estimate the expected times of arrivals of the tsunami and the explosion-induced atmospheric waves at different locations in the ocean. Surprisingly, it was found that the waves in the Pacific and the Atlantic oceans arrived too early for long ocean waves to arrive at these sites but in good agreement with atmospheric sound waves to arrive. Thus, the assessment of Eving and Press (1955) and Garrett (1970) regarding coupling between the ocean surface waves and the explosion-induced atmospheric waves stemmed from the mismatch in time between observed and expected tsunami waves. Thus, the near-field records of the 1883 Krakatau tsunami were apparently related to real surface ocean waves arriving from the source area, whereas the records from intermediately located sea-level gauges included some mixture of directly arriving ocean waves and atmospherically generated waves. Far-field records were associated purely with atmospheric waves.

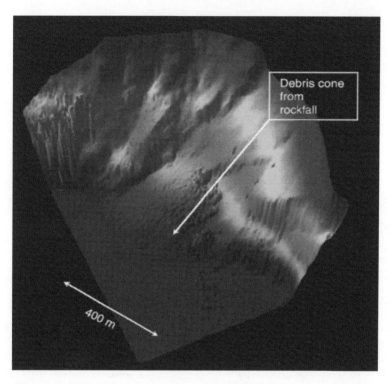

FIGURE 2.7 Multi-beam echo-sounder image of the failure mass cone just north of Adeane Point, Knight Inlet; the source of the Kwalate avalanche-generated tsunami. The maximum water depth (purple) is about 535 m. The slide volume is $\sim 4 \times 10^{6} \, \text{m}^{3}$. *Source: Bornhold et al., 2007; Reproduced with kind permission of Canadian Meteorological and Oceanographic Society (CMOS).*

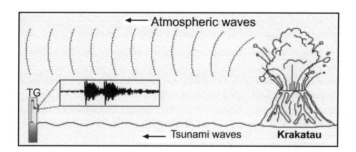

FIGURE 2.8 Sketch illustrating the generation of long-wave sea-level oscillations recorded by remote sea-level gauges after the 1883 Krakatau Volcano explosion. *Source: Monserrat et al., 2006; Natural Hazards and Earth System Sciences; Copernicus Publications.*

A volcano in Anak Krakatau (Child of Krakatau) erupted in 1927 and 1928, and these eruptions generated large local tsunamis in the Sunda Strait. However, the height of the waves attenuated rapidly away from the source region because of very short wave periods and wave lengths; no damages from these tsunamis were reported (George, 2003).

2.4. TSUNAMIS GENERATED BY METEOROLOGICAL DISTURBANCES

Although geophysical tsunamis are the main cause of destructive seiches observed in the World Oceans, long waves generated by atmospheric forcing (passage of cyclones and hurricanes, frontal squalls with associated thunderstorms, atmospheric gravity waves, atmospheric pressure jumps, wind waves) can also be responsible for significant, even devastating, long waves. These waves are called "meteorological tsunamis" (also, meteo-tsunamis) because they have the same temporal and spatial scales as typical tsunami waves (having approximately the same frequencies as seismically generated tsunami waves and look very similar to the latter); and can affect coasts in a similar way, although their catastrophic effects are normally observed only in a limited number of specific bays and inlets (Defant, 1961; Rabinovich and Monseratt, 1996; Gonzalez et al., 2001; Rabinovich and Stephenson, 2004). The term *'meteorological tsunamis'* ('meteo-tsunamis') for such waves has been suggested by Nomitsu (1935), Defant (1961), and Rabinovich and Monserrat (1996, 1998). Because of the pronounced similarity between 'meteo-tsunamis' and seismically-generated tsunamis, it is often difficult to recognize one form of wave from another (Rabinovich et al., 2009). Other mechanisms that may result in a meteo-tsunami include tide-generated internal waves, wave superposition, wind-current interaction, and atmospheric shockwaves from volcanic activity (Rabinovich and Monseratt, 1996; Lowe and de Lange, 2000; Bryant, 2001). Catalogs of tsunamis normally contain references to 'tsunami-like' events of 'unknown origin' that could, in fact, be atmospherically generated ocean waves (See Figure 2.9). Several significant tsunami-like events were recently observed in sea-level records from the coasts of Adriatic Sea, Balearic Islands, Bristol Channel, English Channel, British Columbia (Canada) and the state of Washington (United States). Although similar to seismically generated tsunamis, the events were of nonseismic origin because there were no major or moderate earthquakes at the time of the events. Coincident water-level oscillations were observed at offshore stations, as well as in sheltered bays, inlets, and channels of the Strait of Georgia, Juan de Fuca Strait, and Puget Sound. This involved stations in Victoria, Patricia Bay, Port Angeles, Friday Harbor, Seattle (Washington), and Cherry Point that are all well protected from tsunami waves arriving from the open ocean (Fine et al., 2009). Because these tsunami-like events were of meteorological origin, they have been legitimately termed *meteorological tsunamis.*

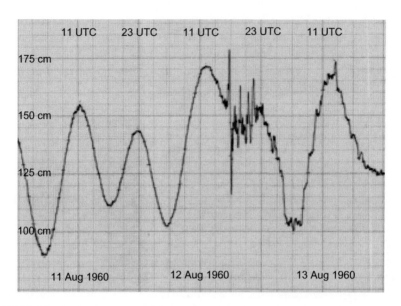

FIGURE 2.9   Sea level curve recorded at the Rovinj tide gauge on August 12, 1960, visualizing the high-frequency oscillations presumably related to meteo-tsunami waves. *Source: Vilibic and Sepic, 2009; Reproduced with kind permission of Elsevier.*

A classical example often cited in connection with meteorological tsunamis is the one occurred on June 21, 1978, in Vela Luka, which is a small town hidden in a bay on Korcula Island in the Adriatic Sea (Figure 2.10). In the early morning of June 21, 1978, the sea suddenly began to rise in the town, over-topping the piers and surging into the streets. The rumble of the incoming wall of water awakened the inhabitants who witnessed a series of destructive ocean waves (Figure 2.11), flooding much of the city and causing devastation and widespread damage (Figure 2.12). Tsunami-like waves with trough-to-crest heights of up to 6 meters appeared without any warning, resulting in the greatest natural disaster in the modern history of Vela Luka. The sea withdrew by about 100 meters leaving the harbor dry and stranding fishes on the beach, and then violently came back. The phenomenon occurred approximately every 18 minutes and lasted for 1 hour. The damage reported in Vela Luka after the event was approximately 23% of the annual income of Korcula Island (Vucetic et al., 2009).

The resemblance of the event with a tsunami was too obvious. This event was felt in the middle and south Adriatic Sea. However, there was no record of an earthquake at that time in the Adriatic Sea. This event is listed even in a number of tsunami catalogs (e.g., Italian Tsunami Catalog, Tinti et al., 2004; Maramai et al., 2007), where the source mechanism is marked as "unknown". Subsequent scientific investigations indicated that the waves were not related to a seismic event or a submarine landslide but to an atmospheric process (a traveling

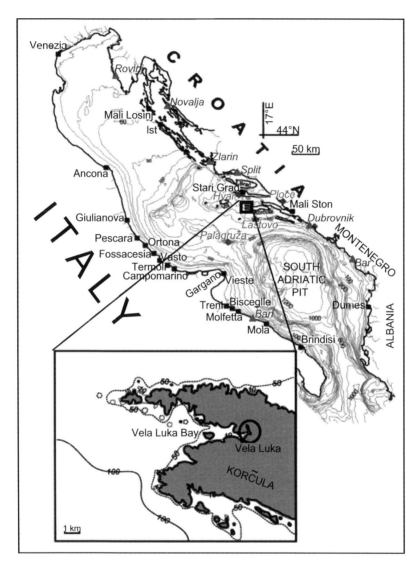

FIGURE 2.10 Map of Vela Luka bay at Korcula Island in the South Adriatic Sea. *Source: Vucetic et al., 2009; Reproduced with kind permission of Elsevier.*

disturbance that had the capability to resonantly transfer energy to the ocean via the Proudman resonance mechanism), thereby identifying this as a meteorological tsunami event. Several destructive meteo-tsunamis affected the eastern Adriatic shore in the last few decades, and therefore attracted much attention of the local researchers in order to explain their source, generation, and propagation mechanism, as well as inundation along the complex coastal topographies.

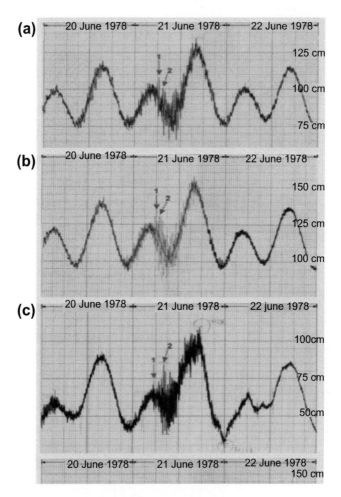

FIGURE 2.11   Sea-level charts recorded at: (a) Bar; (b) Dubrovnik; (c) Split in association with the June 21, 1978 meteo-tsunami event in Vela Luka. The first and the second waves are marked by 1 and 2. *Source: Vucetic et al., 2009; Reproduced with kind permission of Elsevier.*

The Adriatic Sea encompasses a few locations which are particularly vulnerable to destructive meteotsunamis, namely Vela Luka, Mali Ston, Stari Grad, Ist (First), and Mali Losinj (Figure 2.13). After the Vela Luka episode, the next significant meteo-tsunami in the Adriatic Sea occurred on 5 October 1984 on Ist Island (Sepic et al., 2009). In subsequent years, no severe meteo-tsunamis were recorded till 27 June 2003, when a particularly strong meteotsunami was observed in Stari Grad and Mali Ston Bays (Vilibic et al., 2004). The damage in Mali Ston Bay during the 2003 meteo-tsunami was not a result of strong sea-level oscillations, but of severe currents at the bay constrictions. It is note-worthy that Mali Ston Bay is the most important area for production of the

(a)

(b)

FIGURE 2.12 Photos depicting (a) low water and (b) high water in Vela Luka in association with the June 21, 1978 meteo-tsunami event, in which houses and storehouses in the seafront were flooded and some of the boats moored nearby hit the ground and were broken off their moorings by severe currents. As many as 30 yachts were reported to be damaged. *Source: Vucetic et al., 2009; Reproduced with kind permission of Elsevier.*

European flat oyster and Black mussel (Peharda et al., 2007), and the strong turbulent currents swept out the shellfish farms over the 6 km long area, causing considerable financial loss and halting shellfish production for a while. The next event occurred on 22 August 2007, again on Ist Island (Sepic et al., 2009), followed by a flood at Mali Losinj (the largest town of the eastern Adriatic islands) a year later (August 15, 2008). Photographs of the events are shown in Figure 2.14.

At certain places in the World Ocean, hazardous atmospherically-induced waves occur regularly and have specific local names: '*rissaga*' in the Balearic Islands (Gomis et al., 1993), '*marubbio*' ('*marrobio*') in Sicily (Candela et al., 1999), '*sciga*' on the Croatian coast of the East Adriatic Sea (Hodzic, 1979/1980) in the Mediterranean, '*milghuba*' (play of the sea) in Malta (Drago, 2008), '*abiki*' and '*yota*' in Japan (Hibiya and Kajiura, 1982), '*seebar*' in the Baltic Sea (Metzner et al., 2000), '*death waves*' in Western Ireland, and '*inchas*' and '*lavadiads*' in the Azores and Madeira islands (Rabinovich et al., 2009). These waves have also been documented for Greece (Papadopoulos, 1993), Dutch coast (de Jong and Battjes, 2004), the Yellow and Aegean seas, the northwestern Atlantic, English Channel (Douglas, 1929), Florida shelf (Sallenger et al., 1995), the Great Lakes (Donn and Ewing, 1956), coastal areas of Argentina and New Zealand (Dragani, 2007; Goring, 2005), the Yellow Sea (Wang et al, 1987), and in some specific ports such as Port Rotterdam. From

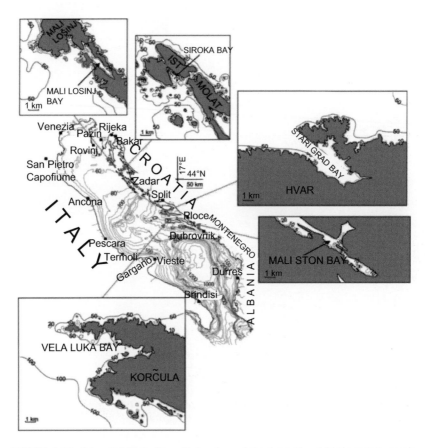

FIGURE 2.13 Map of Adriatic Sea with locations of the destructive Adriatic meteotsunamis, namely Vela Luka, Mali Ston, Stari Grad, Ist, and Mali Losinj. *Source: Vilibic and Sepic, 2009; Reproduced with kind permission of Elsevier.*

time to time, significant long waves up to 1 m high and with periods of a few minutes occur on the New Zealand coast. These waves create a threat to navigation because they cause set-down in vessels navigating to berth, thus reducing their under-keel clearance. These events in New Zealand are strongly correlated with low-pressure systems that originate in the tropics and propagate southwards 1000 km or more to the east coast of New Zealand. The speed of propagation of the weather systems is generally of the order of 20 km/h, which is well below the long wave speed of approximately 800 km/h in the deep sea, but is close to the group velocity of swell waves in deep water which are assumed to be the main source of the observed waves on the New Zealand coast (Goring, 2009). Haslett et al., (2009) examined numerous historical reports on high long waves recorded on the coast of the United Kingdom (Figure 2.15) and concluded that these waves have an atmospheric origin such as storm activity,

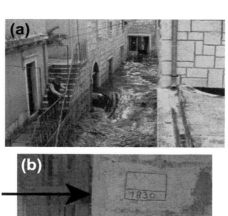

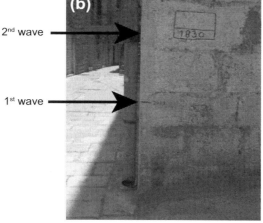

2nd wave

1st wave

FIGURE 2.14 Photographs of the flooding of the houses and the streets at the sea front due to meteotsunamis at (a) Vela Luka in 1978, (b) Stari Grad in 2003 (arrows denote the height of the first and the second wave. It is interesting to notice the sign "1830" on the house wall, which is an indicator of the maximum flood height during a similar event that occurred in that year), (c) Ist in 2007 (maximum wave height is marked), and (d) Mali Losinj in 2008. *Source: Vilibic and Sepic, 2009; Reproduced with kind permission of Elsevier.*

offshore thunderstorms, and squalls. Wave amplification at the coast is supported also by oceanographic mechanisms such as seiching in enclosed basins associated with far-traveled long-period waves from Atlantic depressions. Fatalities have occurred during meteo-tsunami events in the UK, including those of beach users killed by meteo-tsunamis arriving unexpectedly at beaches

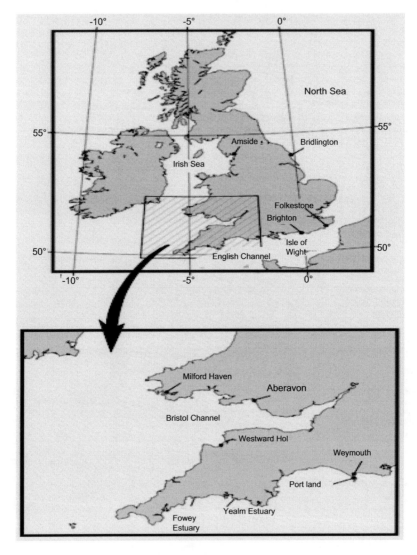

FIGURE 2.15 Locations in the United Kingdom which are sensitive to meteo-tsunami attack. *Source: Haslett et al., 2009; Reproduced with kind permission of Elsevier.*

during summer months (May-September), when beaches attract large numbers of recreational users (Figure 2.16).

For the coastal flooding events triggered by meso-scale forcing and associated with resonance, terms such as *rogue waves, freak waves,* and *giant waves* have been used in the literature to describe them. *Rogue waves* are large relatively short-period meteorological waves that are infamous for sinking ships in the open sea and so differ from tsunami, which are of low-amplitude in

FIGURE 2.16 (a) Folkestone Harbour, Kent, where eight large waves were observed entering on 20th July 1929, washing one boy from the harbor wall to his death; (b) Folkestone Harbour, Kent, where a number of bathers and people paddling were caught by the waves and were in danger of being drowned during the meteo-tsunami event of 20th July 1929; (c) Weymouth Beach, Dorset, where a wave 6 ft high came roaring up the narrow harbor, tearing moorings out of the sea-bed and throwing boats in confusion under the town bridge during the 5th July 1939 meteo-tsunami event; (d) Portland Harbor, Dorset, where the boom defense across the entrance to Portland Roads was moved some distance into Portland Harbor by the pressure of the sea waves during the 5th July 1939 meteo-tsunami event. *Source: Haslett et al., 2009; Reproduced with kind permission of Elsevier.*

the open ocean. However, rogue wave formation in coastal waters may be considered as meteo-tsunami if they take on long-period tsunami-like characteristics (Kharif and Pelinovsky, 2003).

Rissaga is a Spanish word meaning "high-amplitude sea-level oscillation." (Figures 2.17 and 2.18). This was first observed in Spain and was reported in a series of papers. Tintore and colleagues (1988) reported large-scale sea-level

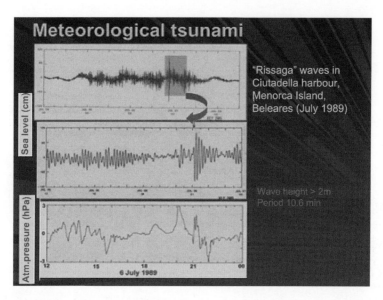

FIGURE 2.17 Meteorological tsunami, known as "Rissaga" waves in Ciutadella Harbour, Menorca Island, Baleares. The top row shows the residual sea-level (i.e., sea-level after removal of tide) just before, during (shown within shaded rectangular box), and just after the rissaga event. The residual sea-level during rissaga event is shown in detail in the middle row. The atmospheric pressure disturbances that lead to the rissaga generation is shown in the bottom row. *Source: http:// www.theplasmaverse.com/pdfs/meteorological-tsunamis-destructive-atmosphere-induced-waves-observed-in-the-worlds-oceans-seas-seiches-abiki-milghuba.pdf*

oscillations (up to 1 m amplitude and about 10 min period) in various bays and harbors of the western part of the Mediterranean Sea. They showed that these oscillations arise due to a three-way resonant coupling among atmospheric gravity waves, coastally trapped edge waves, and the normal modes of a harbor. The energy sources are atmospheric pressure fluctuations with a period of about 10 min and amplitude of about 1.5 mb. Every year from June to September, these oscillations are observed on the northeast coast of Spain and in the Belearic Islands. The largest amplitudes usually occur in Ciutadella, which is an elongated harbor about 1 kilometer long and 90 meters wide, and 5 meter deep that is situated on the west coast of Minorca Island. The fundamental period of the inlet (the Helmholtz mode) is approximately 10.5 min. Due to its particular geometry; Ciutadella Inlet has a high Q-factor which results in significant resonant amplification of long waves arriving from the open sea. Rissaga events (large-amplitude seiches) having wave heights of more than 4 meters, and which have dramatic consequences for the harbor, usually take place once every 5−6 years. One of the largest oscillations observed was about 2 meters on June 21, 1984 (Jansa, 1986), as a result of which about 300 boats were destroyed or severely damaged (Rabinovich and Monserrat, 1996). Jansa and colleagues (2007) reported that on June 15, 2006, a rissaga occurred in

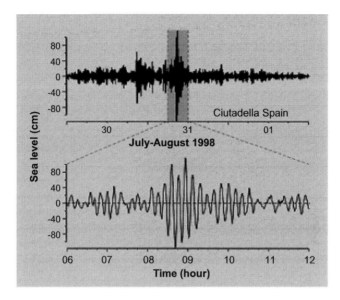

FIGURE 2.18 The strong "rissaga" event recorded in Ciutadella Inlet (Minorca Island, Spain) on 31 July 1998. *Source: Monserrat et al., 2006; Natural Hazards and Earth System Sciences; Copernicus Publications.*

Ciutadella (Menorca) with amplitudes of up to 4 to 5 meters and a period of about 10 minutes. One hundred ships were damaged, 35 of which sank. The damaging effects of rissaga and sciga are shown in Figures 2.19 and 2.20, respectively. Several coastal locations of British Columbia, Canada, also feature relatively large resonant amplification because of the presence of bays, inlets, and fjords and their distinct resonant properties. Rabinovich and Monseratt (1996, 1998) included the Kuril Island (Japan) and China as well in the list of places hit by a rissaga (Figure 2.21). The frequency response at various stations is strong and clear. The corresponding resonant periods are relatively short (from 2.5 to 20 min), demonstrating the major impact for these locations in the event of a relatively high-frequency tsunami. In the case of resonance, the arriving waves at these locations may amplify 10–15 times (Rabinovich and Stephenson, 2004).

Meteorological tsunamis (i.e., destructive atmosphere-generated tsunami-like waves) have been found to be induced by the resonant superposition of external factors such as strong atmospheric disturbances resonantly interacting with open-ocean waves and internal factors such as pronounced resonant properties of a specific bay or harbor. The propagation directions of these atmospheric disturbances have been found to vary among events.

The atmospheric disturbances include abrupt atmospheric pressure changes (isolated pressure jumps; see Hibiya and Kajiura, 1982; Vilibic et al., 2004, 2005), gale winds, moving storm fronts (frontal passage, squalls, trains of

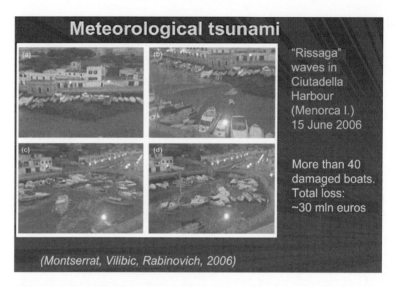

FIGURE 2.19 Illustration of the damaging effect of "rissaga" meteorological tsunami. in Ciutadella Harbour, Menorca Island. *Source: Monserrat et al., 2006; Natural Hazards and Earth System Sciences; Copernicus Publications.*

FIGURE 2.20 Illustration of the damaging effect of "sciga" meteorological tsunami in Vela Luka, Croatia. *Source: http://www.theplasmaverse.com/pdfs/meteorological-tsunamis-destructive-atmosphere-induced-waves-observed-in-the-worlds-oceans-seas-seiches-abiki-milghuba.pdf*

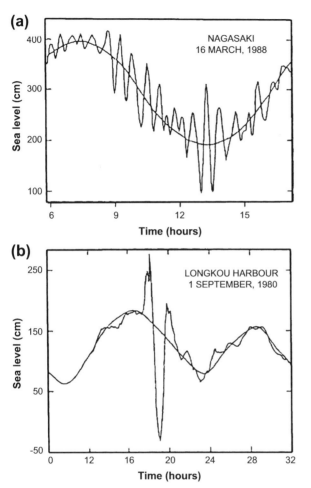

FIGURE 2.21 Rissaga type meteo-tsunami oscillations in (a) Nagasaki, Japan. *(Source: Rabinovich and Monserat, 1996; Reproduced with kind permission of Springer)* and (b) Langkou harbour, China. *Source: Wang et al., 1987; Reproduced with kind permission of American Meteorological Society (AMS).*

atmospheric gravity waves; see Gossard and Munk, 1954; Monserrat et al., 1991; Garcies et al., 1996) (Figure 2.22), and other types of atmospheric disturbances that normally generate barotropic ocean waves in the open ocean and amplify them near the coast through specific resonance mechanisms (Proudman, 1929; Greenspan, 1956). These atmospheric disturbances— atmospheric pressure jumps and trains of atmospheric waves with typical amplitudes of 0.5−1.5 hPa and typical propagation speeds of 20−30 m/s (Fine et al., 2009)—may have different origins, such as dynamic instability, orographic influences, and so on (Gossard and Hooke, 1975). Nevertheless, even during the strongest events, the atmospheric pressure oscillations at these scales typically reach only a few hPa (i.e., mb) which, based on the static sea-level response (inverted barometer effect), correspond to only a few centimeters of sea-level change. Consequently, these atmospheric fluctuations

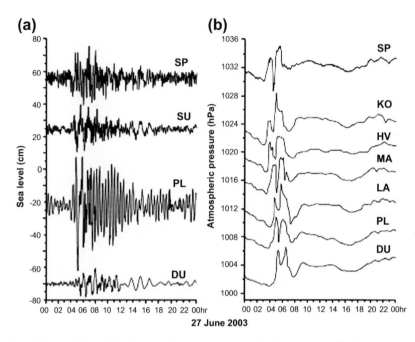

FIGURE 2.22   The 27 June 2003 meteo-tsunami time-series at four locations in the eastern Adriatic coastal areas of Croatia (low-frequency oscillations have been removed by a high-pass digital filter) subsequent to the abrupt air pressure oscillations illustrated in (b). *Source: Monserrat et al., 2006; Natural Hazards and Earth System Sciences; Copernicus Publications.*

can produce a significant sea-level response only when some form of resonance occurs between the ocean and the atmospheric forcing.

In the case of Adriatic meteo-tsunamis, the traveling atmospheric disturbances are commonly generated by a flow over the mountain ridges (Apennines), and keep their energy through the wave-duct mechanism (Lindzen and Tung, 1976) while propagating over a long distance below the unstable layer in the mid-troposphere. However, the Adriatic meteo-tsunamis may also be generated by a moving convective storm or gravity wave system coupled in the wave-CISK (Conditional Instability of the Second Kind) manner (Powers and Reed, 1993), not documented at other world meteo-tsunami hot spots (Vilibic and Sepic, 2009). The wave-duct mechanism represents the propagation of the waves in a stable layer capped by a dynamically unstable steering layer, which prevents vertical energy leakage. The wave-CISK mechanism represents coupling between a gravity wave and convection. It may be noted that wave-associated convergence forces moist convection and this (i.e., moist convection) provides energy for the gravity wave. These waves are visible in surface air pressure series, constantly forcing the ocean during their passage over a certain area (Vilibic and Sepic, 2009).

Proudman (1929) obtained an analytic solution for a situation in which the atmospheric disturbance is traveling over a channel of uniform depth (no friction has been included). According to this solution, if the translational speed (U) of the atmospheric disturbance is equal or nearly equal to the phase speed (c) of the long ocean waves, the wave height gets constantly amplified on their way toward the coast. This is known as Proudman resonance (See Figure 2.23). The degree of resonance is expressed as U/c ratio. Fully resonant conditions happen when $U = c$. Vilibic (2008) observed that the waves are controlled not only by the degree of resonance but also by the friction, Coriolis force and diffusion, as well as by nonlinearity. High-resolution (1-minute) sea-level series recorded at Rovinj during the 2008 Mali Losinj meteo-tsunami support the idea of generation of long ocean waves through the Proudman resonance.

If the alongshore component of the atmospheric disturbance velocity is equal to the phase speed of an edge wave generated along the coastlines, then the amplification of the long ocean waves is driven by Greenspan resonance (Greenspan, 1956). When the atmospheric disturbance and the generated long ocean waves have period or wavelength equal to the resonant period or wavelength of the shelf region, shelf resonance resonantly amplify the long ocean waves.

Finally, the generated long ocean waves hit the harbors, producing stronger oscillations only if the corresponding inner basin has well-defined resonant properties and a large Q-factor (Rabinovich, 2009). The Q-factor, which measures the energy decay of the system, is estimated from the observational data (Miles and Munk, 1961):

$$Q = \frac{f_0}{\Delta f} \tag{2.1}$$

where f_0 is the peak frequency in the spectrum and Δf is the width of this spectrum at its half maximum height. Internal or harbor resonance plays a crucial role in the destructiveness of the meteo-tsunami waves. Effect of strong wind piling up the water inside the basin and triggering the fundamental mode oscillations is also an important factor that aggravates the destructiveness of the meteo-tsunami waves.

During resonance, the atmospheric disturbance propagating over the ocean surface is able to generate significant long ocean waves by continuously pumping energy into these waves. The meteorological tsunami event of December 9, 2005, on the Pacific coast of North America was sufficiently strong to initiate an automatic tsunami alarm, while others generated oscillations in several ports that were strong enough to cause damage to boats (Fine et al., 2009). The Adriatic is characterized by elongated islands with sloping bathymetry (Figure 2.13). The generated long ocean waves hit the funnel-shaped bays or harbors of large amplification factors, resulting in meteo-tsunami waves with heights up to 6 meters at the very end (head) of bays or harbors. Within models the mechanism is

(a)

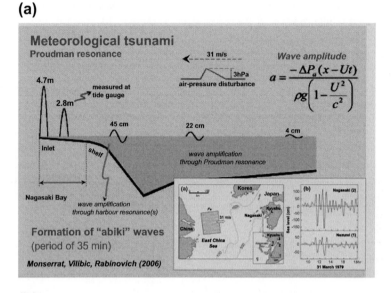

(b)

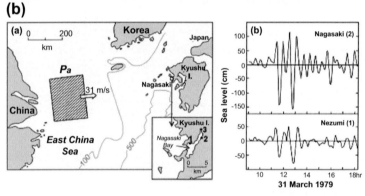

FIGURE 2.23   (a) A sketch illustrating the physical mechanism responsible for formation of the catastrophic meteo-tsunami at Nagasaki Bay (Japan) on 31 March 1979. The initial pressure jump over the western part of the East China Sea was about 3 hPa. The long waves generated by this event first amplified from 3 cm to 16 cm as a result of the Proudman resonance effect, then to 45 cm due to the shelf amplification and finally to 478 cm at the head of the bay due to harbour resonance. U: atmospheric disturbance translational speed; c: phase speed of the oceanic long-wave. (b) Meteorological tsunami: Map showing the location of Nagasaki Bay and the site of the initial atmospheric pressure disturbance (shaded rectangle). Number "3" is at the head of the bay where the maximum wave of 478 cm was observed on 31 March 1979. *Source: Monserrat et al., 2006; Natural Hazards and Earth System Sciences; Copernicus Publications.*

fairly well understood; however, it is extremely difficult to reproduce these events as the meteo-tsunami generating process is highly variable at both temporal and spatial scales (Vilibic and Sepic, 2009). The risks connected with a single meteo-tsunami event in two different bays can be quite different; while in some bays the flooding may be predominantly responsible for the devastation, in other bays the currents may produce substantial damage, or may influence the safety of navigation like in Ploce Harbor (Vilibic and Mihanovic, 2005).

It is a well-recognized fact that periods of extreme events (e.g., normal tsunamis as well as meteorological tsunamis) are mainly related to resonant properties of the local topography rather than to the characteristics of the source, and they are almost the same as those of the background oscillations for the same sites (Honda et al., 1908; Miller, 1972; Rabinovich and Monseratt, 1996). Meteorological tsunamis are regularly observed at the same sites with pronounced local resonant properties. These are called "meteorological tsunamis" (also meteo-tsunamis) because these oscillations have approximately the same frequencies as seismically generated tsunami waves and look very similar to the latter (Defant, 1961; Rabinovich and Monseratt, 1996; Gonzalez et al., 2001; Rabinovich and Stephenson, 2004).

When storm seiches and typical tsunami seiches look very similar, it is unclear whether the source differences could be detected from an analysis of sea-level data alone without additional information on seismic or meteorological events. However, strong visible similarities between normal tsunamis and meteo-tsunamis are related to the same resonance influence of the local topography and do not mean that their sources are the same. In fact, it is the particular properties of the sources that can be used to distinguish between these two phenomena (Rabinovich and Stephenson, 2004).

Rabinovich (1997) and Monserat and colleagues (1998) suggested a simple approach to analyzing tsunamis/meteo-tsunamis and reconstructing their spectral source characteristics. The general idea is that by comparative analysis of the event- and background-spectra, the sources and topography effects can be separated. By eliminating the influence of topography and restoring the source, these phenomena can be identified and an insight into their nature can be gained.

In the long list of historical tsunamis, several involved meteo-tsunami phenomena. For example, an unusual tsunami-like long-period ($\sim 15-60$ min) wave action (with amplitude ~ 0.8 m superimposed on a high tide) affected the entire coastline between southern Namibia and Table Bay on the African coast, beginning on August 20, 2008, and culminating in an extended attack during the morning of August 21, startling boat owners in harbors and causing flood damage at a shoreline factory in St. Helena Bay. This was attributed to a "meteo-tsunami" phenomenon with an atmospheric origin that was unrelated to any submarine geological cause (Hartnady et al., 2009). In another instance, sea-level disturbances along the Japanese coast that were the result of meteorological origin have been reported by Satake and colleagues (2009).

As indicated earlier, the resonant conditions in the open ocean and on the shelf are governed by Proudman resonance (Proudman, 1929) and Greenspan resonance (Greenspan, 1956), respectively. Proudman resonance occurs when the translational speed (U) of the atmospheric disturbance equals the phase speed (c) of the oceanic long wave. Greenspan resonance occurs when the alongshore component (U_l) of the atmospheric disturbance velocity (U) equals the phase speed (c_j) of the j-th mode of edge waves. Shelf resonance occurs when the atmospheric disturbance and the associated atmospherically generated ocean waves have periods and/or wavelengths equal to the resonant periods and/or wavelengths of the shelf region.

Long, narrow inlets; bottle-like bays; and well-protected harbors are suitable candidates for strong resonant amplification (high Q-factor). Harbor resonance occurs when the natural frequency of the harbor and the frequency of the arriving forcing match. V-shaped bays are probable candidates for topographic amplification due to wave geometric focusing (Figure 2.24).

Because meteorological tsunamis have, in general, a multiresonant generation mechanism, the needed coincidence of several resonant factors significantly diminishes the possibility of the occurrence of such events. This is the main reason why these phenomena are rare and restricted to specific locations. The Mediterranean Sea (Figure 2.25) is a region where meteo-tsunamis are quite common. The combination of these factors at such specific locations is like a "ticking time bomb." Sooner or later, when the atmospheric disturbance is intense and the parameters of the disturbance coincide with the resonant parameters of the corresponding shelf topography and embayment geometry, the "bomb" will explode. Vela Luka is a natural bay where a combination of harbor resonance and topographic amplification due to its V shape occur and is therefore probably one such site.

Atmospheric activity generates long waves not only through the nonlinear interaction of wind waves but also directly by dynamic forcing of atmospheric pressure on the sea surface. The destructive meteorological tsunamis are mainly related to atmospheric buoyancy waves or pressure "jumps" and may occur even during calm weather (Rabinovich and Stephenson, 2004).

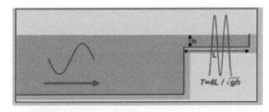

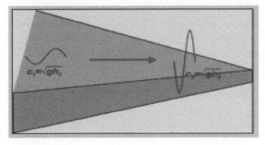

FIGURE 2.24 Meteorological tsunami amplification at harbours and bays through resonance. Vela Luka, Croatia in the Mediterranean Sea is a natural bay where a combination of harbour-resonance and topographic amplification occur because of its geometry as shown above and could therefore be one of "ticking time-bomb" sites. *Source: http://www.theplasmaverse.com/pdfs/meteorological-tsunamis-destructive-atmosphere-induced-waves-observed-in-the-worlds-oceans-seas-seiches-abiki-milghuba.pdf*

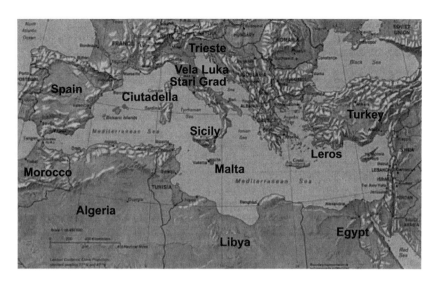

FIGURE 2.25 Mediterranean Sea region where meteo-tsunamis are quite common. Vela Luka Bay in this region is a natural bay where a combination of harbour resonance and topographic amplification due to its V-shape occur and is therefore probably one of "tickling time-bomb" sites. *Source: http://www.theplasmaverse.com/pdfs/meteorological-tsunamis-destructive-atmosphere-induced-waves-observed-in-the-worlds-oceans-seas-seiches-abiki-milghuba.pdf*

Miles and Munk (1961) demonstrated that the relative intensity of seiches oscillations in harbors and bays is determined primarily by the Q-factor of the corresponding basins. Reducing the harbor entrance by wave protection constructions increases the Q-factor, which results in a corresponding increase in the harbor oscillations. From this point of view it is quite understandable why the narrow inlet of Ciutadella experiences much stronger seiches than the open-mouthed inlets and harbors in the Balearic Islands.

Meteo-tsunamis are similar in appearance to ordinary tsunami waves (Figures 2.26 and 2.27) and can affect the coasts in a similar, damaging way, although as indicated earlier, the catastrophic effects related to the former are normally observed only in specific bays and inlets. It has been shown by Monserrat and colleagues (2006) that "normal" tsunamis and meteo-tsunamis have the same periods, the same spatial scales, and similar physical properties. The generation efficiency of both phenomena depends on the Froude number (Fr), with resonance taking place when Fr is approximately 1.0. Meteo-tsunamis are much less energetic than seismic tsunamis, which is why they are generally local, whereas seismic tsunamis can have globally destructive effects.

Rabinovich (1993) proposed that destructive extreme meteo-tsunamis observed in some places are generally the result of a combination of several resonant factors (e.g., the coincidence of resonant frequencies of the shelf and inner basin). The relatively small probability of such coincidences is the main reason why major meteo-tsunamis are infrequent and observed only at some

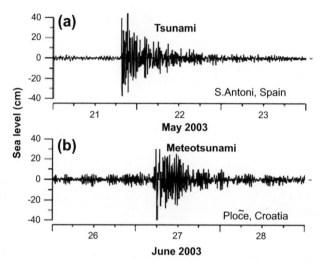

FIGURE 2.26 (a) Tsunami oscillations recorded at Sant Antoni (Ibiza Island, Spain) after the Algerian earthquake of 21 May 2003; and (b) meteo-tsunami recorded at Ploce Harbour (Croatia) on 27 June 2003. *Source: Monserrat et al., 2006; Natural Hazards and Earth System Sciences; Copernicus Publications.*

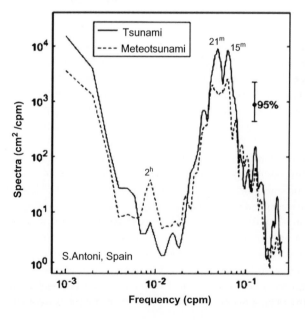

FIGURE   2.27 Sea level spectra for the tsunami of 21 May 2003 and the moderate meteo-tsunami event of 1 May 2003 recorded at Sant   Antoni   (Ibiza Island, Spain). *Source: Monserrat et al., 2006; Natural Hazards and Earth System Sciences; Copernicus Publications.*

FIGURE 2.28 The May-2005 *"Kallakka-dal"* at Poonthura coast of Kerala State, southwestern India, uprooting the coconut palms by large scale removal of soil from the coastal land. *Sources: Courtesy of Dr. N. P. Kurian, Center for Earth Science Studies, Kerala, India.*

specific locations in the ocean (Monserrat et al., 2006). There have been several instances of concurrent arrivals of tsunamis of geophysical and meteorological origins. Such an instance associated with the December 2004 global tsunami at Northwest Atlantic Ocean, and the wavelet analysis of these two waves are given by Thomson and colleagues (2007).

Meteorological tsunamis of a different kind that closely resemble rogue waves, freak waves, or giant waves and are locally known as *kallakkadal* have been reported by Kurian and colleagues (2009). In such cases, the sea suddenly invades the coastal land (Figure 2.28) without any local meteorological disturbances of any kind (either on a synoptic scale or on a meso-scale). *Kallakkadal* is the term used by fishermen in the State of Kerala in southwest India for incidents of this kind, which due to the absence of any precursors or warnings, usually inflict great material loss and physical and mental injuries. In Malayalam (the language spoken predominantly in the State of Kerala) *kallan* means "thief" and *kadal* means "sea." Thus, the two words are combined to produce *kallakkadal,* meaning "sea that arrives like a thief" (unnoticed).

Basically *kallakkadal* occurs on the southern coast of India, mainly during the months of April or May (pre—southwest monsoon season). The swell, which is generated in the southern ocean by storms near Antarctica, propagates northward into the Arabian Sea (and the Bay of Bengal), and when it encounters a coastal current directed southward, the swell is amplified through interaction with the current. Due to the increased wave setup, low-lying areas on the coast get flooded. Flooding becomes more severe when *kallakkadal* occurs on the days of spring tide. The flooding is not continuous all along the coast because the land topography everywhere might not be conducive to flooding. These flooding incidents appear to be more severe and more frequent on the southern Indian coasts than on the northern coasts. The main reason for this is the orientation of the coastline (Kurian et al., 2009). In the southern side of the Indian Peninsula, the coast curves in such a manner that the swell waves can meet the coast in a perpendicular direction.

Although the occurrence of *kallakkadal* is not well documented in the scientific literature, according to fishermen, this occurs almost every year. In recent years, it occurred in 2007, toward the end of April. It also occurred on February 9–10, 2008, although it was of relatively low intensity. The flooding and inundation arising from *kallakkadal* invite the attention of the media and the government only when it severely affects the local population and causes colossal damage to their property. Needless to say, some of the events go unnoticed and undocumented.

The *kallakkadal* of May 2005 (Figure 2.29) is perhaps the most intense one in the recent years and is well documented (Narayana and Tatavarti, 2005; Baba, 2005; Murty and Kurian, 2006). As shocking memories of the December 2004 monstrous tsunami were deeply imprinted in the minds of the people, together with the fact that the impact of the May 2005 *kallakkadal* was of a similar nature, the inundation created much panic and chaos, and therefore the event received considerable multimedia coverage in the State of Kerala. The May 2005 *kallakkadal* marooned almost all of the low-lying areas of Kerala

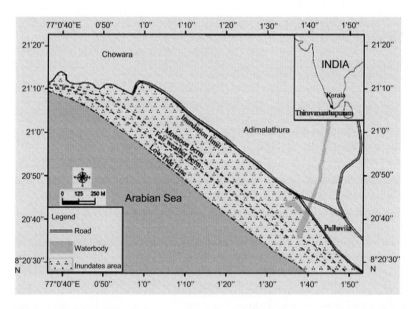

FIGURE 2.29 Inundation-limit of May-2005 *Kallakkadal* at Adimalathura, near Thiruvananthapuram on the coast of Kerala, India. The inundation observed here is a maximum of 435 m from the Low Tide level. *Source: Kurian et al., 2009; Reproduced with kind permission of Springer.*

and the southern Tamil Nadu coasts. The following are the salient features of a *kallakkadal* (Kurian et al., 2009):

- Occurs mostly during premonsoon season and occasionally during post-monsoon season
- Continues for a few days
- Inundates low-lying coasts
- Run-up during high tide can be as much as 3–4 m above MWL
- Associated wave characteristics are typical of swells with moderate heights (2–3 m) and long periods (~ 15 s)
- Occurrence found more on the southwest coast of India than other coasts

To illustrate a *kallakkadal* in more detail, Figure 2.29 compares the inundation limit of the May 2005 event with a monsoon berm, a fair weather berm, and a low-tide line at a specific location on the coast of Kerala. At present, no operational prediction system exists for kallakkadal-type events. Prediction of the events of this type is quite different from the forecasting of storm surges and rissaga because the latter events are triggered by a local weather system, either on a synoptic scale or on a meso-scale. For kallakkadal-type events, the forcing is entirely remote, and a local weather system is absent. Coupled with the fact that wave-current interactions are also involved, predicting kallakkadal-type events becomes all the more difficult. Extreme meteo-tsunamis in various regions of the World Ocean are shown in Table 2.3.

2.4.1. Meteo-tsunami Warning

In the past, little attention has been paid to tsunamis of meteorological origin in comparison to seismically-, volcanically-, or landslide-generated tsunamis. The situation began to change drastically after the devastating megathrust 2004 Sumatra tsunami that killed roughly 230 thousand people (Bernard et al., 2006; Bernard and Robinson, 2009). Firstly, this extreme catastrophic event attracted high public and scientific interest toward tsunamis and other marine natural hazards in general. Secondly, this event initiated a major upgrade of existing sea-level gauge networks around the globe. The new digital instruments were designed to continuously measure sea-level variations with high precision and to store the sea-level record once every 1–2 minutes. According to Rabinovich and colleagues (2009), the newly upgraded coastal sea-level gauges enabled scientists to measure relatively high-frequency local seiches in regions previously inaccessible to such investigations. Moreover at certain sites, the sea-level measurements were accompanied by simultaneous precise observations of atmospheric pressure fluctuations. As a consequence of the perceived increase in global warming, it may be speculated that the number of meteo-tsunamis will increase in the future due to the increased thermal energy in the atmosphere which feeds the atmospheric instabilities and favors the severity of a mesoscale event (Beniston et al., 2007). On the other hand, one

TABLE 2.3 Extreme Meteorological Tsunamis in Various Regions of the World Ocean

Region	Local Name	Typical Period	Maximum Height (m)
Nagasaki Bay, Japan	Abiki	35 min	4.78
Pohang Harbor, Korea	—	25 min	>0.8
Longkou Harbor	—	2 h	2.93
Cuitadella Harbor, Menorca Island, Spain	Rissaga	10.5 min	>4.0
Gulf of Trieste, Italy	—	3.2 h	1.56
West Sicily, Italy	Marubbio	14.6 min	>1.5
Malta, Mediter.	Milghuba	~20 min	~1.0
West Baltic, Finland coast	Seebar	—	~2.0
Vela Luka and Stari Grad Bays, Adriatic	—	10—15 min	~2.5
Southwest coast of India	*Kallakkadal*	~15 s	3.0

Source: http://www.theplasmaverse.com/pdfs/meteorological-tsunamis-destructive-atmosphere-induced-waves-observed-in-the-worlds-oceans-seas-seiches-abiki-milghuba.pdf; Source for last row: Dr. N. P. Kurian, Centre for Earth Science Studies, Kerala, India.

cannot form any conclusion without long-term and continuous monitoring on the process scale.

2.4.1.1. Importance of High-Resolution Atmospheric Pressure Measurements

As indicated earlier, a meteo-tsunami needs to feed its energy from a small-scale air pressure disturbance (gravity wave, pressure jump, frontal passage, squall line, etc.) propagating with certain speed and direction over the region. Two mechanisms that support long-distance propagation of atmospheric gravity waves are wave duct and wave-CISK. The first mechanism was found to be solely responsible for the Balearic rissaga events (e.g., Monserrat and Thorpe, 1992; 1996) whereas wave duct and wave-CISK were found to be responsible for the 2007 Ist and 2003 middle Adriatic events, respectively (Sepic et al., 2009; Belusic et al., 2007). If one of these mechanisms is present, then a disturbance can be sustained over a few hundred kilometers, and it may be captured by the surface air pressure measurements. Atmospheric pressure gradient (pressure temporal changes) plays an important role in strengthening

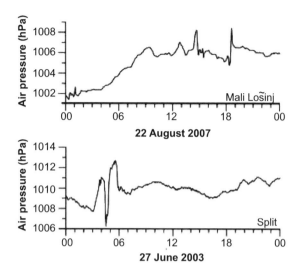

FIGURE 2.30 Surface air pressure series measured with a 2-minute resolution during (a) the Ist meteo-tsunami in 2007 at Mali Losinj, and (b) the middle Adriatic meteo-tsunami in 2003 at Split. These time-series measurements were used for high-frequency air pressure data analysis. *Source: Vilibic and Sepic, 2009; Reproduced with kind permission of Elsevier.*

the meteo-tsunami waves (Vilibic, 2005). However, to observe these disturbances and to properly resolve the high-frequency pressure oscillations, one needs non-standard meteorological measurements on 1-minute timescale (Vilibic and Sepic, 2009). In the absence of such digital measurements, the required high-resolution data can be extracted by careful digitization of analog air pressure charts. Figure 2.30 displays such time series obtained for the 2007 Ist meteo-tsunami as well as the series measured with a 2-minute resolution during the 2003 middle Adriatic meteo-tsunami. In the case of meteo-tsunami at the Ist Island, the atmospheric gravity wave of a narrow width (less than 100 km) was capable of provoking the strongest long ocean wave just over the major axis of the disturbance (Vilibic and Sepic, 2009). As expected, sea-level high-frequency oscillations exceeding 40 cm (induced by the air pressure disturbance) were found to be superimposed on the long meteo-tsunami ocean waves measured at coastal sea-level gauges.

As a result of the development of new digital instruments having the capacity to continuously measure sea level variations with high precision and to store the sea-level record once every 1–2 minutes, high-frequency records of meteorological tsunamis now exist for many areas of the World Oceans. For example, analysis of coincident 1-min sea level data and high-frequency atmospheric pressure data confirmed that the tsunami event of 9 December 2005 along the coast of British Columbia (Canada) and Washington State (USA), which was sufficiently strong to trigger an automatic tsunami alarm, originated with atmospheric pressure jumps and trains of atmospheric gravity waves with amplitudes of 1.5 to 3 hPa. Also, stimulating interest in this problem are the several recent destructive events which occurred in the vicinity of the Balearic Islands. It is believed that a combination of high-resolution in-situ measurements from a network of sea-level and surface-meteorological stations, satellite images of meteorological features favorable for meteo-tsunami generation, together with thoroughly validated meteotsunami detection algorithms can be effectively used in future meteo-tsunami warning systems. Such centers will have the capacity to provide coastal managers with data to better inform local hazard risk assessment for beach users and coastal communities.

Meteo-tsunami warning has been operational for the Balearic rissagas for more than 20 years, having been established after the 1984 destructive meteo-tsunami (Jansa et al., 2007). The rissaga warning system that exists in the Balearic Islands is primarily based on the synoptic-scale data. The conditions favorable for rissaga development are found to be the intrusion of warm air from Sahara in the levels near the surface and strong middle-level south-westerly wind (Monserrat et al., 1991). These conditions may lead to the formation of atmospheric gravity waves, which, if propagating with a certain phase speed, can generate meteotsunamis. According to Belusic and Mahovic (2009), due to a very complex orography of the eastern Adriatic coast such a set of favorable conditions should be defined for every bay or inlet in order to be able to detect these conditions and to produce a warning.

The strength of a meteo-tsunami is largely dependant on the topographical characteristics of the affected area. The appearance of a destructive meteo-tsunami is a multi-resonant phenomenon in which a number of specific condi-tions have to be satisfied (Monserrat et al., 2006). Prior to developing a method for meteo-tsunami prediction, it is necessary to investigate the common prop-erties related to meteo-tsunamis in general. Based on several previous studies, the conditions required for the occurrence of a meteo-tsunami are as follows (Belusic and Mahovic, 2009; Vilibic and Sepic, 2009):

- A sudden small to mesoscale atmospheric pressure- or wind- disturbance propagating over the sea (a small-scale atmospheric pressure jump visible in surface air pressure series).
- The propagation speed of the atmospheric disturbance needs to be close to the phase speed of the barotropic open sea waves.
- Occurrence of external resonance between the atmospheric disturbance and long ocean waves
- Propagation of the disturbance needs to be directed toward the mouth (entrance) of the harbor/bay/inlet
- Eigenfrequency of the harbor/bay/inlet matches the frequency of the incoming open sea waves so that internal resonance can occur between the arriving long ocean waves and the harbor/bay eigenmodes
- The harbor/bay/inlet has a large amplification (Q) factor.

The spectra of the atmospheric disturbances in the case of the Adriatic meteo-tsunami reveals a substantial energy at the harbor or bay eigen-frequencies (e.g., Sepic et al., 2009), this being a favorable condition for the intense amplification of the arriving long ocean waves through harbor resonance. The heavy monetary impact inflicted by meteotsunamis opens a question of how to assess future risk and how to mitigate the impact. However, building of a meteo-tsunami warning system in the Mediterranean Sea requires latency (i.e., time between the detection of the potential tsunami waves and alerting the population) to be lower than 5 minutes (ICG-NEAMTWS, 2008). The future warning system should therefore include (Vucetic et al., 2009):

- A real-time assessment and watch service for meteo-tsunami threats
- A warning service for alerting the coastal population
- Education activities to inform the population on how to mitigate meteo-tsunami threats

According to Vucetic and colleagues (2009), the first part of such a service is particularly hard to develop as it demands non-standard ocean and atmospheric measurements which are not easy to upgrade to the real-time level. The efficiency of applicable numerical models in research, warning, and mitigation activities should also be improved to an operational level, starting with the production of better bathymetry charts in the most affected coastal regions. Meteo-tsunami researchers are optimistic that such an approach may ultimately be developed on a basin scale (e.g., for the Adriatic Sea) and may become a part of the Mediterranean tsunami warning system in the near future.

According to Vilibic and Sepic (2009), apart from the assessment and forecasting of synoptic conditions, the meteo-tsunami research should include high-resolution air pressure measurements (sampling interval 1-min or lower) at a number of microbarograph stations close to the affected areas. Such real-time measurements may be useful to provide an indication of a possible impending meteo-tsunami. For example, atmospheric pressure measurements in Brazil from three locations, separated by a few hundred kilometers, during the September 2002 meteo-tsunami event (Figure 2.31) provided a clear indication of a strong pressure gradient (pressure jump of up to 10 hPa/2 h) traveling with a speed of approximately 31 m/s. In a similar study, passage of atmospheric pressure perturbations at two Balearic group of islands have been used to examine the external forcing responsible for the above-normal seiches that are potentially dangerous to the infrastructure and boats in Ciutadella Harbor (Menorca Island, Western Mediterranean). From this study, particularly based on the direction of propagation of atmospheric disturbances, it was found that sea-level measurements at Cala Ratjada (CRA) could be used to forecast destructive events in Ciutadella Harbor (Figure 2.32) as a part of a Mediterranean Tsunami Warning System ICG/NEAMTWS (Marcos et al., 2009). During specific synoptic meteorological situations, trains of atmospheric pressure gravity waves travel from southwest to northeast across the Mediterranean (Monserrat et al., 1991). When these disturbances propagate with a phase speed of about 22–30 m/s, resonant conditions occur on the southeastern shelf of Mallorca Island and kinetic energy is efficiently transferred from the

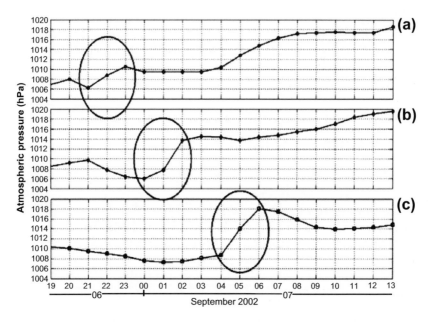

FIGURE 2.31 Atmospheric pressure records for (a) Cananeia, (b) Ubatuba, and (c) Arraial do Cabo in Brazil leading to the September 6 and 7, 2002 meteo-tsunami event. The pressure jumps are marked on each record indicating that there has been a deepening of the system along the coast. *Sources: Candella 2009; Reproduced with kind permission of Elsevier.*

atmosphere into the ocean (Garcies et al., 1996). Arriving at the coast these open ocean long waves may significantly amplify seiche oscillations inside bays and inlets due to harbor resonance (Figure 2.33). It was found that the destructive seiches (rissaga waves) in Ciutadella Harbor (Menorca Island) are normally accompanied by above-normal seiches on the Coast of Mallorca Island (Rabinovich and Monserrat, 1996, 1998). Moreover, these seiches commonly begin 1 to 1.5 hour earlier than in Ciutadella, thereby hypothetically enabling the use of the former as a predictor for rissaga phenomenon in Ciutadella Harbor (Marcos et al., 2009).

At least three stations should be distributed in a triangle in order to estimate properly the speed and the direction of an atmospheric disturbance. Measurements from such multiple stations are necessary for the assessment of meteo-tsunami generation potential on the spatial and temporal scales of the process; i.e., for the assessment of the gravity waves and various atmospheric disturbances capable of provoking a meteo-tsunami (Monserrat and Thorpe, 1992). The gravity waves and the associated atmospheric disturbances are highly variable both in space and time, with a spatial scale of a few hundred kilometers at the most (Vilibic and Sepic, 2009). From a meteorological perspective, the tracking of convective clouds in real time may also be used for detecting the favorable conditions for a meteo-tsunami (Belusic and Strelec-Mahovic, 2009).

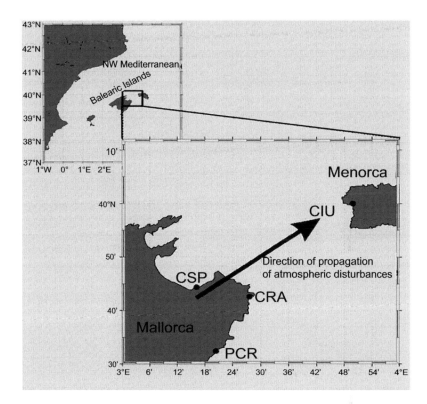

FIGURE 2.32   Map of Mallorca and Menorca islands in the Western Mediterranean, where the direction of propagation of atmospheric disturbances influences the sea-level elevations at these two islands and therefore the sea-level measurements at Mallorca have the potential to be used to forecast destructive meteo-tsunami events at Menorca. *Sources: Marcos et al., (2009); Reproduced with kind permission of Elsevier.*

2.5. TSUNAMIS GENERATED BY UNDERWATER GAS EMISSION

In old documents describing the 1854 Ansei Nankai earthquake and the 1946 Showa Nankai earthquakes, Tsuji and Tachibana (2009) learned that there appeared "columns of fire" or huge burning columns of clouds above the sea surface at several villages on the coast of Kii Peninsula and Shikoku Island. Some of the descriptions, which Drs. Tsuji and Tachibana presented at the 24th International Tsunami Symposium at Novosibrisk, Siberia, Russia, are shown in Table 2.4. The eyewitness accounts of the fire columns that appeared just before the tsunami, the gas bubbles that rose to the surface from the bottom of the sea on the days after the tsunami, the presence of hot water in association with the tsunami, and the dead fish that appeared to have been "boiled" provide a credible indication of tsunamis causing the emission of inflammable gases from the seafloor. Places where fire columns were observed in association

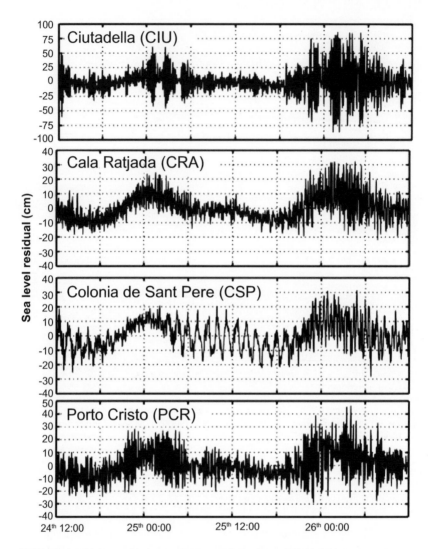

FIGURE 2.33 Sea level residual oscillations at CIU, CRA, CSP and PCR at Menorca and Mallorca islands in the Western Mediterranean during the rissaga meteo-tsunami events of May 2008. Note the different vertical scales for CIU. *Sources: Marcos et al., (2009); Reproduced with kind permission of Elsevier.*

with the 1854 Ansei-Nankai earthquake and the 1946 Nankai earthquake are shown in Figures 2.34 and 2.35, respectively. Posttsunami field survey notes (Table 2.5) also provide credible indications of tsunami generation as a result of intense gas emission from the seafloor.

According to eyewitness accounts, the fire columns (Figure 2.36) were accompanied by a sound like the roaring of a canon. Geological surveys have

TABLE 2.4 Eye-witness Account of Fire Columns That Appeared Before the Tsunami Attacked the Coast

Source	Eye-witness Description
Tekeuchi Denshichi's diary	After shaking, a roaring like a canon firing was heard from offshore, a bright column looking like a fire column appeared, and after a while the tsunami wave came.
Record of earthquake and a tsunami in 1854 at Minabe Town, Wakayama Prefecture	After shaking, at about 16 o'clock, a fire column appeared on the southwest horizon, just after that a huge tsunami came.
Hamaguchi Goryo's memo, Hiro, Wakayama Prefecture	A fire column, about three meters high and diameter one foot appeared on the sea surface around Karimo Island.
History of Yuasa Town	In southward offshore, a large fire vomited from the sea surface with a big noise, and 5 ~ 6 black balls jumped up from the sea surface. Due to the dark smoke of the fire, Karimo Island became invisible.
Letter of Kaheiji Ichien, Kashiwajima, Kochi Prefecture	Above the sea near Okinoshima Island, a black cumulonimbus cloud fired up; one side of it, devouring flames fired up.
Chronicle of Fukuchido temple, Tenricity, Nara Prefecture	A large shaking happened in the evening. At that time a large black cloud rimmed with red color appeared in the southwest sky.
A fisherman in Tsubaki hot spring, Tanabe town	At 3 o'clock in the morning, two fire columns appeared offshore Shirahama town and Susami town. The sea surface at the root of the latter column looked like a hollow plate.
Inami Town, Wakayama Prefecture	Two bright fire columns appeared before the tsunami attacked the coast.
A fisherman	Bubbles were coming up from the bottom of the sea after the day of the tsunami till the end of September at a point 10 kilometers away from the lagoon mouth.
Another fisherman	A light flashed from the sea bottom at the time of tsunami generation.
Saveral people in Arop 1 and 2 villages	Roaring of sound like from an airplane or a drum was heard.
Tsunami survivors	The seawater was hot, and a few people had traces of scalds on their bodies. In the next morning after the tsunami, many dead fishes (looking like boiled ones) were noticed in the sea.

Source: Courtesy of Drs. Yoshinobu Tsuji, Earthquake Research Institute, University of Tokyo; and Toru Tachibana, Soil Engineering Corporation, Okayama, Japan.

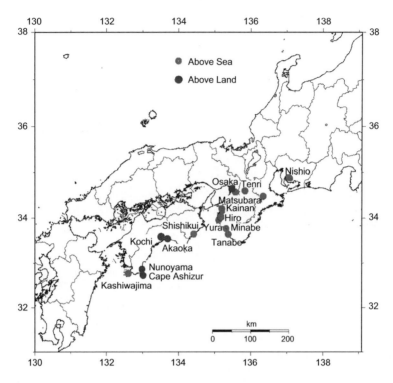

FIGURE 2.34   Places where fire columns were observed in association with the 1854 Ansei-Nankai earthquake. *Source: Courtesy of Drs. Yoshinobu Tsuji, Earthquake Research Institute, University of Tokyo; and Toru Tachibana, Soil Engineering Corporation, Okayama, Japan.*

indicated the presence of a layer of methane hydrate that is distributed offshore of the coast of those districts where fire columns and tsunamis have been observed. Based on this finding, it has been inferred that the probable cause of formation of the fire columns could have been a burst of inflammable methane hydrate contained in the seabed layer and the gushing up of the associated methane gas from the seabed to the sea surface. There have been instances in which the dissociated gas hydrate played a role in inducing landslides. The mechanism of landslide inducement by dissociated gas hydrate and methane hydrate bursting caused by a submarine landslide is shown in Figure 2.37. Tsunami generation by methane seeping bubbles arises from a pressure balancing mechanism. Pressure balancing on the seafloor inside and outside of the gas-charged water column demands the following:

$$P = h_{gas} \times \rho_{gas} \times g = h_{water} \times \rho_{water} \times g \qquad (2.2)$$

where:

P: hydrostatic pressure at the seafloor
h_{gas}: instantaneous height of the gas-charged water column above the seafloor

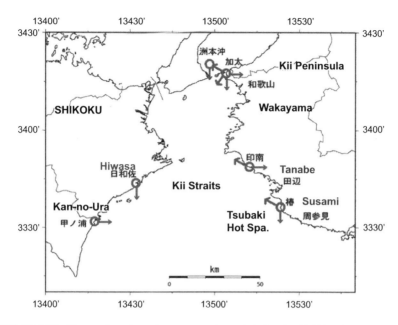

FIGURE 2.35 Places where fire columns were observed during the 1946 Nankai earthquake. *Source: Courtesy of Drs. Yoshinobu Tsuji, Earthquake Research Institute, University of Tokyo, and Toru Tachibana, Soil Engineering Corporation, Okayama, Japan.*

h_{water}: instantaneous height of the seawater column (exterior to the gas-water mixture) above the seafloor

ρ_{gas}: density of the gas-charged water column

ρ_{water}: density of the water column outside the gas-water mixture

g: gravitational acceleration

Because $(\rho_{gas}) < (\rho_{water})$, the fluid dynamical requirement of maintaining pressure balancing at the seafloor (i.e., maintaining the same seafloor pressure P inside and outside of the gas-charged water body) demands that $(h_{gas}) > (h_{water})$. The impulsive increase in water height locally over the gas-emitting region by this mechanism gives rise to the generation of an impulse, which is modified as a smooth bump on the sea surface and transmitted as a tsunami wave. In the light of the preceding observations, it has been argued that strong shaking (say, due to an earthquake or a landslide) of the seabed layer, which is impregnated with rich methane hydrate and associated gas, can result in the generation of a burst and exhalation of a large volume of inflammable gas, which in turn can trigger tsunami or tsunami-like waves. There are instances where methane plumes and epicenters of large earthquakes coexist or are located relatively close together spatially. In such places, an earthquake can trigger strong emissions of methane gas plumes, which in turn can give rise to powerful landslides.

TABLE 2.5 Field Notes of Posttsunami Surveys

Location	Field Note Description
Pantai Congot, Lawang, Temon Town (Lat: 7° 53′ 59.5″ S, Long: 110° 02′ 02.3″ E)	Tsunami height=5.2 m. No earthquake felt. Vertical yellow-colored smoke akin to atomic bomb explosion appeared.
Pantai Nam Prejo, Jatimalang Town (7° 52′ 44.7″ S, 109° 58′ 59.9″ E)	Tsunami height =5.7 m (from marks on concrete wall). No shaking, nobody killed. Lights flahsed 3 times akin to an atomic bomb explosion. A pencil-like vertical smoke column was observed at the horizon.
Pantai Keburuhan, Kaptaijaya (7° 51′ 15.0″ S, 109° 54′ 47.2″ E)	Tsunami height = 4.5 m. Seawater went upstream 3 kilometers from the shoreline. No shaking of earth, nobody killed. A vertical smoke column was observed.

Source: Courtesy of Drs. Yoshinobu Tsuji, Earthquake Research Institute, University of Tokyo; and Toru Tachibana, Soil Engineering Corporation, Okayama, Japan.

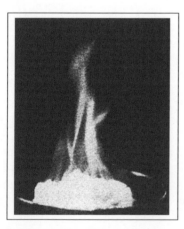 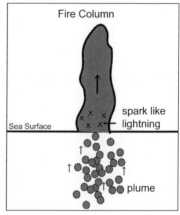

FIGURE 2.36 Illustration of fire column generated from the burning of methane gas gushed up to the sea surface from the methane hydrate below the sea bed. *Source: Courtesy of Drs. Yoshinobu Tsuji, Earthquake Research Institute, University of Tokyo, Japan; and Toru Tachibana, Soil Engineering Corporation, Okayama, Japan.*

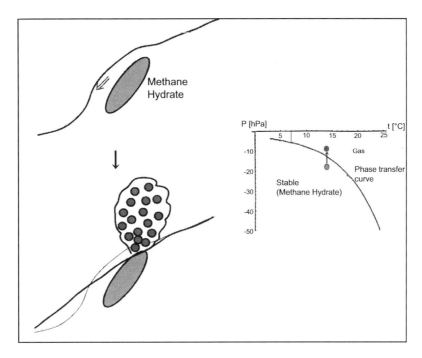

FIGURE 2.37 Bursting of methane hydrate induced by a submarine landslide. *Source: Courtesy of Dr. Yoshinobu Tsuji, Earthquake Research Institute, University of Tokyo, Japan; and Toru Tachibana, Soil Engineering Corporation, Okayama, Japan.*

Whereas methane gas emissions have been found to have the potential for tsunami generation, it has been discovered that a mixture of carbon dioxide, nitrogen, and methane also emits from some locations on the seafloor. For example, natural seepage of such a mixture of gases of volcanic origin at the rate of approximately 48,500 ton/year has been identified from the seafloor in Wakamiko Caldera in the Kagoshima Bay off the Japanese coast (Harata et al., 2010). The Japanese coast is vulnerable to earthquakes, and an earthquake event occurring at the seafloor at or in the immediate vicinity of such natural gas reserves may lead to tsunamis. In such cases, the generated tsunami wave height can be the sum of the contributions from all three sources: earthquake, gas emission, and landslide.

On July 17, 2006, an earthquake took place in the south of Central Java, Indonesia. After the earthquake, a large tsunami hit the southern coast of the central part of Java Island, and about 660 people were killed. Seawater rose to a height of 7.7 meters above the mean sea level (MSL) at Panbandaran. It is curious that the tsunami was too large for the earthquake magnitude (7.2), and the fault model obtained by sea-level records could not explain such a large run-up height. Moreover, a mushroom-shaped cloud (probably gas emitted from the seafloor) was observed in the offshore sea area from the coast south of Jogyakarta.

Professor Viacheslav K. Gusiakov, an active tsunami scientist who spearheaded tsunami research at the Siberian Branch of the Russian Academy of Sciences and was primarily responsible for holding the two landmark International Tsunami Symposia at Novosibirsk, Russia, in 1989 and 2009, pointed out, "Magnitudes of several tsunamis are too large for their earthquake magnitude, and they cannot be explained based on tectonic crustal motion. Many of them are suggested to be secondary tsunamis induced by submarine landslides." Tsuji and Tachibana (2009) have suggested the possibility that a methane hydrate burst makes the magnitude of a submarine landslide tsunami additionally larger.

2.6. TSUNAMIS GENERATED BY ASTEROID IMPACTS

In addition to oceanic seismic activities, submarine landslides, volcanic eruptions, and meteorological sources, evidences of substantive impacts into the ocean by sizable comets and the physical and environmental effects resulting from such asteroid impacts have been reported in the literature. Although an asteroid impact is beyond human response, examining such events that have occurred on our planet in the past would satiate the human "curiosities" and are therefore worthy of mention, at least in academic interest.

Impact of a large asteroid (estimated to be approximately 10 km in diameter) at Chicxulub, on the Yucatan peninsula, Mexico (which occurred within the time of Deccan flood basalt volcanism in India) generated a crater (called Chicxulub crater) about 180−200 km in diameter. This impact triggered mass flows and tsunamis in the Gulf of Mexico and adjacent areas over 65.5 million years ago (Ma). Convincing evidence that reinforces the occurrence of the Chicxulub event continues to emerge even today. For example, in 1980, a team of researchers lead by the Nobel prize-winning physicist Luis Alvarez discovered that sedimentary layers found all over the world at the Cretaceous-Tertiary boundary (called K-T boundary, which is a geological signature dated to approximately 65.5 Ma) contain a concentration of Iridium (a rare metal) 30−160 times greater than normal. As Iridium is extremely rare in the earth's crust, but remains abundant in most asteroids and comets, the Alvarez team suggested that an asteroid struck the earth at the time of the K-T boundary. Till date, the ejecta-rich deposits (which are compositionally linked to the Chicxulub impact), have been found to be globally distributed over more than 350 marine and terrestrial sites. The observational evidence suggesting that the ejecta distribution pattern is distinctly related to distance from the Chicxulub crater points to an impact event in the Gulf of Mexico−Caribbean region (Schulte et al., 2010).

Experimental data about tsunamis generated from large-scale collisions is not available. For this reason, one of the main tools used is mathematical modeling. Asteroid impact models have predicted that an impact large enough to generate the Chicxulub crater would induce earthquakes (magnitude > 11), shelf collapse around the Yucatan platform, and widespread tsunamis sweeping the coastal zones of the surrounding oceans (Schulte et al., 2010). Moreover, models have suggested that the Chicxulub impact had sufficient energy to eject and distribute materials around the globe, possibly enhanced by decomposition of the volatile-rich carbonate and sulfate sediments. Near-surface target material was ejected ballistically at velocities of up to a few km/s as part of the ejecta curtain. This yielded the thick impact-derived ejecta materials layer at several spatially distributed geographical sites. Parts of the impact-derived ejecta would be entrained within the impact plume, which is a complex mixture of hot air; projectile material; and impact-vaporized, shock-melted, and fragmented target rocks. Because of the presence of sulfur-bearing sediments in Chicxulub, about 100–500 Gt of sulfur was presumably injected into the atmosphere within minutes after the Chicxulub impact. The sulfur in the ejecta was probably rapidly transformed to sunlight-absorbing sulfur aerosols with the capacity to cool Earth's surface for years to decades by up to 10 °C. The sulfur release also generated acid rain, which would have severely affected marine surface waters, continental catchments, and watersheds. It has been suggested that the extremely rapid injection rate of dust and climate-forcing gases that emanated from the Chicxulub impact would have magnified the environmental consequences.

Subsequent to the estimated period of occurrence of the impact episode, several proximal areas around the northwestern Gulf of Mexico from 500 to 1000 km from Chicxulub are characterized by a series of cm- to m-thick impact-derived ejecta. These ejecta are indicative of high-energy sediment transport and are believed to be from tsunamis and high-gravity flows from the impact site. Analysis of Chicxulub impact ejecta materials suggests that the impact episode caused a collapse of the Yucatan carbonate platform and triggered mass flows and tsunamis in the Gulf of Mexico and adjacent areas. Sedimentological and petrological (i.e., pertaining to hard rock, including volcanic rock) data suggest that the lens-like ejecta deposits in Mexico were generated by impact-related liquefaction and slumping, consistent with the single very-high-energy Chicxulub impact. Synthesis of recent stratigraphic, micropaleontological, petrological, and geochemical event-deposits data from the globally distributed sites suggested that the size of the crater and the release of climatically sensitive gases from the carbonate- and sulfate-rich target rocks could have caused catastrophic environmental effects such as extended darkness, global cooling, and acid rain. These effects have provided an array of potential mechanisms for the ecologically diverse but selective abrupt extinctions. The temporal match between the ejecta layer and the onset of the extinctions and the agreement of ecological patterns in the fossil record with modeled environmental perturbations (for example, darkness and cooling) led Schulte and colleagues (2010) to conclude that the Chicxulub impact triggered the most devastating mass-extinction event in Earth's history and abruptly ended the age of the dinosaurs. Analyses of fossil records reveal that several other major groups suffered considerable, although not complete, species-level extinction. For marine phytoplankton (which are the major drivers of ocean productivity), darkness and suppression of photosynthesis were likely major killing mechanisms. On land, the Chicxulub impact resulted in the loss of diverse vegetation, destruction of diverse forest communities, abrupt mass extinction, and ecosystem disruption.

It has been suggested that the only "wet" crater for which a tsunami has definitely been proven is Chicxulub. This tsunami was most likely caused by multiple slumps on the continental shelf that was made unstable by the violent shaking produced in the very large impact. However, recent studies indicate the possible occurrences of asteroid-impact-induced tsunamis subsequent to the Chicxulub episode.

It is believed that two sets of oceanic impact events have occurred within the last 5000 years, one in the Indian Ocean about 2800 BC and the other in the Gulf of Carpentaria (Australia) about AD 536, causing the most energetic natural catastrophes during the late Holocene, with large-scale environmental and historic human effects and consequences. In southern Madagascar there is evidence for oceanic water run-up reaching 205 meters with inland penetration up to 45 km, which is far beyond the run-ups of any historical tsunami. In field studies, a number of features have been found that are well explained by flooding that resulted from mega-tsunami waves caused by asteroid impacts (Gusiakov et al., 2009; Roberts, 2009). Incidences of a few other asteroid-induced tsunamis have also been reported in the literature. For example, tsunamis of asteroid origin have occurred in the Barents Sea (Shuvalov et al., 2002) and at Kaluga (Masaitis, 2002), which was covered with sea at that time. At present, much attention is focused on tsunami occurrences of asteroid origin (Kharif and Pelinovsky, 2005).

Recent model studies have provided more useful insights on tsunami generation by asteroid impacts. For example, Simonenko and colleagues (2009) considered a 1-km-diameter stone asteroid falling into a 4-km-deep ocean at a speed of 22 km/s at angles 30, 60, and 90 degrees, as well as axisymmetrically into the ocean. Assessment of a cylindrically diverging surface wave and its impact on the shore indicated that the aftereffects of the falling of a stone asteroid of about 1 km diameter are likely to be destructive to the ocean shore. Calculations showed that the wave height on the shelf may increase from 60 to 100 m followed by a decrease of wave height on the shallow water.

In another study by Gisler and colleagues (2009), hydrodynamic calculations of impacts into both deep and shallow seas indicated that contrary to popular belief, ocean impacts of asteroids do not produce tsunamis that lead to worldwide devastation. According to these researchers, the most dangerous features of ocean impacts are the atmospheric effects, just as for land impacts.

Chapter 3

Tsunami Databases

Tsunami hazards have widespread social consequences as a result of the vulnerability of populated areas, especially coastal communities. In contrast to many other natural disasters, tsunamis may produce catastrophic effects thousands of kilometers away from the source area. With nearly 230,000 fatalities, the December 26, 2004, Indian Ocean tsunami was the deadliest tsunami in history, illustrating the importance of developing basinwide warning systems. The global reach of this tsunami generated a significant increase in scientific and public interest in this natural hazard. Collecting the data on observed tsunamis and cataloguing of historical information on tsunamis is the key problem for scientific examination of tsunami waves, estimation and mitigation of tsunami risk for the coastal areas, and improvement of the existing tsunami warning systems. Further, key to creating basinwide warning systems is easy access to quality-controlled and verified data on past tsunamis. This is because warning centers, emergency managers, and modelers need to know if, and when, similar events have occurred in the past (Dunbar, 2009).

Historical tsunami sea-level records help us chronicle the nature of a tsunami starting soon after it is first generated, often by an earthquake, and as it impacts coastal shores nearby and then far away across oceans and sometimes into other basins. Historical data on tsunami occurrence and coastal run-up are important for basic understanding of the tsunami phenomenon, its generation, propagation and run-up processes, and its damaging effects. Such data are widely used for evaluating the tsunami potential of coastal areas and for determining the degree of tsunami hazard and risk for use in coastal-zone management and disaster

Tsunamis: Detection, Monitoring, and Early-Warning Technologies. DOI: 10.1016/B978-0-12-385053-9.10003-1

preparedness. Also, historical data are of critical importance for real-time evaluation of underwater earthquakes by the operational tsunami warning centers, for the establishment of thresholds for issuing tsunami warnings and for design-criteria for any tsunami-protective engineering construction. Thanks to significant efforts by Sergey Soloviev, James Lander, Patricia Lockridge, Viacheslav Gusiakov, and their colleagues, large databases and reliable catalogues exist for almost the entire coast of the Pacific Ocean, where about 85% of all tsunamis have occurred. The oldest and the longest databases of historical tsunamis and related natural hazards are maintained by the United States and Russia. The Canadian coast was probably the only region without a comprehensive database of historical tsunami data. In an effort to fill this "white spot" and to complete the existing tsunami database, S. O. Wigen began the process of cataloguing tsunamis for British Columbia, Canada, but he died in 2000, before it was completed. However, under the mutual cooperation of the Canadian Hydrographic Service and the Russian Academy of Sciences, Stephenson and colleagues (2007) finished the job. Recently, UNESCO's Intergovernmental Oceanographic Commission and the International Tsunami Information Center (UNESCO/IOC/ITIC) and Japan Nuclear Energy Safety Organization and the Disaster Control Research Center of Tohoku University have also begun to create a tsunami database by collecting historical materials and documents concerning tsunamis pertaining to the regions under their purview. Greece has also developed a database of tsunamis, but it is limited to the Mediterranean Sea.

3.1. THE UNITED STATES TSUNAMI DATABASE

One important source of information on significant earthquakes, volcanic eruptions, and global historical tsunamis are the natural hazard catalogues produced by the National Oceanic and Atmospheric Administration (NOAA) National Geophysical Data Center (NGDC). Dunbar and Stroker (2009) provided an account of the general features of the NGDC database. Historical databases do not cover long enough periods of time to reveal a region's full tsunami hazard, so a global database of citations to articles on tsunami deposits was added to the archive. The way to obtain long-term data to develop a predictive chronology of events is to study paleo-tsunami deposits. These deposits provide a proxy record of large earthquakes. Paleo-tsunami sediments can be chronologically dated using radiocarbon, optically stimulated luminescence, and thermoluminiscence dating methods. Darienzo and Peterson (1990) dated a series of tsunami deposits and determined the recurrence interval for subduction zone earthquakes and associated tsunamis. Studies of paleo-tsunami sediment sheets that rise in altitude inland as tapering sediment wedges (Dawson, 1994) have revealed important information about several past tsunamis (Clague et al., 1994; Atwater et al., 2005). At present, paleo-tsunami studies are being carried out in several sites along the east coast of India (Kumar et al., 2007).

NGDC further expanded the archive to include high-resolution tide gauge data, deep-ocean sensor data, and digital elevation models used for propagation and inundation. NGDC continuously reviews data for accuracy, making modifications as new information is obtained. The added databases allow NGDC to provide the tsunami data necessary for hazard assessments, mitigation efforts, and warning guidance. The data is also useful to scientists in other fields, such as storm surge modeling.

The NGDC global historical tsunami database includes over 2000 events, dating from 2000 BC to the present. The first database table contains information on the source, date, time, location, magnitude, and intensity (e.g. earthquake magnitude or volcanic explosivity index), and the tsunami intensity, magnitude, validity, and maximum water height. This table also includes detailed socioeconomic information such as the number of fatalities and injuries, the number of houses destroyed, and total financial loss. Where possible, the effects caused by the source are differentiated from the effects caused by the tsunami. The second database table includes information on the locations where tsunami waves were observed (e.g. run-ups, eyewitness observations, reconnaissance field surveys, sea-level gauges, or deep-ocean sensors) such as the water height, horizontal inundation, time of arrival, distance from the source, and socioeconomic information at the run-up location. Since 2004, NGDC has been carefully examining and verifying the information in the database with the original source references. As a result, accurate statistics can now be collected on these tsunamis. For example, an examination of the database reveals that over 400 tsunamis have caused damage, and 200 caused fatalities. Although the total effects are dominated by the tragic 2004 Indian Ocean tsunami, every ocean in the world has experienced fatal and damaging tsunamis. The majority of these effects resulted from regional or local tsunamis; in fact, only nine tsunamis have caused fatalities 1000 km from the source. All of these originated in the Pacific Ocean, except the 2004 Indian Ocean tsunami event. A comparison of fatal tsunamis and water heights reveals that 35% of tsunamis with over 1 meter maximum water height resulted in fatalities; 60% of tsunamis with over 5 meters water height caused fatalities (Dunbar and Stroker, 2009).

Long-term data from these events can be used to establish the past record of natural hazard event occurrences, which is important for planning, response, and mitigation of future events. Efforts such as NGDC's are transforming the perception of data centers as "dusty old archives" as scientists embrace new technologies for data preservation and access. Aside from archiving old analog records, NGDC is at the forefront of standards-based Web delivery of integrated science data through a variety of tools, from Web-form interfaces to interactive maps. Further, the data download options support a variety of desktop tools, including spreadsheets, ESRI ArcGIS, and Google Earth™. The majority of the data in the tsunami archive are available online at http://www. ngdc.noaa.gov/hazard/tsu.shtml. Scientists, journalists, educators, planners, and emergency managers are among the many users of this public domain data, which may be used without restriction provided that users cite data sources.

In order to meet the needs of local authorities and tsunami warning centers (TWCs) from countries around the world, Varner and Dunbar (2009) have reported a stand-alone geographic information system (GIS) application to interact with a local copy of the historical hazards database. The software is based on uDig (User-friendly Desktop Internet GIS), an open-source GIS framework written in Java that is built upon the well-established Eclipse Rich Client Platform (RCP). They have customized the behavior of uDig by developing plug-ins that allow the user to query the database using many different search parameters and to display information about events on the map and in a table format. The many ways to access data from the NGDC database are reported by Dunbar (2009).

3.2. THE RUSSIA TSUNAMI DATABASES

The Russian Novosibirsk Tsunami Laboratory also maintains a Global Historical Tsunami Database (GTDB). This database is available through http://tsun.sscc.ru/On_line_Cat.htm. The basic historical event information in such databases generally includes the date, time, and location of the event; the magnitude of the phenomenon (e.g. tsunami intensity, seaquake magnitude, volcanic explosivity index); and socioeconomic information such as the number of fatalities and degree of financial loss. The tsunami database includes an additional table with information on the locations where tsunami waves were observed (e.g., run-ups, eyewitness observations, reconnaissance field surveys, tide gauges, or deep-ocean sensors). The present version of the global catalogue on tsunami and tsunami-like events covers the period from 2000 BC to the present and currently contains 2139 historical events that have occurred in the world oceans during the last 4000 years, including 1206 of them in the Pacific, 263 in the Atlantic, 125 in the Indian Ocean, and 545 in the Mediterranean region. These events are responsible for nearly 700,000 lives lost in tsunami waves during the whole historical period of available observations (Gusiakov, 2009). This large bulk of historical data, although obviously incomplete for many areas, provides a good basis for assessment of tsunami hazards in the main tsunami-prone areas of the world oceans, as well as for retrieval of empirical studies of tsunami intensity with the earthquake source parameters.

3.3. THE UNESCO TSUNAMI DATABASE

Understanding how to read and interpret sea-level mareograms to confirm a tsunami's destructiveness is essential and is a part of the necessary standard operating procedures of every national TWC. The observed sea-level record is an indication of not only the strength of the tsunami but also the local conditions where the sea-level station was situated. Understanding how to interpret the tsunami and untangle and identify local resonance and the resulting wave amplification is especially important for TWC staff, who must be able to quickly interpret the records for tsunami potential and then decide whether the threat level is high enough to issue a warning alert. The duty staff therefore needs to be familiar with the characteristics of tsunamis, as recorded instrumentally, in order to consistently and correctly evaluate a tsunami's destructiveness and at the same time recognize other nontsunami signals, including those related to signal integrity and instrument failure. To build a learning database, UNESCO/IOC/ITIC have compiled historical sea-level records from past tsunamis in order to chronicle the anatomy of tsunamis in the near, regional, and far-field areas and from a variety of source zones around the world. The compilation is intended to provide an interpreted reference on historical mareograms from destructive tsunamis, and it is accompanied by timelines of actions and reactions by TWCs and affected countries. Until July 2009, compilations have been completed for the 1946 Aleutian Islands, 1960 Chile, 1975 Hawaii, 1998 Papua New Guinea, 2004 Sumatra, 2006 Kuril Islands, and 2007 Solomon Islands and Peru events (Igarashi et al., 2009).

3.4. THE JAPAN TSUNAMI DATABASE

Since 2007, the Japan Nuclear Energy Safety Organization and the Disaster Control Research Center of Tohoku University have been working on the development of a Tsunami Trace Database by collecting historical materials and documents concerning tsunamis that have hit Japan. The sources of the database are the two conventional historical tsunami databases: NGDC/NOAA and Novosibirsk Tsunami Laboratory. The database is proposed to be released over the Internet around the year 2011. To enable Internet browsing of trace information placed over a base of seamless map images, a Web-GIS function has been mounted on the system. Users will be able to download the PDF file of the original document via the Internet. This database is expected to be used effectively in tsunami assessment and also in the field of nuclear seismic safety and general tsunami disaster prevention (Iwabuchi et al., 2009).

3.5. THE GREECE TSUNAMI DATABASE

The Institute of Geodynamics, National Observatory of Athens, Greece, has compiled a catalogue of tsunamis in the Mediterranean Sea from antiquity up to the present, quantified them in terms of tsunami intensity, and performed statistics to assess repeat occurrences of tsunami events of certain intensity ranges (Papadopoulos et al., 2009). Intensity is by definition a measure of the event impact, and, therefore, tsunami intensity maps or other intensity attributes may provide tools for the description of tsunami risk that is of particular value for risk managers.

Geophysical Tsunami Hydrodynamics

Just as an earthquake is quantified in terms of Mw (formerly, in terms of the Richter scale), a tsunami is quantified in terms of two parameters: *Imamura-Iida tsunami magnitude scale*, m and *Soloviev-Imamura tsunami intensity, I*. These parameters are expressed as:

$$m = \log_2 H_{max} \qquad (4.1)$$

where H_{max} is the maximum observed trough-to-crest tsunami height, and:

$$I = 1/2 + \log_2 H \qquad (4.2)$$

where H is the mean tsunami height at the nearest coast.

Since the midnineteenth century, the strength of a natural event traditionally has been measured by intensity, a parameter that measures the event impact in a particular location. This implies that the event intensity is spatially variable and that only maximum intensity may be a measure of the event size. The first tsunami intensity scale was introduced by Sieberg (1927), and later several other attempts were made to quantify tsunamis with different or modified scales. However, some scales are more for measuring magnitude than intensity, since they do not describe a tsunami's impact but only the physical parameter of the event—for example, the tsunami's height. The most recent and detailed intensity scale proposed is that of Papadopoulos and Imamura (2001). Efforts to quantify tsunami size in terms of physical magnitude scales

Tsunamis: Detection, Monitoring, and Early-Warning Technologies. DOI: 10.1016/B978-0-12-385053-9.10004-3

(Abe, 1979; Murty and Loomis, 1980) have had limited application so far. Papadopoulos and colleagues (2009) showed that tsunami intensity is a valuable parameter for a number of reasons. The first is the need to quantify tsunami events and to perform statistics for hazard assessment purposes. Tsunami quantification through intensity renders the investigation of relations between tsunami intensity and earthquake size realistic, and this may be useful in several applications.

Tsunami magnitude, M_t, is a fundamental parameter that measures the overall physical size of a tsunami at the source. According to the empirical relation put forward by Abe (1979, 1981), the tsunami magnitude M_t, which is determined using tsunami data to match the moment magnitude M_w is expressed as:

$$M_t = \log H + B \qquad (4.3)$$

which is applicable to far-field tsunamis, and

$$M_t = \log H + \log \Delta + 5.80 \qquad (4.4)$$

which is applicable to near-field tsunamis, where H is the maximum single amplitude (crest or trough) of tsunami waves measured in meters by coastal tide gauges, B is a constant that depends on a combination of source region and the station, and Δ is the distance measured in kilometers from an earthquake epicenter to a coastal tide gauge station along the shortest oceanic path. Abe (1985) found that Eqs. (4.3) and (4.4) are applicable to far-field and Japanese near-field ($100 \text{ km} < \Delta < 3500 \text{ km}$) tsunamis, respectively.

In tsunami wave analysis, a parameter customarily identified is the *maximum* (trough-to-crest) *wave height* (H_{max}), known as the "queen wave." However, some researchers (Candella et al., 2008) also examine the *significant wave height* ($H_{1/3}$) parameter, defined as the average height of the highest one-third of all the waves (LeBlond and Mysak, 1978). Whereas H_{max} is a unique feature of the tsunami wave field, related to one specific peak value, $H_{1/3}$ is an integral measure of the tsunami wave field during the observational period.

The ratio $r = \frac{H_{1/3}}{H_{max}}$ is representative of the rate of tsunami energy decay, which is slow for large r and fast for small r. In Brazil, where tsunami heights as large as 1.2 m were recorded in the case of tsunami waves that arrived from the source region in Sumatra in December 2004, $r \sim 0.3{-}0.5$ for most records. The largest r was ~ 0.61 for Santos and Santa Teresta, indicating slow energy decay at these sites, while the smallest $r \sim 0.30$ was for Arraial do Cabo, where the tsunami energy decayed rapidly (Candella et al., 2008). The wave energy usually has a general exponential decay with time (Van Dorn, 1984, 1987). However, in some instances, broad "bursts" of wave energy have been found superimposed on this basic structure due to additional pumping of tsunami energy from tsunami waves reflected from ridges, continental shelves, and so forth (Kowalik et al., 2007; Candella et al., 2008).

It is well known that the catastrophic Indian Ocean tsunami generated off the coast of Sumatra on 26 December 2004 was recorded by a large number of sea-level gauges throughout the World Ocean. A study by Rabinovich and colleagues (2009a) used gauge records from 55 sites to examine the energy decay of tsunami waves from this event in the Indian and Atlantic Oceans. Their findings revealed that the *e*-folding decay time of the tsunami wave energy within a given oceanic basin is not uniform, as previously supposed, but depends on two independent time scales: (1) the tsunami travel time for the waves to reach the basin from the source region, and (2) the time required for the waves to cross the shelf adjacent to the coastal observation site. The observed decay time increased with distance, and tsunami travel time ranged from 9.3 hours for Male (Maldives) to 34.7 hours for Trident Pier (Florida). A close examination of the data revealed that the broader and shallower the shelf, the more incoming tsunami wave energy it can retain. This retention leads, in turn, to more protracted tsunami ringing and to slower energy decay at adjacent coastal sites. Tsunami energy flux is given by (Kowalik et al., 2008):

$$E = \rho g H \varsigma u \tag{4.5}$$

where E is the energy flux vector, ρ is the water density, g is gravitational acceleration, H is the water depth, ς is the sea surface elevation, and u is the water velocity.

4.1. PROPAGATION

The tsunami wave front (i.e., phase) radiates away from the source in almost all directions, depending on the nature and geometric configuration of the source region. By virtue of the tsunami wave length being enormously longer than even the greatest ocean depths, it behaves as a *shallow-water* wave regardless of the depth of the ocean. The speed of propagation of a very long wave (shallow-water wave), such as a tsunami, is expressed approximately as:

$$c = \sqrt{gH} \tag{4.6}$$

where c is the phase velocity, g is the acceleration due to gravity, and

$$H = (D + \eta)$$

where D is the water depth below the mean sea level and η is the wave height. Figure 4.1 shows the relationship between water depth and the speed of propagation of tsunami waves, based on Eq. (4.6). A more accurate expression for c is given by (Nirupama et al., 2007a):

$$c = \sqrt{gD} \left(1 - \frac{2\pi^2 \mu}{3}\right) \tag{4.7}$$

where

$$\mu \equiv \frac{D^2}{\lambda^2} \tag{4.8}$$

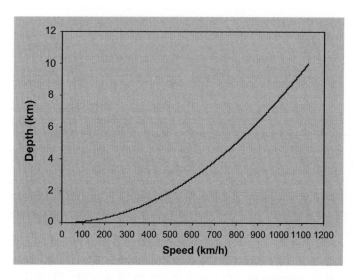

FIGURE 4.1   Relationship between water depth and the speed of propagation of tsunami waves.

in which λ is the wave length. From Eq. (4.7), it can be seen that long waves travel with a speed mainly determined by water depth but subject to a small negative correction proportional to μ. The implication of Eq. (4.7) is that two wave components with a slightly different value of μ will tend to separate as they propagate (by virtue of different phase velocities). Thus, μ is considered as a measure of "frequency dispersion." Frequency dispersion mainly happens during the propagation of the tsunami wave. Changes in ocean bathymetry (changes in D) encountered by the tsunami wave during its propagation, among other things, contribute to the separation and spreading of the tsunami wave into a multiwave event. Frequency dispersion was one of the salient features of the December 2004 Indian Ocean tsunami (Kowalik et al., 2005a, b; Kowalik et al., 2007a; Murty et al., 2005; Murty et al., 2006; Nirupama et al., 2005; Nirupama et al., 2006). Just as wind-generated waves have both phase and group velocities, so is the case with tsunami waves. The group velocity of tsunami waves arriving at any point can be expressed as (Nirupama et al., 2007b):

$$c_g = \sqrt{gD} \left(1 - \frac{K^2 D^2}{2} \right) \qquad (4.9)$$

where K is the wave number. During an earthquake, the fault motion is elongated to one direction (length), but the tsunami amplitude is larger in the direction perpendicular to the fault strike (or in the short axis direction) (Prof. Kenji Satake, Earthquake Research Institute, University of Tokyo, Japan, Personal communication, 2009). Accordingly, the period T of the tsunami wave is related to the source width W and the depth H, and is given by the relation

$T \sim \frac{W}{c}$, where c is the speed of the tsunami wave at the source. If the tsunami source is similar to an ellipse with axis-length a_1 and a_2 ($a_1 < a_2$), then the full wave length (elevation and the following depression of sea level) will be equal to a_1 for the wave propagating in the short-axis direction and a_2 for the wave propagating in the long-axis direction (Dr. Andrei Marchuk, Institute of Computational Mathematics and Mathematical Geophysics, Siberian Branch of the Russian Academy of Sciences, Novosibirsk, Russia, Personal communication, 2009). The salient features of tsunami wave propagation are shown in Table 4.1. Although the period of the primary tsunami wave (i.e., the initial wave generated by the earthquake) is defined by the width of the bottom deformation and the ocean depth, as just indicated, impinging of the tsunami wave on the shelf break results in its splitting into two waves due to partial reflection. Consequently, the periods of the secondary waves are defined by reflection and generation of the new modes of oscillation through an interaction of the tsunami waves with the shelf/shelf-break geometry. The evidence of both tsunami transformations and trappings have been achieved both theoretically and in observations (Loomis, 1966; Nekrasov, 1970; Abe and Ishii, 1980; Yanuma and Tsuji, 1998; Mofjeld et al., 1999a, b).

In the Pacific Ocean, where the typical mean water depth is about 4 kilometers, the tsunami waves propagate at a mean velocity of 713 km/h (approximately the speed of a jet plane) and can traverse the Pacific Ocean basin in less than 24 hours. Tsunami waves can have wavelengths (i.e., distance covered between successive crests or successive troughs) in excess of 100 km and periods (i.e., time elapsed between successive crests or successive troughs) in the range of a few minutes to about an hour. For example, the most common spectral peaks that were observed at several sites during the 2004 Sumatra tsunami were 13–14 minutes, 37–38 minutes, 40–43 minutes, and 48–55 minutes. Low-frequency peaks (i.e., larger period waves) were observed at some sites—specifically, 1.8 h at Cape Town and Port Nolloth, 2.1 h at Port Elizabeth, and 2.6 h at Saldanha in the South African coastlines (Rabinovich

---

TABLE 4.1 Salient Features of Tsunami Wave Propagation

- Waves propagate at *long-wave speed,* $c = \sqrt{(gH)}$; ($H = water\ depth + wave\ height$). $L \sim 100\ km$. (a) waves "long" when $L \gg H$; (b) wave period (T) is determined by source width (W) and depth (H): $T \sim W/c$.
- Wave heights $\eta < 1$ m in open ocean but can exceed tens of meters at the coast.
- Wave *propagation* modified by refraction, reflection, diffraction, scattering, and dispersion (wave groups).
- Wave dynamics are *highly* linear except near the source region and in shallow water.

Source: Courtesy of Dr. Alexander B. Rabinovich, P.P. Shirshov Institute of Oceanology, RAS, Moscow.

and Thomson, 2007). Similar periods were observed at the Indian coastal stations as well. Larger-period waves with periods greater than 1 hour (approximately 3.2, 1.9, and 1.4 hours) were observed from the Takoradi (Ghana, West Africa) spectrum as well (Joseph et al., 2006). However, these peaks can be considered to be the "birthmarks" of the measurement location and are primarily related to the local topographic resonance at these sites.

In terms of wave length L and wave period T, the tsunami wave propagation speed c can be expressed as:

$$c = \frac{L}{T} \tag{4.10}$$

From Eqs. (4.6) and (4.10):

$$T = \frac{L}{\sqrt{gH}} \tag{4.11}$$

The period, T, of a given tsunami wave depends on several factors related to the tsunami generation source characteristics. The expected periods of arriving tsunami waves may be roughly estimated based on the source dimensions and the ocean depth in the source area (Rabinovich, 1997). It has been found both theoretically and through observations (e.g., the Chilean tsunami of November 10, 1922; the Kamchatka tsunami of April 13, 1923; and the Aleutian tsunami of April 1, 1946) that the period, T, of a tsunami wave and that of a seismic surface wave traveling over large distances increases during propagation but decreases with time at a given station (Munk, 1947). However, such deviations in the wave period are not considerably large, so for all practical purposes, the period of a given tsunami wave constituent is assumed to remain invariant all along its propagation path. Thus, $\frac{L}{\sqrt{gH}}$ remains constant during tsunami propagation. Also, from Eq. (4.11), $L = (T)\sqrt{gH}$. Thus, for a given wave of period T (i.e., for T remaining constant), the wavelength L is directly proportional to $\sqrt{H}$. This scenario forces the tsunami wave length, L, to decrease as the local depth, H, decreases as the wave propagates over shallow coastal waters. Figure 4.2 illustrates the reduction of tsunami wave lengths of different periods with a decrease in local water depth.

The expression $c = \sqrt{gH}$ is a simplified relationship. Actually, the velocity of the wave also depends on its wavelength, as embodied in the expression:

$$c = \sqrt{\frac{g \times \tanh(kH)}{k}} \tag{4.12}$$

where $k = \frac{2\pi}{L}$. The shorter the wave, the slower it propagates. Due to the effect of linear dispersion, the real front is usually behind this theoretical estimate (Kulikov et al., 2005). The longer the route of propagation, the stronger the effect of dispersion. The wave front can be even more strongly distorted in the shallow water and during its propagation along the straits. However, the

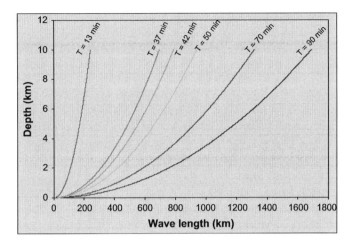

FIGURE 4.2 Reduction of tsunami wave lengths of different periods with decrease in water depth.

estimate of the position of the tsunami wave front calculated from relation $c = \sqrt{gH}$ is sufficiently exact, and the error does not usually exceed the size of the tsunami source ($\sim 50–100$ km). The evolution of the wave train is caused by the dependence of the group velocity on the frequency (Kulikov et al., 2005):

$$c_g = \left(\frac{\omega}{2k}\right)\left(1 + \frac{2kH}{\sinh 2kH}\right) \qquad (4.13)$$

where ω is angular frequency, k is the wave number, and H is the depth of the ocean. In order to analyze the effect of the linear dispersion of tsunami waves, an accepted method is to construct a diagram of the dependence of the spectral amplitude of the signal on time and the wave number. This method is widely practiced in seismology (Dzienovski et al., 1969) as Multiple Filter Analysis. In recent times, methods of wavelet analysis (which are closely related to this method) have been widespread in tsunami wave analysis.

Because the tsunami wave has a very large wavelength in the open ocean, it loses little energy as it propagates through the vast expanse of the open ocean (the rate at which the wave loses its energy is inversely related to its wavelength). As a result, the tsunami waves travel great transoceanic distances with limited energy loss. In the open ocean, the tsunami waves are nearly undetectable (typically <1 m). However, as the tsunami wavetrain approaches the shallow water regions, its speed diminishes (i.e., its kinetic energy decreases) and the wave energy is compressed into a much shorter distance as a result of decrease in wave length. This effect results in a corresponding increase in potential energy (manifested as a corresponding increase in wave height). For example, satellite-recorded tsunami waves in the open part of the Indian Ocean during the December 2004 Sumatra tsunami were about 0.7–0.8 m, but when these waves arrived at the coast of India, they were 3–5 m (Rabinovich and Thomson, 2007; Chadha et al., 2005; Gower, 2005). Because of this effect, a tsunami that is imperceptible in deep water grows to dangerous heights in shallow coastal waters posing a threat to life in coastal communities (Figure 4.3).

Tsunami amplitude at an open coast is estimated using the energy conservation law known as Green's law. Green's law states that the tsunami amplitude (h_0) at the coast is represented by the fourth root of the ratio of sea depth (d_1) at the forecast point and the depth (d_0) at the coast (Figure 4.4) and is given by the expression:

$$h_0 = \sqrt[4]{\frac{d_1}{d_0}}\, h_1 \qquad (4.14)$$

Because the forecast points are set close enough to the coast, horizontal convergence and divergence of the rays can be ignored; that is, the wave front is regarded as almost parallel to the coast due to refraction toward the direction of the largest sea-depth gradient. It must be noted that if the tsunami enters restricted water bodies such as bays, inlets, harbors, estuaries, and so forth where resonant- and/or geometrical- amplifications occur (see Chapter 2), the observed tsunami amplification will be considerably larger than that estimated from Eq. (4.14).

The relatively large spatial steps (about 1 nautical mile) generally used in models does not allow resolving small-scale features of the coastal and bottoming topography. Second, the length of the spatial grid step determines the minimum wave length and, therefore, the maximum wave frequency that can be simulated numerically. More precisely, the model grid step plays the role of a low-frequency filter that effectively suppresses high-frequency waves

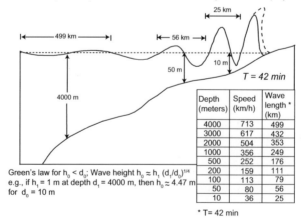

Arrival at the Coast

Tsunami speed is reduced in shallow water as wave height increases rapidly.

Green's law for $h_0 < d_0$; Wave height $h_0 \approx h_1\,(d_1/d_0)^{1/4}$
e.g., if $h_1 = 1$ m at depth $d_1 = 4000$ m, then $h_0 \approx 4.47$ m for $d_0 = 10$ m

Depth (meters)	Speed (km/h)	Wave length * (km)
4000	713	499
3000	617	432
2000	504	353
1000	356	249
500	252	176
200	159	111
100	113	79
50	80	56
10	36	25

* T= 42 min

FIGURE 4.3 Illustration of depth-dependent height amplification and wave-length shortening of tsunami during its approach to the shallow coastal waters.

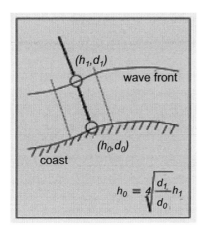

FIGURE 4.4 Schematic sketch illustrating the application of energy conservation law (Green's law) for tsunami amplitude calculation in near coastal areas. Ray convergence and divergence are ignored. *Source: Kamigaichi, 2009; Reproduced with kind permission of Springer Science+Business Media.*

(Rabinovich et al., 2008). Thus, numerical resolution of tsunami-related information such as run-up requires a very detailed fine-grid model with precise bathymetry (Titov and Synolakis, 1997).

It has been noted that tsunami amplitude at the coast is modified by several factors. The first factor is amplification due to shoaling near the coastline, which is assumed to be governed by Green's law, in which the maximum amplitude is proportional to $h^{-1/4}$, where h is the water depth (Abe, 1973; Satake, 2002). The second factor is amplification due to reflection at a fixed edge (coastline), in which wave amplitude is doubled (Baba et al., 2004). In reality, the reflection coefficient would be strongly dependent on a wide variety of run-up effects. By taking these two factors into consideration, deep-ocean tsunami amplitudes can be corrected so they can be treated the same way as tsunami amplitudes measured at the coastline. The corrected maximum single amplitude (crest or trough) of the tsunami is defined as (Baba et al., 2004):

$$H_a = \frac{2H_p}{h^{-\frac{1}{4}}} \tag{4.15}$$

where H_p is the original maximum single amplitude (crest or trough) recorded in the deep ocean. Replacement of Eqs. (4.3) and (4.4) with H_a of Eq. (4.15) provides the modified empirical relations for determining M_{tp} using tsunami amplitudes measured with deep-ocean tsunameters (Baba et al., 2004):

$$M_{tp} = \log\left(\frac{2H_p}{h^{-\frac{1}{4}}}\right) + B \tag{4.16}$$

which is applicable to a far-field tsunami, and:

$$M_{tp} = \log\left(\frac{2H_p}{h^{-\frac{1}{4}}}\right) + \log\Delta + 5.80 \tag{4.17}$$

which is applicable to a near-field tsunami.

Equations (4.16) and (4.17) were, in practice, found to be applicable to far-field and near-field tsunamis, respectively, measured at the Japanese coasts (Baba et al., 2004). A typical tsunami is much shorter than a tide wave. For this reason, tides are usually neglected in tsunami modeling. The computed sea-level dynamics are superimposed on the tidal motion nondynamically. However, in coastal areas with strong tidal activity, dynamic nonlinear interaction of tidal and tsunami waves can amplify the magnitude of inundation (Androsov et al., 2009). Because shallow-water approximations cannot provide accurate estimates of complex tsunami behavior, such as breaking, splashing, flowing around structures, and so on, it often becomes necessary to use efficient numerical modeling techniques such as the Lattice Boltzmann Method (LBM). LBM is a new technique in computational fluid dynamics (CFD), and it is completely different from traditional methods that are based on a direct solution of flow equations. LBM has been primarily developed in physics and computational science and applied in the field of fundamental fluid dynamics. Recently, its research area has been expanded to application problems including tsunami hazards and to simulate the complex behavior in tsunami wave fronts (Araki and Koshimura, 2009).

4.2. FEATURES OF GEOPHYSICAL TSUNAMIS

Until very recently, the true features and structures of tsunami waves were the least understood subjects in the realm of tsunami research. Because of the clearly manifested "smoothness" of the geophysical tsunami waves (resulting primarily from their overall "unidirectionality") in contrast to the seeming "roughness" and chaotic nature of the frequently encountered wind-waves (resulting primarily from their "multidirectionality"), there was unanimity among tsunami researchers that tsunami waves were distinctly different from wind-waves. If so, then what are they? There was no definite answer to this question, primarily because closely sampled sea-level records were not available. There were no instruments designed exclusively for tsunami measurements. In the absence of such instruments (tsunameters), tsunami waves were studied based on "tide gauge" records. Unfortunately, all types of tide gauges were designed to filter out most of the short-period waves such as wind-waves and swell (periods in the range of a few seconds to a few minutes) so the desired long period ($\sim$12 h) semidiurnal tides (that occur twice a day) are clearly separated out from the background noise (i.e., wind-waves and swell). This universal feature of the tide gauge technology precluded a clear understanding of the true structure of the tsunami wave.

When tsunami researchers were still looking for an answer regarding the true structure of tsunami waves, John Scott Russell, in his experiments in hydrodynamics, discovered for the first time in the middle of the nineteenth century a special type of well-defined and fast-moving shallow-water wave. This fast-moving wave, which was considerably different from all other forms of waves that are frequently encountered in the ocean, was called a *solitary wave* because of the presence of *soliton(s)* (heap of water wave(s)) on the wave crest. It is interesting to note that long waves, which are dynamically similar to the hazardous tsunami waves, are frequently generated by high-speed ferries. For example, the Tallinn Bay, a semienclosed body of water approximately 10×20 km in size in the Baltic Sea, is one of the few places in the world where several high-speed ferries operate at depths ranging from 20 m to 40 m in near-critical high-speed conditions, generating packets of large, solitonic, long-crested waves. In many aspects, these waves can be used as a dynamically similar input, allowing modeling and measurements of the shoaling and run-up properties of extreme, long, large-scale tsunami waves. Didenkulova and colleagues (2009) have suggested that with the use of geometric similarity, the Tallinn Bay can be used as a natural laboratory for tsunami wave studies.

Soliton theory was given a sound mathematical footing primarily by Korteweg and de Vries (Zabusky and Galvin, 1971; Hammack and Segur, 1974). Solitons possess some special properties: They are practically frictionless; do not break up, spread out, or lose strength easily; and are very robust against perturbations. The well-defined and fast-moving heap of water approaching the shore one after the other in a queue during a tsunami event (Figure 4.5) was traditionally considered to be a train of solitons, and thus a tsunami wave has been described as a *solitary wave train* in shallow water. From practical experience, it is well known that as a tsunami approaches the coast, ultimately the shore halts further progress, which results in a considerable rise in the vertical height of the coastal water level, called surge. In contrast to wind-waves that come and go without flooding higher areas, tsunami waves run quickly over the land as a wall of water (Figure 4.6). This specialty of tsunami, featuring various stages of wave breaking and bore formation (Figure 4.7), could be due to its directionality and depth penetration. It was

Sumatra tsunami hitting Koh Pu, Thailand

© Anders Grawin

FIGURE 4.5 Well-defined and fast-moving heaps of water approaching the shore one after the other in a queue during tsunami propagation. *Source: Courtesy of NOAA Center for Tsunami Research.*

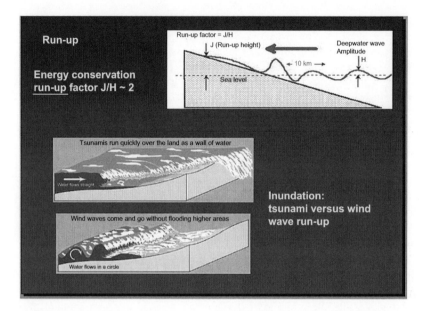

FIGURE 4.6 Illustration of coastal flooding scenario, tsunami versus wind-wave run-up. *Source: Courtesy of R.E. Thomson.*

believed that the tsunami waves swept into the seacoast like a huge wall of water (See Figure 4.8 for glimpses of damage and destruction at Phuket, Thailand, caused by the December 26, 2004 Sumatra tsunami) primarily because of its solitary wave nature, and had it not been for the presence of solitons, the tsunami waves would have been as harmless as the tide waves (another shallow-water wave) that occurs in the ocean twice daily in most ocean basins.

According to Madsen and Mei (1969), a long wave propagating over the continental shelf can fission into a series of solitary waves (single crests propagating over the ocean surface), depending on the wave steepness and the bottom slope. These series of solitary-type waves are generated only at the leading edge of the original carrier long wave and ride on the gradual variation of the main wave. Such solitary waves may break, but the energy of the carrier long wave is approximately conserved. Therefore, local wave breaking in a very long wave does not influence the maximum wave run-up height significantly.

In the study of tsunami waves traveling over the continental shelf, both linear and nonlinear long-wave equations are usually inadequate; rather, an intermediate-type wave equation known as a solitary wave equation (e.g., Boussinesq or Kortweg-de Vries [KdV] type equation) is considered to be more appropriate. The phase velocity of a solitary wave is given by (Nirupama et al., 2007a):

$$c \cong \sqrt{gD}\left(1 + \frac{\varepsilon}{2}\right) \tag{4.18}$$

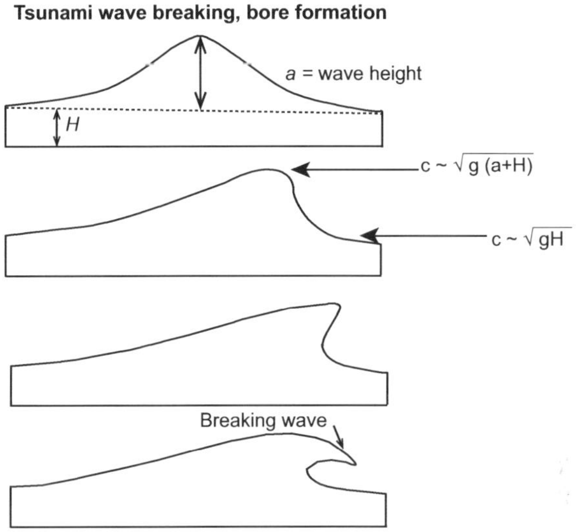

Tsunami wave breaking, bore formation

a = wave height

H

$c \sim \sqrt{g\,(a{+}H)}$

$c \sim \sqrt{gH}$

Breaking wave

FIGURE 4.7 Illustration of various stages of tsunami wave breaking and bore formation.

.... *death and destruction*

Phuket, Thailand (photos by Helmut Issels)

FIGURE 4.8 Glimpses of damage and destruction at Phuket, Thailand, caused by the December 26, 2004 Sumatra tsunami. *Source: http://www.theplasmaverse.com/pdfs/meteorological-tsunamis-destructive-atmosphere-induced-waves-observed-in-the-worlds-oceans-seas-seiches-abiki-milghuba.pdf. Photos by Helmut Issels.*

where ε is the relative amplitude and is given by $\varepsilon \equiv \frac{\eta}{D}$, in which η is the wave amplitude. Thus, the phase velocity of a solitary wave is approximately equal to that of a shallow-water wave but is subject to a small positive correction proportional to ε. Thus, ε is a measure of the amplitude dispersion. Both amplitude and phase dispersions tend to distort the wave forms. In deep water, especially in the near-field of tsunami generation, the linear wave theory is considered to be adequate, and phase dispersion alone is relevant. However, on the continental shelf, both phase and amplitude dispersions are important, and therefore the solitary wave concept is quite often employed in the mathematical analysis. In very shallow coastal areas (bays, harbors, and inlets), the amplitude dispersion dominates. Both amplitude and phase dispersions tend to cause a gradual distortion of the tsunami waves. The theoretical basis for the choice of a particular wave theory (e.g., linear long-wave theory, Boussinesq equations, nonlinear long-wave theory, etc.) was discussed by Nirupama and colleagues (2007a).

Ramsden and Raichlen (1990) conducted experiments to examine the impact forces on a vertical wall by a broken wave evolved from a solitary wave offshore. It was found that the maximum run-up of the water on the wall occurs prior to the maximum force exerted on the wall. Zelt and Raichlen (1991) conducted laboratory experiments to examine the overland flows on a horizontal bed generated by an incident solitary wave. The water depth in the experiments varied gradually and smoothly from a constant-depth section (a horizontal bottom) to a zero-depth section. By comparing numerical solutions of the one-dimensional Boussinesq equation (i.e., solitary wave equation), it was found that the bottom friction is not important during the initial collapse of solitary waves on the horizontal bed, but that it becomes a dominating factor in the overland flow.

From considerations of practical usefulness to save lives, it is not the prediction of a tsunami event but the run-up heights on the coasts that are more important. Because the coastal water bodies are highly complex from a fluid dynamics point of view, the same tsunami may behave very differently in different coasts due to differences in bathymetry and the shape of the land-water interface. A tsunami that is harmless in one coastal location can be very destructive in another location. Hence, a clear estimation of the beach run-up heights at different coastal locations that are vulnerable to tsunami attacks is an important consideration.

Solitary wave theory has been applied to study and solve various tsunami-related problems. For example, Synolakis (1987, 1991) applied the solitary wave concept to understand and quantify tsunami evolution far from the shoreline, maximum tsunami run-up, and tsunami breaking criterion. According to the solitary wave concept, maximum run-up (R) can be expressed as:

$$\frac{R}{d} = 2.83 \left(\frac{H}{d}\right)^{5/4} \left(\sqrt{\cot \beta}\right) \tag{4.19}$$

In Eq. (4.19) d denotes the constant water depth offshore, H is the offshore wave height at the constant-depth region, and (cot β) is the beach slope. The breaking criterion during run-down is given by the expression:

$$\frac{H}{d} < (\cot \beta)^{10/9} \tag{4.20}$$

The maximum height evolution is given by:

$$\left(\eta_{max}/H \right) = \left(d \cot \beta / X \right)^{1/4} \tag{4.21}$$

In Eq. (4.21) X is the distance from the shoreline. Based on the preceding mathematical relationships, it was possible to study through computer animation the climbing of a solitary wave up a sloping beach (Liu et al., 1991).

Subsequent to the December 2004 global tsunami, deep-sea as well as coastal sea-level measurements were made at very close sampling intervals ($\frac{1}{2}$ -1 min). Such closely sampled and precise deep-sea and coastal sea-level measurements obtained during weak, as well as relatively strong, tsunami events began to remove the prevailing obscurity regarding the structure of tsunami waves. Based on such measurements, it is now broadly agreed that tsunami waves are "*a series of waves that penetrate the entire water depth, and propagating at high speed with a coherent elevation and depression of large expanses of the ocean surface in the wave*", rather than solitary waves. However, this new definition does not dilute the fact that tsunamis are potentially hazardous waves.

Despite the preceding definition, the *solitary wave* concept is still used for tsunami modeling, inundation mapping, and so on. For example, Nicolsky and colleagues (2009) conducted analytical and laboratory benchmarking for the cases of solitary wave run-ups on simple and composite beaches, a run-up of a solitary wave on a conical island, and the extreme run-up in the Monai Valley, Okushiri Island, Japan, during the 1993 Hokkaido-Nansei-Oki tsunami. Additionally, they field-tested the developed model to simulate the November 15, 2006, Kuril Islands tsunami and compared the simulated water height to field observations at several DART (Deep-ocean Assessment and Reporting of Tsunamis) buoys. In all the tests conducted, they calculated a numerical solution with an accuracy recommended by NOAA standards. By choosing several scenarios of potential earthquakes, they applied the developed model (near-shore bathymetry and inland topography on a grid of 15-meter resolution) to estimate the maximal aerial extent of potential inundation of the city of Seward located in Resurrection Bay, Alaska, and obtained good agreement between the computed and observed datasets with regard to the 1964 earthquake in Alaska.

In another engineering study, Komarov and colleagues (2009) applied the solitary wave theory (Boussinesq model) to study the interaction of tsunami waves with very large floating platforms and structures that are designed to be used for the ocean space utilization (e.g., floating airports, storage facilities, ice fields in polar regions, etc.).

In some shallow coastal locations, the tsunami waves focus energy against the coastline as a result of the bending of wave crests during its propagation; this refraction process depends on the shape and topography of the seafloor. Reefs, bays, entrances to rivers, undersea features, and the slopes of beaches all modify the tsunami wave train as it strikes the coastline. The bathymetry and the shape of the beach heavily influence the focusing characteristics and the height of the tsunami waves on a beach. Consequently, whereas one coastal area may see no damaging tsunami wave activity, another area only a few km away may witness destructively large and violent activity. For example, in 1952, even though no abnormal rise in water level was observed at most of the coastal sites in Hawaii, the adjacent Hilo bay recorded the highest level of run-up. Thus, the striking capacity of tsunami waves may vary widely from place to place.

Tsunami waves impose dynamic water pressures on coastal structures as well as buildings and bridges near the coastline, inflicting serious damage to the surrounding infrastructure. For instance, the damage observed in Thailand in association with the December 26, 2004, Indian Ocean tsunami was almost entirely due to water pressures that varied from impulsive pressures of breaking waves at the shore to reduced dynamic pressures inland as water velocity decreased due to surface friction (Saatcioglu et al., 2007). An empirical expression used in engineering practice for the estimation of impulsive water pressure of breaking waves, developed by Goda (1995) is given by:

$$P_{\max} = \frac{\pi \gamma c H_w}{4 g \tau} \tag{4.22}$$

where γ is the specific gravity of seawater (10.3 kN/m^3), c is the celerity of the wave in m/s, τ is the impact duration (in seconds), and H_w is the wave height in m. Equation (4.22) was developed to compute the effects of breaking waves on coastal structures at the shore. Because the hydrodynamic pressure associated with wave celerity is reduced progressively with increasing surface friction, an expression that is more appropriate to the mechanism of tsunami wave impact within some distance from the shore is (Hiroi, 1979):

$$P = 1.5 \gamma H_w \tag{4.23}$$

Although the basic equations for tsunami analysis have been known for decades, the progress toward the development of benchmarked models for forecasting tsunami inundation took considerably longer. This is primarily because only the largest tsunamis were being reported for several decades, with the exception of the 1960 and 1964 events, and only qualitative information was available on tsunami inundation. Due to the lack of reliable quantitative information, tsunami hydrodynamics has progressed slower than research in other natural hazards. Tsunami hydrodynamic modeling had several facets by necessity and was driven by milestone scientific meetings and posttsunami surveys that kept identifying novel problem geometries and previously unrecognized phenomena (Synolakis and Kânoğlu, 2009). The existing synthesis leading to real-time forecasts that are currently available had to await the large-scale laboratory experiments of the 1980s to 1990s, deep-sea tsunameter measurements, and the development of sophisticated modeling tools. Computational efforts were developing rapidly toward modeling the two-plus-one (two horizontal space dimensions and time) propagation of long waves from the source to the target, but coastal evolution and inundation remained unexplored. Advances in tsunami hydrodynamics were jumpstarted in the 1990s. State-of-the-art inundation and forecasting codes have evolved through a painstaking process of careful validation and verification (Liu et al., 1991). The field survey results in the 1990s served as crude proxies to free-field tsunami recordings and allowed for the validation and verification of numerical procedures. Operational tsunami forecasting was only made possible through the availability of deep ocean measurements.

Combinations of analytical and numerical modeling research have begun to shed more light on many hitherto unknown effects of tsunamis on harbors and coasts. It is now understood that an abnormal amplification of tsunami waves in some harbors can be associated with the processes of wave shoaling and resonance in the bay. The shoaling effects are especially important on gentle beaches where the wave reflection is weak. Traveling "nonreflecting" waves approach the coast without loss of energy, and their amplitude significantly increases. However, tsunami amplification is influenced by beach profile as well. For example, Didenkulova and Pelinovsky (2009) showed that the run-up of a tsunami wave to a beach with a bottom profile of the kind $h \sim x^{4/3}$ (where h is water depth and x is a coordinate) may be considerably higher than for a beach with a linear profile.

Hydrodynamic forces of a tsunami current may cause additional damage from drifting ships or vessels in a harbor area. Hashimoto and colleagues (2009) examined the damage mechanisms of drifting ships and vessels, and they developed a model to analyze the ship drifting motion by a tsunami current. The model consists of a set of equations of motion of a floating body driven by a tsunami current considering surge, sway, and yaw (Sakakibara et al., 2005; Kubo et al., 2005; Kobayashi et al., 2008), with additional terms to formulate the grounding effect. The equations are solved numerically in temporal and spatial domain under the tsunami current field estimated by a tsunami inundation model based on the nonlinear shallow-water wave theory. The model was implemented to simulate a true event involving the hydrodynamic response of a 2600-ton power plant barge 63 m long, which was berthed in Ulee Lheue ferry port (Banda Aceh, Indonesia) before the 2004 Indian Ocean tsunami and drifted approximately 3 km inland from the tsunami inundation flow. The modeled hydrodynamic features (water levels and current velocities) that were input to the ship drifting model beautifully simulated the sequence of drifting motion and grounding. At the ITS-2009 tsunami symposium in Novosibirsk, Russia, Hashimoto and colleagues (2009) presented a film of the model-simulated drifting of the ship several kilometers inland from its original position. It was stated that the simulated grounding position was consistent with the true position of the drifted ship in the middle of Banda Aceh City, thereby suggesting the model's applicability in understanding the damaging effects of tsunamis in harbors and beaches if validated models incorporating high-resolution topography/bathymetry data (23 m spatial resolution in the present case) are used to obtain the hydrodynamic features of tsunami inundation.

The preceding observations regarding the fairly good success of simulating several observed tsunami effects at the coasts based on solitary wave theory, as well as nonlinear shallow-water wave theory, suggest the operational utility of both of these theories. Thus, it appears that the recently acquired knowledge on the structure of a tsunami wave as a series of waves that penetrate the entire water depth, and propagating at high speed with a coherent elevation and depression of large expanses of the ocean surface in the wave over the older concept of its solitary nature do not make much difference in a practical sense.

4.3. INFLUENCE OF MIDOCEAN RIDGES, STRAITS, AND CONTINENTAL SHELVES

The world oceans abound in several ridges, straits, basins, and continental shelves. The Indian Ocean, which was a victim of the December 2004 Sumatra tsunami, consists of two major ridges and a few minor ones (Figure 4.9). These ridges function as wave guides and exercise considerable influence in determining the direction of tsunami propagation. The large number of sea-level observations available for the December 2004 tsunami event enabled scientists to investigate the propagation and transformation of the tsunami in the open ocean and to examine tsunami characteristics in terms of the relative influence of the source and submarine topographic features on the arriving waves.

Like any other waves, tsunami waves are subject to reflection, refraction, and amplification by topographical influences. For example, a noteworthy feature found in all tsunami records for the coast of South Africa was the very long (>4 days) ringing time. Several distinct wave trains with typical durations of 12 to 24 hours appear in the South Africa tsunami records, apparently due to multiple wave reflections from continental boundaries, including the Antarctic coast, the coasts of south and southeast Asia, and western Australia (Rabinovich and Candella, 2009).

Global tsunami propagation models (Titov et al., 2005a, b; Kowalik et al., 2007a, b) demonstrate that the Mid-Atlantic Ridge served as a wave guide, efficiently transmitting the December 2004 tsunami's energy from the source area to far-field regions of the Atlantic Ocean. According to their simulation

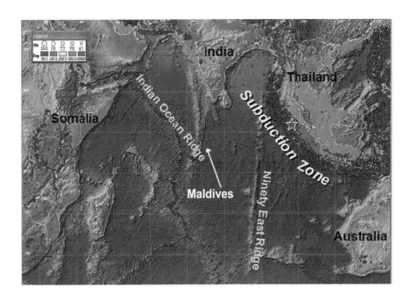

FIGURE 4.9 Ridges and islands in the Indian Ocean. *Source: Gusiakov. V.K., ITDB/WLD: Integrated Tsunami Database for the World Ocean, Version 6.52 of October 31, 2007, CD-ROM, Tsunami Laboratory, ICMMG SD RAS, Novosibirsk, Russia.*

results, upon reaching the Tropic of Capricon, a branch of the ridge-trapped tsunami energy flux split from the topographic wave guide and headed toward the coast of Brazil (striking the coast exactly where the maximum waves were observed), while another branch headed toward Nova Scotia (Geist et al., 2006). Candella and colleagues (2008) observed that as a result of such a wave-guide effect, a "tongue" of high tsunami energy that extended from the Mid-Atlantic Ridge to the central Brazilian coast resulted in strong tsunami oscillations at locations such as Imbituba, Umbatuba, Arraial do Cabo, and Santos. As revealed by the same modeling results, the north coast of Brazil was in a "shadow" zone for the tsunami energy flux, explaining why low wave heights were recorded at Natal (Candella et al., 2008).

Likewise, numerical simulation studies by Lobkovsky and colleagues (2009) indicated that the propagation of tsunami waves generated by the November 2006 and January 2007 Kuril earthquakes at Simushir, off Japan, was clearly affected by the Emperor and Hawaiian submarine ridges. As emphasized by Kowalik and colleagues (2008) and Tanioka and colleagues (2008), the instance of maximum waves in these two tsunami episodes occurring a long time (~ 5 h) after the first wave arrivals is attributable to the influence of large-scale bottom features such as the Emperor Ridge. According to these researchers, the maximum observed oscillations constitute "secondary" tsunami waves reflected from the Emperor Ridge. Kowalik and colleagues (2008) also suggested that the Emperor Ridge, Koko Guyot, and the Hess Ridge in the northwestern Pacific played a primary role in the reflection of the 2006 tsunami and the subsequent formation of new sets of high-amplitude waves.

Rabinovich and colleagues (2009b) examined high-resolution (15 sec and 1 min) open-ocean records of the 2004 Sumatra tsunami detected in the observations from two DART and two ODP CORK (Ocean Drilling Program, Circulation Obviation Retrofit Kit) stations at depths of 1500 to 3500 m (located within 200 to 300 km of each other) in Cascadia Basin (northeast Pacific) and found that tsunami waves, following their 20,000 km journey from the source region, arrived at these two deep-sea stations approximately 7 hours later than predicted; this suggested that the energy flux was strongly determined by the midocean ridge wave guiding.

Satake and colleagues (2009) observed that the long duration of the Simushir tsunamis was due to reflected waves at Emperor Seamount Chain. Midocean ridges and continental shelves play important roles also in propagating the tsunami to significantly greater distances by the reflection mechanism. For example, global tsunami propagation models (Titov et al., 2005a, b; Kowalik et al., 2005a, b) demonstrate that midocean ridges served as wave guides to the 2004 tsunami, efficiently transmitting tsunami energy from the source area (Indonesia) to far-field regions of the Pacific and Atlantic coasts of North America.

Evidence of the influence of the ridge on tsunami propagation has been found also in the tsunami signals obtained from Takoradi (Ghana) in the central west coast of the African continent in the Atlantic Ocean (Joseph et al., 2006); this included the presence of two distinct bursts at discrete times, the peaks of whose envelopes were separated by approximately 14 hours, and the second burst being larger in height than the first burst. Because only one distinct burst (with rapid amplitude decay thereafter) has been reported from the sea-level record at Male and other sites that are close to the source region with no shallow scatterers in between (Merrifield et al., 2005a; Titov et al., 2005a), the observed second burst at Takoradi can be interpreted as evidence of reflected tsunami waves. The reason for the larger height of the second burst is believed to be the arrival of tsunami waves from certain regions of the continental shelves or from the Mid-Atlantic Ridge (see also Woodworth et al., 2005).

Interestingly, by running numerical simulations on the high seas, Okal and Hartnady (2009) determined that the South Atlantic Ridge, the Southwest Indian Ocean Ridge, and the Agulhas Rise caused strong focusing of tsunami waves, resulting in increased tsunami heights in Ghana, southern Mozambique, and certain parts of the coast of South Africa.

Kowalik and colleagues (2007) showed through numerical modeling that the reflected waves from Sri Lanka and the Maldives Island chain were often larger than the primary wave. The sea-level data obtained from the west and south coasts of Sri Lanka (Colombo and Kirinda) during the December 2004 tsunami confirmed this finding and indicated the importance of tsunami wave reflections from the Maldives Island chain, which produced the maximum wave at the Sri Lanka locations 2 to 3 hours after the arrival of the first wave (Pattiaratchi and Wijeratne, 2009). Similarly, along the west coast of Australia, the highest waves during the December 2004 tsunami occurred 15 hours after the arrival of the first wave. Kowalik and colleagues (2007) attributed the source of this higher energy to reflected waves from the Maldives and Seychelles (Mascarene Ridge). However, the time lag of 13−15 hours suggested that the Island of Madagascar could also have been responsible for the reflection (Pattiaratchi and Wijeratne, 2009).

The continuous presence of the tsunami wave field at several locations with gradually reducing energy levels has also been found at some locations, which is attributable to dispersed secondary-scattered/reflected signals arriving from far-off continental shelves or midocean ridges. González and Kulikov (1993) reported incidences of spatial dispersion of tsunami wave fields observed by the precise deep-ocean bottom pressure recorders in the North Pacific.

Tsunami reflections from sea mounts have in some instances been sufficiently strong enough to the extent of affecting the credibility of tsunami forecasting unless the presence/influence of sea mounts was included in the tsunami forecast model simulation estimates. For example, when a large earthquake with $M_w = 8.3$ occurred in the Kuril Island Region off Japan on November 15, 2006, which caused a Pacific-wide tsunami, the tsunami forecasting agency in Japan (i.e., Japan Meteorological Agency [JMA]) disseminated tsunami forecasts for Japan based on the estimated tsunami amplitudes in the database. About 4 hours after the first detection of the tsunami on the coast of Japan, the JMA cancelled all the forecasts, judging from the amplitude decay in the real-time sea-level record at a station that no larger tsunami would be observed. Surprisingly, after the cancellation, larger amplitudes (than the first detected tsunami) arrived on the coast of Japan and were clearly recorded at many of the sea-level stations along the Pacific coast of Japan up to 5 hours later (Kamigaichi, 2009). The JMA conducted a close examination of this case and found that these large amplitudes were tsunami waves reflected from the Emperor Sea Mount Chain in the middle of the Pacific (Figure 4.10). Because the computational area for the tsunami simulation database of the JMA until that time was limited around Japan, such reflected waves could not be represented by the simulation. Therefore, today, the JMA examines the scenarios for which the computational area should be expanded (Kamigaichi, 2009).

The influence of ridges on tsunami propagation and amplification has been felt also during the September 2007 weak tsunami, generated due to the $M_w = 8.4$ earthquake in Sumatra on September 12, 2007. Except for Padang (00°57' S, 100°22' E) in Indonesia and Salalah on the Arabian coast in the Persian Gulf, which reported trough-to-crest tsunami wave heights of 2.40 m and 1.45 m, respectively, all other locations recorded less than a meter. The relatively larger wave height reported at Salalah is believed to be due to focusing or interference from the Carlsberg Ridge. As noted earlier, tsunami wave properties

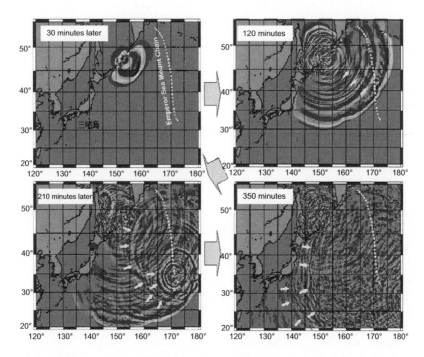

FIGURE 4.10 Snapshots of numerical tsunami simulation for the November 15, 2006, Kuril earthquake event ($M_w = 8.3$) demonstrating tsunami wave reflection from the Emperor Sea Mount Chain in the middle of the Pacific. Coriolis force has been considered for this case. The *white dotted line* marks the position of the Emperor Sea Mount Chain. Broken red curves with yellow arrows denote reflected tsunami waves from the Emperor Sea Mount Chain. The estimated arrival times of the reflected waves at the coast of Japan are consistent with the actual tide records. *Source: Kamigaichi, 2009. Reproduced with kind permission of Springer Science+Business Media.*

(particularly wave height and period) in continental coastal regions are strongly affected by the resonant characteristics of the shelf and coastline.

Straits also play a considerable part in directing the tsunami energy. According to Rabinovich and colleagues (2008), the two deepest straits in the Kuril Islands—Bussol Strait and Kruzensterna Strait—provided the main "gateways" for the tsunami energy flux during the 2006 and 2007 Kuril Islands tsunami events. For both events, the tsunami energy "tongue" protruded into the Sea of Okhotsk (which was a little stronger for the 2006 event) and propagated northwestward toward the northern shore of Sakhalin Island.

4.4. TSUNAMIS ON ISLANDS

Tsunamis on isolated, open-ocean islands are much less affected by topography. For example, at St. Helena Island in the Atlantic Ocean, the largest signal in association with the December 2004 Sumatra tsunami was just 3 cm (Woodworth et al., 2005). However, the prominent tsunami signal at Takoradi

(Ghana), which is the closest coastal location to St. Helena Island, was almost 13-fold larger than the tsunami signal at St. Helena Island (Joseph et al., 2006). Reduced tsunami amplitudes at island stations have been reported also in the case of the September 2007 tsunami caused by the $M_w = 8.4$ earthquake in southern Sumatra (Indonesia) in the Indian Ocean (4.517° S, 101.382° E), which was a relatively weaker one compared to the powerful December 2004 tsunami. Whereas the maximum trough-to-crest tsunami wave heights at Colombo and Trinconmalee in Sri Lanka were 50 cm and 60 cm, respectively (Pattiaratchi and Wijeratne, 2009), those at Male (Maldives archipelago) and Kavaratti Island (Lakshadweep archipelago in the eastern Arabian Sea) were only 24 and 5 cm, respectively (Prabhudesai et al., 2008). The reason for the observed relatively small height of the tsunami wave at these islands is their open-ocean position in contrast to the other two locations.

Whereas tsunami signals at open-ocean islands are relatively weak in general, exceptions have been reported. For example, while the December 2004 tsunami at St. Paul Island located in the South Indian Ocean recorded 15 cm height, that at neighboring Kerguelen Island recorded 68 cm height, which is a factor of 4.5 greater than the tsunami wave height at St. Paul Island. According to the numerical model of the 2004 Sumatra tsunami by Titov and colleagues (2005a, b), this effect is due to local amplification of tsunami waves that impinged on the Kerguelen Ridge (Rabinovich and Thomson, 2007).

There have been instances where islands have trapped tsunami waves. For example, it has been suggested that part of the energy of the tsunami waves generated by the November 2006 and January 2007 Kuril earthquakes at Simushir, off Japan, was trapped by the shelf around the Kuril Islands. Due to wave trapping (LeBlond and Mysak, 1978), part of the tsunami energy propagated as trapped waves (Stokes edge waves; see Efimov et al., 1985) southwestward toward Japan and northeastward along the Pacific shelf of the Kuril Islands in the direction of Kamchatka. This effect plays a crucial role in tsunami energy conservation, wave transmission distance, and long-term tsunami "ringing" (Miller et al., 1962; Kajiura, 1972). More energy gets trapped when the source area is closer to the island. For example, it has been observed that the 2006 Kuril tsunami source area (which was closer to the Kuril Island shore than was the 2007 source area) resulted in more of the energy from the 2006 event becoming "trapped" over the Kuril shelf where it propagated as edge waves along the Kuril Islands (Rabinovich et al., 2008). It has also been suggested that the late arrival of the highest waves at Malo-kurilsk and at some sites on the coast of Japan is related to the "trapping" effect (Lobkovsky et al., 2009). Edge waves at the shelf "conserve" tsunami wave energy and transport it with minimum loss, but their speed is much lower than the speed of tsunami waves in the open ocean, which determines the propagation of the leading tsunami front and first arrival of the waves at coastal sites.

4.5. TSUNAMI-INDUCED SEICHES IN HARBORS

In a tsunami event, the most important happenings occur at the coastline, mainly because it is here that loss of lives and material damages occur due to land inundation at high speeds. The destructive effects of tsunamis in coastal regions are usually limited to several hours. The coastal water body is the landward region of the continental shelf, which behaves as a semienclosed body of water and is therefore subject to resonance phenomenon. Free oscillations (the so-called normal modes of oscillations) of the coastal gulfs, bays, estuaries, inlets, lagoons, and backwaters play an important role in determining the coastal behavior of tsunamis. Free oscillation in a semienclosed water body is similar to the oscillation of a pendulum where the oscillation continues after the initial force has ceased to exist. Several factors cause the initial displacement of water from a level surface, and the restoring force is gravity, which tends to maintain a level surface. Several different physical phenomena can set a water body into oscillation (i.e., excite its normal modes). Once formed, the oscillations are characteristic only of the geometry (length, width, and depth) of the water body and may persist for many cycles before decaying under the influence of friction. Thus, the frequencies of the fundamental normal mode and its higher harmonics can be determined solely from knowledge of the geometry of the water body. The standing wave, or seiche, develops due to repeated reflections (assuming negligible dissipation) from the physical boundaries of the water body (i.e., the boundary walls).

A much neglected tsunami effect is the setup of coastal oscillations (seiches) that can last for three to five days after the impact of the tsunami and can affect port operations and coastal sea levels (Pattiaratchi and Wijeratne, 2009). The fundamental period of oscillation of a continental shelf is given by Merian's formula for an open system (Pugh, 1987):

$$T_n = \frac{4L}{n \times \sqrt{gH}} \tag{4.24}$$

where L is the width of the continental shelf, H is the mean water depth, g is acceleration due to gravity, and n is the mode number ($=1$ for the fundamental mode). Tsunamis enhance the existing natural oscillation of the continental shelf. For example, Pattiaratchi and Wijeratne (2009) found that after tsunami events, enhanced (i.e., higher-energy) oscillations were predominant in all of the sea-level time-series from the continental shelf adjacent to Colombo (Sri Lanka), as well as Western Australia (Figure 4.11). The time-frequency plot (Figure 4.12) indicates higher-frequency energy around the 75-minute band (which is the fundamental period of oscillation of the continental shelf adjacent to Colombo) before the tsunami. Although spectral energy increased across all frequencies during the tsunami, the energy at the higher frequencies decreased subsequent to the tsunami. Nonetheless, the energy at the 75-minute frequency band continued for almost 10 days. Persistence of tsunamis in harbors, bays,

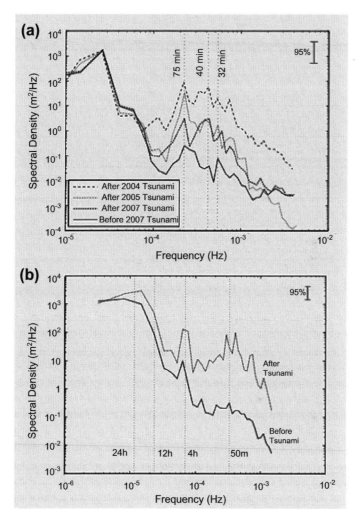

FIGURE 4.11 Enhanced oscillations predominant in the sea-level time-series from the continental shelf adjacent to (a) Colombo (Sri Lanka) and (b) western Australia. *Source: Pattiaratchi and Wijeratne, 2009. Reproduced with kind permission of Springer Science and Business Media.*

and estuaries is determined largely by the fundamental period of oscillation of the respective water bodies and therefore is decided by their geometry.

Free oscillations of coastal water bodies play an important role in the coastal behavior of tsunamis. Not only do these oscillations contribute to the transformation of the tsunami waves in the coastal region, but they also play a very important role in the secondary undulation that persists in the coastal water bodies up to several days after the main tsunami activity has died down. A glaring example of high-water-level persistence due to free oscillations of

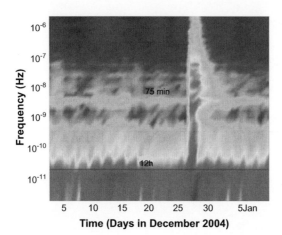

FIGURE 4.12 *Time-frequency plot of the sea-level record from Colombo pertaining to the December 2004 tsunami, indicating the persistence of the fundamental period of oscillation of the continental shelf (75 min) before and after the tsunami. Source: Pattiaratchi and Wijeratne, 2009. Reproduced with kind permission of Springer Science and Business Media.*

coastal water bodies is that which occurred in the State of Kerala in India during, and even after, the dissipation of the main Indian Ocean tsunami of December 2004. According to Nirupama and colleagues (2007c), the reason for the tsunami persistence in the Kerala coastal water bodies is the so-called Helmholtz mode of resonance (also known as co-oscillating mode (Platzman, 1972) and pumping mode (Lee and Raichlen, 1972)), whose natural frequency of oscillation is given by the expression:

$$\omega_n \equiv c\sqrt{\frac{A}{lV}} \tag{4.25}$$

where c is the wave velocity, A is the area of cross section, l is the length of the narrow channel, and V is the volume of the water body.

In this case, the long gravity wave energy entered the wide water bodies in Kerala through narrow entrance channel(s). The tsunami energy, once it has entered the water body, cannot easily get out because each successive reflection from its physical boundaries (i.e., walls) leaks out only a small amount of energy from the resonating water body to the external open water body. Basically, Helmholtz resonance represents the balance between the kinetic energy of water flowing in through a narrow connecting channel and the potential energy from the rise in the mean water level within the water body (Freeman et al., 1974).

One of the physical oceanographic processes that contribute to the amplification of tsunamis is the coupling of tsunami waves to internal waves generated at density interfaces. Using the analytical numerical model developed by Imamura and Imteaz (1995), Nirupama and colleagues (2007d) showed that the internal wave coupling could add at least 1 m to the overall tsunami amplitude at some locations on the Tamil Nadu coast of India, as well as in Sri Lanka, during the December 2004 tsunami. They have also suggested that since the Bay of Bengal is subject to strong stratification, it is quite conceivable that the generation of internal waves at the density interfaces is quite routine.

4.6. TSUNAMIS IN INLAND WATER BODIES

Tsunamis and tsunami-like water excitations have occurred from time to time in internal seas and inland basins such as lakes, rivers, and even artificial water supply reservoirs, although such excitations were not as intensive as those in oceans. Through proven geological methods, prominent evidences of paleo-tsunamis have been revealed on the shores of these water bodies. Natural environments can preserve tsunami records that can be extended beyond the historical record in several ways. Identification and analysis of tsunamigenic deposits provide insight into their characteristics, including their recurrence interval. Low-elevation lakes less than 5 m above mean sea level and situated close to the shore have only recently been explored for paleo-tsunami deposits (Hutchinson et al., 2000). Tsunami deposits commonly exhibit evidence of rapid deposition such as grading or massive structure. Storm-surge deposits most closely resemble tsunami deposits. However, storm-surge waves do not penetrate the distance of a considerably longer wave such as a tsunami wave. A tsunami deposit is usually identified by the sedimentary context (e.g., deposited on soil associated with coseismic subsidence), a larger grain size than surrounding sediments (indicating higher-energy depositional conditions), the spatial distribution of the deposit, and ruling out other high-energy depositional modes (e.g., storm surges or floods). Additional information that indicates a marine source of the sediments, such as microfossils (Hemphil-Haley, 1995) or geochemical signature, is also useful for determining that a deposit was formed by a tsunami. Foraminiferal analyses are considered to be the best method to differentiate a storm surge from the tsunami. Whereas the foraminifera found in storm-surge deposits are characteristic of the beach environment, those found in tsunami deposit are deep-sea foraminifers.

The earliest instance of reported tsunami-like waves in lakes is perhaps of volcanic origin (Didenkulova and Pelinovsky, 2006). For example, in 1305, a series of pyroclastic flows descended from Mount Tarawera to Lake Tarawera and created tsunami waves that measured 6–7 m on the opposite coast of the lake (De Lange et al., 2002). Eruption of a volcano near Lake Taal in the Philippines generated a 5-m tsunami (De Lange and Healy, 2001). Similar events occurred also in Russia, and Didenkulova (2005, 2006) compiled the first list of such events.

The reason behind the tsunami phenomenon at inland water bodies must be predominantly sought in the earthquakes, although in some cases, it was found to be due to volcanic eruption and underwater landslides. Seismic causes are quite understandable for such seismic-prone regions as basins of the Black Sea and Lake Baikal. But now it is also true of the Scandinavian area and its vicinity, including the territory of Russia as well, because relatively intense seismic events have occurred in the past (Nikonov, 2009). Thus, tsunami danger is a genuine problem to be thoroughly studied in the above-mentioned intra-continental water basins.

For about two decades, Professor Nikonov of the Institute of Physics of the Earth, Moscow, Russia, has been involved in the acquisition of primary written sources concerning tsunamis and tsunami-like water disturbances in relatively small water bodies such as the Black and the Caspian Seas, the Baltic and the White Seas, Lake Ladoga in the western part of Russia, and Lake Baikal in east Siberia. It was found that the shores of each of them were affected by tsunamis in historical periods many times, including events recorded in the twentieth century. His investigations have brought to light tens of tsunamis and tsunami-like events in areas of ancient civilizations such as the shores of the Black and the Baltic Seas. Soloviev (1978) was the first to report tsunamis in rivers originated by slides.

There have been instances where large buildings collapsing have generated tsunamis. For example, a tsunami wave was observed in Nizhny Novgorod (Russia) on July 18, 1597, when the whole Pechersky monastery (200 m to 300 m) with all of its buildings slid into the Volga River. The resulting tsunami caused a wave run-up up to 25 m (Didenkulova et al., 2002; Didenkulova and Pelinovsky, 2006). This event occurred in the vicinity of Nizhny Novgorod, which was one of the major ancient Russian cities, situated about 400 km from Moscow at the junction of two large Russian rivers: the Volga and the Oka. This landslide-generated tsunami event became well known primarily because of the destruction of this monastery (at that time, a famous center of cultural and literary activities) in the vicinity of Nizhny Novgorod. Under the tsunami force, boats anchored in the river under the monastery were thrown 43 m inland (Didenkulova and Pelinovsky, 2002). A chronicle about this incident is provided by Gatsysky (2001). During a period of over 400 years (1597−2006), nine tsunami events have taken place in the internal basins (rivers, lakes, and artificial water supply reservoirs) of different regions of Russia.

Historical documents contain several reports on tsunami-like events that occurred in the African lakes, with the most recent case reported at Lake Nyos, Cameroon, on August 21, 1986, when a giant "bubble" of CO_2 gas came from the depths of the lake to the surface and generated destructive water waves that killed many people (Gusiakov, 2009).

Chapter 5

Impact of a Tsunami on Coastal and Island Habitats

Tsunami waves that hit coastal- and island-lagoon water bodies can cause inundation and considerably modify their topography and destroy shallow water habitats. The marine ecosystem (a unit of unique biological communities consisting of a variety of coexistent flora and fauna species) plays a major role in regulating atmospheric oxygen, carbon dioxide, and filtering the pollutants in the water (Cohen and Tilman, 1996; Cotanza et al., 1997; Daily, 1997). Post-tsunami offshore, as well as near-shore, field surveys to assess the impact of tsunamis on offshore, coastal, and island habitats revealed extensive ecological damage. The devastation of these vast stretches of coral reef beds, mangroves, sea grass, and seaweed beds that function as feeding and nursery grounds for a myriad of species of fish and shellfish populations would adversely affect the fishing industry and fish farms for long periods (Aswathanarayana, 2007). Several aquaculture units located close to the shore zone were badly affected by the tsunami; the farms were damaged significantly, and cultured shrimp were either killed or washed into the sea.

Perhaps the most serious ecological damage was the trail of destruction of coral reefs, which are limestone structures built by soft corals and other calcifying algae through secretion of calcium carbonate (Sridhar et al., 2007). The corals act as a sink for over 2% of the present output of anthropogenic CO_2 (Kinsey and Hopley, 1991), and their reef-building process helps to reduce global warming. Corals also supply raw materials for life-saving drugs (Krishnakumar, 1997). Coral reefs are well known not only for their fascinating beauty and ecological uniqueness but also for the livelihood of millions of people in and around the coral reef area. Coral reef ecosystems can be considered the ecosystem most likely to be damaged by environmental stressors.

The December 2004 tsunami partially or fully destroyed coral reefs from several areas in the Indian Ocean (Chadha et al., 2005). Many coral reefs lost both their structure and biota and were reduced to rubble due to mechanical damage. Corals and coral reef organisms were dislocated and washed ashore in the north Andaman Islands. Several staghorn corals were broken off by the force of tsunami waves. This tsunami event caused two kinds of damages to the coral reefs of the Andaman and Nicobar islands (Rao et al., 2007): The reefs

Tsunamis: Detection, Monitoring, and Early-Warning Technologies. DOI: 10.1016/B978-0-12-385053-9.10005-5

were either completely eroded and/or sand and mud were deposited on them. The giant tsunami waves smashed and crushed the reefs, while the backwash caused the deposition of debris on the coral reefs. In Southern Andaman and Nicobar Islands, the large volume of sand and debris swept by the tsunami and deposited over corals and other marine species suffocated and choked them (Sarang, 2005). Generally, persistent stress can trigger bleaching in corals. However, they can recover from bleaching when the environment is favorable. Sedimentation is one of the major threats to the coral reefs. When the sedimentation persists, it cuts off the light that is essential for photosynthesis by their symbiotic partner, from which it receives nearly 98% of its nutritional requirements. As a result, the corals are deprived of energy and stressed. Further, it is possible that the sediments may carry pathogens that can cause coral diseases (Szmant, 2002). Thus, sedimentation is a matter of much concern, since it interferes with the corals' recovery and results in the permanent loss of coral colonies. Excessive sedimentation can kill live coral colonies by suffocating the corals and preventing them from feeding, reducing the light levels reaching the corals that are essential for the coral symbionts *zooxanthellae* to perform photosynthesis, and interrupting the food supply from the symbionts to the corals. Although some coral species are reported to be sediment-tolerant (Solandt et al., 2001), most of the species are highly intolerant to even small amount of sediment. The permanent loss of corals could unbalance the intricate relationship among dependent species and therefore affect the biodiversity and functioning of the entire ecosystem. Estimates made from satellite imagery (SAC, 2005) indicated loss of coral reefs over 23,000 and 17,000 ha in the Andaman and Nicobar Islands, respectively, in the Indian Ocean. About 6740 ha of reef in the Andaman Islands and 6140 ha of reef in the Nicobar Islands are presently covered by sand and mud.

The coral reefs of the Gulf of Mannar off the southeast coast of India have been protected to a large extent by the Island of Sri Lanka from direct impact of the Sumatra tsunami (SDMRI, 2005). In this region, fragments of seaweed and sea grass were washed ashore by the tsunami, and in some areas the fragments were entangled with branching corals. There was, however, no major deposition of sand and debris on the seaweed and sea grass beds (SDMRI, 2005). However, this tsunami reduced the live coral cover significantly in the adjacent Palk Bay.

It was found that the uprush of the tsunami surge brought enormous amount of sediments to the coast and altered the beach profile, especially at the intertidal zone. Intense mixing by the tsunami, as indicated by an increase in surface water salinity, a decrease in subsurface water salinity, a decrease in surface water temperature, an increase in subsurface water temperature, and an increase in suspended particles, was reported by several investigators (Sadhuram et al., 2007; Rao et al., 2007). A decrease in sea surface temperature (SST) by about 1°C on the day of the December 2004 tsunami was reported from the Andaman and Nicobar Islands region of the Indian Ocean (DOD,

2005). Disturbance to the seafloor mostly affects the benthos because of their bottom-dwelling nature. The immediate impact of the tsunami was found to be the draining of nutrients with the consequent lowering in chlorophyll concentration and thus of primary productivity. For five days after the December 2004 tsunami, the chlorophyll levels (an indicator of phytoplankton biomass) decreased abruptly in several places, including northeast of Sumatra Island and the coastal waters adjacent to some southeast Asian countries such as India, Sri Lanka, Myanmar, and Thailand (Rao et al., 2007). This affected the food web and resulted in the depletion in fish production. However, with time, there was perceptible recovery in the marine system, as evidenced from the improvement in biological productivity a few months after the tsunami (Iyer, 2007).

Evaluation of posttsunami effects on different coastal and island habitats has been found useful in planning the restoration of the affected habitats, understanding the socioeconomic implications to coastal and island communities, planning rehabilitation measures needed by local communities, and future preparedness. When a catastrophic natural disaster struck the Norwegian coast of Kvennavatnet, it took between 80 and 120 years for a comparatively diverse invertebrate community to develop. However, in the tropics, where the biological wheels turn faster (unlike in the subarctic), recovery of the marine environment from the tsunami impact could be much faster (Aswathanarayana, 2007).

Chapter 6

The Protective Role of Coastal Ecosystems

In some coastal zones the tsunami waves might take the form of a bore (i.e., step-like wave with a steep breaking front). A bore may form if the tsunami moves suddenly into a shallow-water shelf, bay, estuary, or river. The height of such tsunami-related onshore flooding frequently exceeds 10 m. This water-swell propagates shoreward with a large velocity, accumulating enormous kinetic energy and destroying targets on its way. In the majority of cases, the enormous volumes of seawater that break onto the shore during a tsunami event cannot be stopped by artificial barriers.

Mangroves are known to possess a multiplicity of features that allow them to sustain substantial environmental challenges such as sea-level rise and storm damage. Based on observational evidence, these features are considered to effectively function to ameliorate the effects of catastrophes such as hurricanes, tidal bores, cyclones, and tsunamis (Chapman, 1976; UNEP-WCMC, 2006; Alongi, 2008), although such observations have rarely been empirically tested or adequately assessed in the past (Ewel et al., 1998; Valiela and Cole, 2002). However, in recent experimental and model studies, the role of mangroves in attenuating the tsunami wave height has been demonstrated scientifically. Some models using realistic mangrove-forest variables suggest significant reduction in tsunami wave flow pressure for forests at least 100 m in width (Alongi, 2008). Recently, other supporting studies have indicated that mangroves and other coastal tree vegetation have the capacity to considerably reduce the damaging effects of a tsunami. Harada and colleagues (2002) conducted a hydraulic experiment to study the reduction of the tsunami effect on various coastal permeable structures such as mangroves, coastal forests, wave-dissipating blocks, rock breakwaters, and houses. Model simulations using data from hydrological experiments to predict the attenuation of tsunami energy by mangroves were generated by Hiraishi and Harada (2003) based on the 1998 tsunami that destroyed parts of the north coast of Papua New Guinea. The model output (Figure 6.1) suggests a 90% reduction in maximum tsunami flow pressure for a 100-m-wide forest belt at a density of 3000 trees ha^{-1}. Alongi (2008) examined model results obtained by Hamzah and colleagues (1999), Harada and Imamura (2005), Latief and Hadi (2007), and Tanaka and

Tsunamis: Detection, Monitoring, and Early-Warning Technologies. DOI: 10.1016/B978-0-12-385053-9.10006-7

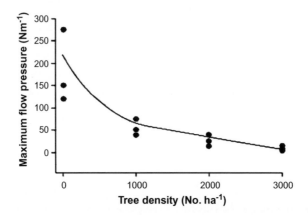

FIGURE 6.1   Model simulation of the decline in maximum flow pressure of a tsunami with increasing mangrove tree density, based on data collected after the 1998 tsunami on the north coast of Papua New Guinea. *Source: Modified from Hiraishi and Harada (2003). Adapted from Alongi (2008); Reproduced with kind permission of Elsevier.*

colleagues (2007) for various types of coastal vegetation, including mangroves, and found that the results were very similar. Tanaka and colleagues (2007) modeled the relationship of species-specific differences in drag coefficient and vegetation thickness with tsunami height and found that species differed in their drag force in relation to tsunami height, with the palms *Pandanus odoratissimus* and *Rhizophora apiculata* being more effective than other common vegetation, including the mangrove *Avicennia alba* (Figure 6.2). Thus, coastal vegetation plays a distinctive role in shore-front dynamics. Indeed, it was observed during the Sumatra tsunami episode that the damage to casuarinas plantations and coconut groves by the tsunami onslaught was minimal.

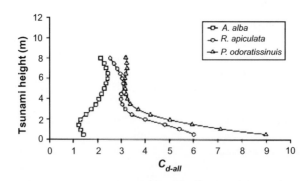

FIGURE 6.2   Changes in the relative drag coefficient for a vertical vegetation structure with increased height of a tsunami for mangroves Avicennia alba and Rhizophora apiculata and the palm Pandanus odoratissimus, estimated from modeling data from Sri Lanka and Thailand. *Source: Modified from Tanaka et al. (2007). Adapted from Alongi (2008); Reproduced with kind permission of Elsevier.*

Although the frontal casuarinas strips were attacked, bent, and stripped of their leaves by waves washing ashore, they remained largely intact and healthy (Mascarenhas and Jaykumar, 2007). These findings indicate the importance of preserving or selecting appropriate species to act as tsunami wave barriers to offer sufficient shoreline protection. Bioshields are useful for various reasons (Mascarenhas and Jaykumar, 2007):

- Shrubs control erosion and stabilize the shore.
- Green belts significantly alleviate wind energy, thus protecting the hinterland from oceanic forces.
- Trees are beneficial for biodiversity and can induce habitats for wildlife.
- The inhabitants of hazard-prone coasts would benefit from green belts in terms of security, access to food, materials, shelter, and income.

Various modeling and mathematical studies carried out thus far to understand the role of coastal mangrove forests and other vegetation in attenuating (absorbing) tsunami wave energy and thereby determining the extent of protection from tsunamis include several factors, such as the following (Brinkman et al., 1997; Mazda et al., 1997, 2006; Massel et al., 1999; Quartel et al., 2007):

- Bathymetry
- The width of the forest
- The slope of the forest floor
- Tree density
- Tree diameter
- Tree height
- The proportion of above-ground biomass vested in roots
- The diameter of stems and roots
- Soil texture
- Forest location (i.e., open coast vs. lagoon)
- Type of adjacent lowland vegetation and cover
- Presence of foreshore habitats (i.e., sea grass meadows, coral reefs, dunes)
- Size and speed of the tsunami
- Spectral characteristics of the incident tsunami waves (i.e., height, period, etc.)
- Distance from the tectonic event
- Angle of tsunami incursion relative to the coastline
- Tidal stage at which the tsunami wave enters the forest

Various studies reveal that mangroves are nearly as effective as concrete seawall structures for reduction of tsunami effects on houses behind the forest. Measurement of wave forces and modeling of fluid dynamics suggest that tree vegetation may shield coastlines from tsunami damage by reducing wave amplitude and energy (Massel et al., 1999). Several other studies indicate that the wave attenuation is achieved primarily because the mangrove forests function as sinks for the suspended sediments (Woodroffe, 1992; Wolanski et al., 1992; Wolanski, 1995; Furukawa et al., 1997). In numerous cases, the annual sedimentation rate, ranging between 1 and 8 mm, has been reported in mangrove areas with expansion of land (Bird and Barson, 1977). The density of mangrove species and their complexity and flexibility of aerial root systems also determine the sedimentation process and the wave reduction process (Kathiresan, 2003).

The closest scientific evidence for the protective buffering function of trees and foliage comes from socioeconomic and ethnobiological surveys, according to which these vegetations act as living dykes in coastal zones, protecting against disastrous ocean wave actions and the fury of a wide variety of cyclones and storm events (Pearce, 1996; Walters, 2003; Walters, 2004). Mangroves function as physical barriers against ocean influences, such as waves and storm surges, by means of their large above-ground aerial root systems and standing crops. In general, judging from the December 2004 tsunami, the coastal communities living behind the mangrove forests largely escaped from the fury of the tsunami. Although little could have prevented catastrophic coastal destruction in areas with the maximum tsunami intensity during this event, it has been found that further away, areas with thick coastal tree vegetation, such as mangrove forests and *Casuarina* plantations, were markedly less damaged by tsunami-induced waves than those areas without such lush vegetation (Danielsen et al., 2005). In this regard, a case study from India on the mitigating effects of mangroves on human lives against tsunamis has been reported by Kathiresan and Rajendran (2005). It has been found that in the southeast coast of India, there were only a few human casualties, and there was less economic damage in places where dense mangrove forests were present. Strangely, most of the deaths were a result of injuries from the thorny plant species *Prosopis spicifera*, which caused serious wounds eventually led to death. In another instance, the mangroves of Bhitarkanika in Orissa, India (which serve as the breeding ground for the Olive Ridley turtles), greatly reduced the impact of the "October 1999 Orissa super cyclone" that had a wind speed of 310 km/h, whereas in the areas without mangroves, over 10,000 people were killed, and millions were made homeless. Likewise, in and around the Yale National Park (Sri Lanka), where the waves were very strong and there were no physical barriers present, such as sand dunes, the mangroves and scrub jungle bore the full fury of the December 2004 tsunami. Also, a wide stretch of nearly impenetrable mangrove forest (covering 4262 square kilometers and 56 islands) in Sundarbans saved West Bengal and Bangladesh from the killer forces of the December 2004 tsunami. Sundarbans is the world's largest estuarine forest and part of the world's largest delta formed by the rivers Ganga, Brahmaputra, and Meghna. It is also a UNESCO World Heritage Site and home to the Tiger Reserve and National Park in India, with a larger portion of it spreading into Bangladesh.

Mangroves have also been observed to have the capacity to mitigate recent tsunami effects in other places. For example, in Indonesia, even though the epicenter of the tsunami was closer to Simeuleu Island, the death toll on this island was significantly lower because of the abundance of mangroves. In another example, in Thailand, the island chain of Surin off the west coast escaped heavy destruction because the ring of coral reefs and mangroves that surrounded the island helped to break the lethal power of the tsunami.

In general, the *Rhizophora* species, which occur seaward, have been found to be a more suitable species to mitigate the effects of tsunamis than the *Avicennia* species that exist generally landward (Kathiresan et al., 2005). This is because the aerial stilt roots of the former are more tolerant than the pneumatophores of the latter to long periods of submergence by floodwaters (Kathiresan and Bingham, 2001). These findings were corroborated in some field studies (e.g., at Cuddalore in southeast India) through ground surveys and analyses of cloud-free QuickBird pretsunami and IKONOS posttsunami satellite images from vegetated as well as unvegetated areas. It was found that areas with mangroves and tree shelterbelts were significantly less damaged than other areas (Danielsen et al., 2005).

In another study, multivariate analysis of mangrove field data (Dahdouh-Guebas et al., 2005) covering the entire Tamil Nadu coast of India suggests less destruction of man-made structures located directly behind the most extensive mangroves. Several studies have suggested that conserving or replanting coastal mangroves and greenbelts should buffer communities from future tsunami events. Mangroves also enhance fisheries and forestry production. These benefits are not found in artificial coastal protection structures. Regardless of their monetary value, mangroves offer protection from waves, tidal bores, and tsunamis, and they can dampen shoreline erosion (Mazda et al., 2007). Mangrove forests are fairly robust and highly adaptable (or tolerant) to waterlogged saline soils in warm subtropical and tropical environments. They are also capable of recovering quickly from catastrophic disturbances (Ward et al., 2006). However, they are suitable for planting only on coastal mudflats and lagoons. Elsewhere, the conservation of dune ecosystems or greenbelts of other tree species, such as *Casuarina,* would fulfill the same protective role. In coastal Asia, the resilience of the ecosystems has been degraded by activities such as deforestation of mangroves for intensive shrimp farming, coral mining, land clearing, and so on. Consequently, the adverse effects of the tsunami were felt heavily in some places in Sri Lanka and the Tamil Nadu coast of India during the Sumatra tsunami episode.

Sand dune complexes also function as nature's protection and are generally recognized as the last line of defense between oceanic forces and inland property. As natural geomorphic barriers, dune ridges protect the hinterland from wave run-up due to extreme oceanographic conditions such as those prevailing during tsunamis and storm surges. Wide beaches and high dunes serve as efficient dissipaters of wave energy. Posttsunami field surveys conducted by Mascarenhas and Jaykumar (2007) immediately after the December 2004 tsunami provided strong indications of the natural buffering capacity of littoral sand dunes during this tsunami. Although the sand dunes suffered various degrees of overwash, erosion, and even flattening during this tsunami, they exhibited their capacity to withstand high waves. Most of the assets behind dunes remained intact due to the natural buffer protection offered by the sand dunes. Globally, dune fields have offered protection in the wake of extreme events (Clark, 1996; Nordstrom, 2000; Pilkey et al., 2000).

According to Aswathanarayana (2007), an important component of preparedness systems is the application of dual-use technology and practices—that is, giving preference to the use of the technology/practice that will serve the dual purpose of protecting the coast against natural hazards, such as tsunamis and storm surges, and providing social, economic, and environmental benefits. In this sense, bioshields deserve more weight than a seawall because the latter provides only protection, whereas the former provides protection plus other benefits. The healthier the ecosystem, the more capable it is of regenerating itself and continuing to deliver resources and ecosystem services that are essential for human livelihoods and societal development.

Chapter 7

Earthquake Detection and Monitoring for Early Warnings of Seismogenic Tsunamis

Because undersea earthquakes (ruptures within the Earth caused by stress) have been responsible for approximately 82% of past tsunamis, early detection of such earthquakes has been found to be an effective method to evacuate from seismic tsunamis. Some methods have already been successfully employed, and others are still in experimental stages. These methods rely on a variety of signals that are emanated before the occurrence of an earthquake. Such pre-earthquake signals are known as earthquake precursors. In many cases, such as the Sumatra event, a warning of a tsunami is provided by a seismic signal.

7.1. EARTHQUAKE PRECURSORS

Some of the earthquake precursors have been used to advantage by the observing communities to evacuate areas before seismogenic tsunamis approach. Some other precursors, discussed in the following sections, are currently being analyzed in depth and should be of practical use in the coming years.

7.1.1. P-, S-, Love, and Raleigh Waves

An earthquake is associated with the release of strained energy that has been building over many years. The released energy propagates radially in all directions in the form of elastic waves or seismic waves. There are three basic

Tsunamis: Detection, Monitoring, and Early-Warning Technologies. DOI: 10.1016/B978-0-12-385053-9.10007-9

types of elastic waves associated with earthquakes, both before and during their genesis (Rao, 2005). The two types of body waves (because they propagate through the earth's body) are primary waves (known as *P*-waves) and secondary waves (known as *S*-waves). The two types of surface waves are Love waves and Raleigh waves.

P-waves propagate with tremendous velocities of the order of 19,800 to 28,800 km/h, and the motion is characterized by compressions and rarefactions (i.e., longitudinal waves). *S*-waves are relatively slower (velocities of the order of 11,880 to 17,280 km/h) and are transverse waves. Love waves are generally faster than the Raleigh waves. Whereas Love waves have horizontal particle motion (transverse to the direction of propagation), Raleigh waves have elliptical particle motion.

P-waves travel at a speed at least 22 to 40 times faster than the fastest tsunami waves (~713 km/h). *S*-waves travel at a speed that is at least 17 to 24 times faster than the fastest tsunami waves. The *P*-waves in the December 2004 Sumatra earthquake took only 4 minutes to get to the southeastern tip of Sri Lanka, 12 minutes to get to Antarctica, and 21 minutes to get to Europe. The *P*-waves arrived at every location on the globe (US Geological Survey Website). It is this considerable speed of seismic waves compared with tsunami waves that allows the applicability of seismic waves as precursors of seismogenic tsunamis. Detection of *P*- and *S*-waves using a network of seismometers is a practical method used in early warning systems. Raleigh waves have been detected by seafloor-mounted pressure gauges deployed in the deep oceans.

7.1.2. Microelectric Discharges and Electromagnetic Emissions

It is an accepted fact that a few days before a major earthquake occurs, when the plates begin to crack, microelectric and electromagnetic discharges begin to emanate from the preparation zone of the earthquake (Gufeld et al., 1994; Hayakawa et al., 2004). There are slow electrotelluric and magnetic field variations that could be useful tools for earthquake prediction. Molchanov and Hayakawa (1994) examined a mechanism of stochastic microcurrent activity due to the microfracturing process and explained the temporal behavior of Ultra Low Frequency (ULF) precursors, which were similar to those observed. The earthquake precursors can be observed on both electric and magnetic fields, but one of the two fields can be stronger than the other, at least on some occasions. ULF magnetic field precursors can be observed only for strong ($M_s > 5-6$) and low-depth ($z_o < 30$ km) earthquakes and at distances less than 100 to 200 km from the epicenter (Molchanov et al., 1995). Only a magnetic-type underground seismic source can be effective for the penetration of fields into the upper ionosphere and magnetosphere, and penetration is possible only in the ULF range.

It has been found that in addition to electric and magnetic fields, electromagnetic fields are also induced by seismic activity, and they can penetrate into the space plasma. The microelectric discharges ionize the lower part of the ionosphere that lie over the earthquake preparation zone. A number of electromagnetic phenomena associated with earthquakes have been reported in different frequency ranges: ULF ($f = 10-20$ Hz), VLF ($f = 0.1-100$ kHz), and HF ($f \geq 1$ MHz). The field variations depend on the dimensions of the source and its depth. The density of electromagnetic energy is small compared with seismic energy in the ground or the kinetic energy of the particles in the ionosphere. Also, polarization characteristics and the intensities of electromagnetic fields on the ground are considerably dependent on the configuration of the source current. The low efficiency of source radiation is connected with the inductive behavior of the fields and their dissipation inside the ground medium (Molchanov et al., 1995).

Thus, geophysical events leading to an imminent earthquake can give rise to ionospheric disturbances, and these can occur from several minutes to several days prior to the earthquake strike. It is believed that ingenious application of this effect would provide warning (strength, location of the earthquake) several days ahead of time, as opposed to several hours in other methods. Murty and colleagues (2007) suggested that the linkage between the troposphere and the ionosphere mainly occurs through the possible amplification of the acoustic and internal gravity waves in the atmosphere.

The probability of observation of seismic VLF or ULF emissions on board satellites traversing in the upper ionosphere and low magnetosphere of Earth have been predicted by several investigators (Parrot and Lefeuvre, 1985; Larkina et al., 1989; Bilichenko et al., 1990; Serebryakova et al., 1992; Molchanov et al., 1993). A review of satellite seismic observations is given by Molchanov (1993). Bilichenko and colleagues (1990) observed the pre-earthquake ULF emission bursts above the epicenter at altitudes of 850–900 km, using magnetic detectors in the frequency range 0.1–8.3 Hz on board the *Bulgaria*-1300 satellite. Molchanov and colleagues (1993) observed ULF emission bursts above earthquake epicenters, using an electric detector in the frequency band 8–1000 Hz on board the *Intercosmos*-24 satellite at altitudes of 500–2000 km. The efficiency of the ULF energy penetration into the ionosphere depends on the source configuration in the same way as on the ground. Gwal and colleagues (2007) discussed seismic precursors in the ionosphere as registered by the *DEMETER* (*Detection of Electro-Magnetic Emissions Transmitted From Earthquake Region*) satellite when it was flying over seismic regions, and they concluded that analysis and interpretation of ionospheric anomaly data would be of considerable application in the possible prediction of earthquakes and perhaps even seismogenic tsunamis.

7.1.3. Pressure Signals in the Troposphere

Even small earthquakes can generate pressure signals in the troposphere (the layer of atmospheric air extending about 7 miles above Earth's surface in which the temperature falls with increasing height) in the form of a train of waves (Benioff et al., 1951). Rayleigh waves emanating from the epicentral area can generate large pressure fluctuations in the atmosphere (acoustic waves in the gaseous envelope surrounding Earth) by impulsive effect, provided that their vertical displacement is large. Some regions exhibit resonant ground-air coupling. Such ground-coupled air waves (micropressure fluctuations) generated prior to the Alaskan earthquake of March 1964 were recorded at Honolulu, Berkeley, and Palisades (Donn and Posmentier, 1964).

7.1.4. Rise in Surface Temperature Near the Earthquake Zone

Accumulation of stress in the fault plane of the tectonic plate margins prior to the actual occurrence of the earthquake gives rise to an increase in subsurface pressure, which in turn results in an increase in temperature. Thus, Earth's crustal deformation prior to an earthquake affects the thermal regime near the ground surface. An increase in stress might lead to the release of greenhouse gases such as CO_2, CH_4, N_2, and so on, which become trapped in the pore spaces of the rocks. These gases escape to the lower atmosphere and create a localized greenhouse effect and thus augment the land surface temperature (LST) of the region (Saraf et al., 2007). Thus, Earth's emissivity can unfold many natural processes associated with earthquakes. Any thermal anomaly in tectonically active regions occurring on the land surface can be monitored regularly, and any abnormality observed under normal meteorological conditions could be an indication of an impending earthquake (Choudhury, 2005; Dasgupta, 2005). Consequently, close monitoring of LST in the vicinity of earthquake zones can yield valuable clues about possible imminent earthquakes and possibly seismogenic tsunamis.

Several recent earthquakes have indeed exhibited increased LST. Saraf and Choudhury (2003, 2004, 2005) examined several major earthquakes in different parts of the world and observed a rise in the LST around the epicentral region before each of these earthquakes, primarily induced by strain buildup over an extended area. Such strain buildups preceding an earthquake are known to be the cause of a number of other effects (earthquake precursors) such as the electric, magnetic, and electromagnetic fields described earlier. Identification of thermal anomalies associated with impending earthquakes can lead to the identification of areas of increased seismic risk.

Already available and evolving techniques of remote sensing have the potential to aid in prior warning of earthquakes. Satellite-based remote sensing provides repetitive coverage of multispectral information about events on the Earth's surface, which could also be manifestations of what happens in the Earth's interior. The use of remote sensing in earthquake studies has made great strides in the last few years in identifying signs (indicators) of impending danger. Thermal remote sensing in seismic studies was first used in Russia in 1985 (Gorny et al., 1988; Tronin, 2000). Subsequently, case studies on earthquakes were carried out in Russia, Japan, and China. The realization that the Earth's surface temperature is significantly related to the Earth's physical processes led to the development of an interesting trend of earthquake studies. For example, Saraf and colleagues (2007) analyzed the thermal data from the *NOAA-AVHRR* (*Advanced Very High Resolution Radiometer*) satellite (the temperature accuracy of *AVHRR* is 0.5°C) for the region around Sumatra and Myanmar and observed a sharp rise in temperature prior to the powerful earthquake in Banda-Aceh, Sumatra. They found that thermal anomalies started appearing about 5 days before the December 26, 2004, Sumatra earthquake, and the LST anomaly reached a maximum of 4−6°C on December 25, 2004. The anomaly (spread over a vast region, extending as far as Thailand and Myanmar) disappeared after the earthquake.

7.1.5. Increases in Sea Surface Temperature Near the Epicenter

It has been observed that the sea surface temperature (SST) on the west coast of northern Sumatra (close to the epicenter of the earthquake that generated the notorious Indian Ocean tsunami) was appreciably high (~30−31°C), with a variation of 0.5−1.0°C on December 26, 2004 (Rao et al., 2007). This is probably due to the release of thermal energy (midinfrared emission) from the genesis area of the strong earthquake prior to the earthquake's actual occurrence (Ouzounov and Freund, 2004). The spatial extent of the elevated SST region diminished, probably because of the subsequent incursion of subsurface cold waters due to the vertical mixing caused by the tsunami wave propagation (Rao et al., 2007).

7.2. EARTHQUAKE DETECTION THROUGH MONITORING THE BEHAVIOR OF ANIMALS, REPTILES, AND BIRDS

Although humans are believed to possess superior intellectual capabilities relative to other animals, certain land and marine animals, reptiles, and birds are found to be more sensitive to vision, hearing, vibrations, touch, temperature changes, and smell. For example, whereas the human ear can hear only the audible part of the acoustic spectrum, certain animals can hear parts of the ultrasonics and infrasonics. It has been observed for quite some time that some of these creatures are sensitive to earthquake and tsunami signals. It is inferred that they can recognize and react to any, or at least some, of the earthquake precursors just mentioned. It is interesting to note that on December 25, 2004 (i.e., the day before the great Indian Ocean tsunami), snakes in several parts of India came out from their hiding places and acted as if they were under stress (the author observed firsthand the hysterical movements by a cobra in his garden that day).

An interesting observation about the Andaman and Nicobar Islands (the only places in the world where Paleolithic tribes live in their native reserved habitat with primitive culture) in the Indian Ocean during the December 2004 tsunami was that members of the aboriginal tribal population in these islands safely retreated to the elevated areas because of premonition and their gift of native intelligence and thus escaped the wrath of this tsunami (Shaw, 2005). Also, no animals died because of the tsunami on these islands (Mohan, 2005). This tells us that some type of instinctive knowledge warned them to escape from the tragedy. It is extremely ironic that while many civilized, educated individuals perished during the tsunami, these aboriginal tribes with no formal education were able to predict and avoid danger, probably by observing animal behavior and responding to nature's danger signs quickly and effectively. We can still learn a lot from studying these groups and the behavior of animals in these situations (Murty et al., 2007).

Chapter 8

Numerical Models for Forecasting

Once the generation of a seismic tsunami has been predicted, the most basic information an early tsunami warning center (TWC) requires to provide to the people at imminent risk is the expected time of arrival (ETA) of the first wave in the tsunami wave train at selected coastal locations from the tsunami source location. Judging from past experience, the first wave in a tsunami event is not generally the wave with the greatest amplitude. In fact, one of the most dangerous phenomena that occurs at some locations (which is proved to be highly deceptive in several instances in the history of tsunami events) is the withdrawal of the ocean before the first tsunami wave crest arrives at that location. This phenomenon (generally referred to as initial withdrawal of the ocean, or IWO) neither occurs at the same location for every tsunami nor at every location for the same tsunami (Nirupama et al., 2007f). Sometimes, IWO leads to tragic circumstances because curious onlookers walk right up to the edge of the water, which is usually very far away from the normal beach because of the IWO, to watch a plethora of fascinating exposed fauna and precious marine treasures of archeological significance, which under normal circumstances are never exposed. During the crest phase of a tsunami wave train, seawater penetrates the coast at a high speed and causes extensive inundation (known as run-up, which is expressed in meters either above the mean sea level or above the normal tide). The 1775 Portugal tsunami episode is a classic example of the loss of several lives with the arrival of the succeeding crest of an IWO type of tsunami. Although the IWO process can partially be explained on the basis of the so-called N-waves (waves with a leading depression), such an explanation is relevant only in those cases where the epicenter of the earthquake is about 100 km from the coastline where the IWO occurs (Tadepalli and Synolakis, 1994).

Tsunami travel time charts (i.e., charts that indicate the travel times of the tsunami wave front from the tsunami source location in hourly contours) that are relevant to the earthquake that is responsible for the tsunami generation are helpful for issuing early tsunami warnings for evacuation purposes. For the problem of real-time tsunami forecasting by a warning center, there is too little time to compute such synthetic data for a scenario as it is playing out.

Tsunamis: Detection, Monitoring, and Early-Warning Technologies. DOI: 10.1016/B978-0-12-385053-9.10008-0
115

Therefore, it is necessary to use precomputed model runs to prepare tsunami travel time charts and detailed information about likely tsunami scenarios for preselected coastal and island locations. Once a reasonable number of charts have been prepared, taking into account all historical tsunami events, it is quite reasonable to assume that for any future tsunami events, the travel time information that is required can be quickly and effortlessly obtained from these charts.

The precomputed synthetics that best fit the seismic parameters and sea-level measurements available at the time can form the basis for estimating impacts further afield. The success of this forecasting scheme depends on several factors, including the appropriateness and accuracy of the precomputed model runs, the accuracy of the seismic parameters, the availability and accuracy of sea-level measurements, and the uniqueness of the fit (McCreery, 2005).

Although the first wave in a tsunami event is not generally the wave with the greatest amplitude, tsunami travel time charts are generally constructed for the first wave rather than the wave with the highest amplitude. The reason for this is that advance knowledge of travel times for the first wave (in the tsunami wave train) provides some additional valuable time for the evacuation of people if and when evacuation is needed. Also, tsunami travel times can be precomputed, independent of the seismic moment magnitude of the earthquake, only for the first wave. The heights of the tsunami waves depend on a combination of several factors, and, therefore, predetermination of the travel time of the highest wave remains to be a Herculean task at present.

Tsunami travel times from the source area are estimated from knowledge of the bathymetry of the ocean at successive locations. A variety of numerical tsunami modeling has taken place in recent years. A model known as MOST (**M**ethod **O**f **S**plitting **T**sunamis), which was developed by Dr. Vasily V. Titov (a chief scientist at the NOAA Center for Tsunami Research, the joint operation of NOAA's Pacific Marine Environmental Laboratory and the Joint Institute for the Study of the Atmosphere and Ocean of the University of Washington), is now being widely used for tsunami forecasting and the study of tsunamis in many countries. The MOST model that was run for a comprehensive suite of Alaska-Aleutian sources (Titov and Gonzalez, 1997a, b; Titov et al., 1999), the WC/ATWC model (Whitmore and Sokolowski, 1996) that has now been run for a variety of historical and hypothetical sources around the Pacific, and the Cheung model (Wei et al., 2003) that has been run for historical and hypo-thetical events in the Alaska-Aleutian region are among some popular tsunami models. For all of these models, synthetic deep-sea records have been computed for comparison with tsunameter data as a key constraint for fitting an actual teletsunami scenario (i.e., a tsunami that traverses across an entire ocean basin) to a synthetic one. The National Tsunami Hazard Mitigation Program (NTHMP) in the United States is presently providing support for the devel-opment of an automated forecasting tool that will incorporate all three modeling approaches. The tool will ingest the seismic parameters and sea-level data from coastal stations and deep-ocean tsunameters as it becomes available. Based on these constraints, a program is underway to produce forecasts of near-shore tsunami waveforms, run-ups, and inundations at selected locations (Titov, 2009a, b). Titov and colleagues (2005a) provides the global chart of energy propagation (hourly isochrones, indicated by the thin solid line) of the December 26, 2004, Sumatra tsunami from the source area in the Indian Ocean following the $M_w = 9.3$ megathrust Sumatra earthquake (indicated by the star), which was prepared based on the results of the MOST model estimates. Spatial distribution of the pattern of tsunami wave fronts observed in this figure indicates that the influence of bathymetry and ridges on tsunami propagation is clearly brought out by the model.

Tsunami modeling methods have matured into a robust technology that has proven to be capable of accurate simulations of past tsunamis. The NOAA's Pacific Marine Environmental Laboratory (PMEL) has developed the method-ology that combines real-time deep-ocean measurements with tested and veri-fied model estimates to produce real-time tsunami forecasts for coastal communities. This method (also known as a *Short-Term Inundation Forecast*, or SIFT) is currently being implemented at the NOAA's TWCs. Tsunami forecasts should provide site- and event-specific information about tsunamis before the first wave arrives at a threatened community. The next generation of tsunami forecasts provides estimates of all critical tsunami parameters (amplitudes, inundation distances, current velocities, etc.) based on direct tsunami observa-tion and model predictions. There are significant challenges in meeting a tsunami warning system's (TWS's) operational requirements: speed, accuracy, and user interfaces that provide guidance that is easy to interpret (Titov, 2009a, b).

As indicated earlier, one of the most important requirements for a TWS is generating simulations of expected travel times (i.e., time taken by the tsunami wave to reach the particular coast) and run-up heights for tsunamigenic earthquakes. Bhaskaran and colleagues (2005) developed the first tsunami travel time atlas for the Indian Ocean, covering the 37 countries that have a coastline on the Indian Ocean. For this purpose they used Geoware Ltd. software. In this package, the ETOPO-2 ocean bathymetry (www.gfdl.noaa. gov) with a resolution of 2 minutes are the input data, and the output data are the tsunami travel time contours in hours (Bhaskaran et al., 2007). This atlas provides the travel times for over 250 coastal locations in the Indian Ocean, for tsunamis generated in the potentially seismic locations in the Indian Ocean, and for some events in the western part of the Pacific Ocean that could have an impact on the Indian Ocean. These charts are expected to be accurate to ± 1 minute for each hour of tsunami travel. It is possible to compute travel times independent of the size of the earthquake because tsunami travel times depend primarily on ocean depth.

Another popular tsunami model is N2. The software for the *TSUNAMI N2* model was originally developed by Fumihiko Imamura in Tohoku University,

Japan, modified by Ahmet Cevdet Yalciner and his group at the Middle East Technical University, and then further modified by Costas Synolakis at the University of Southern California. This model is used in the TWCs of several IOC-UNESCO member nations. In the Indian TWC system, the *Tsunami N2* model has been used for predicting surges for different scenarios of earthquakes. This model uses the available earthquake parameters and assumes the worst slip rate. Because the model takes about 40 minutes to provide the output (i.e., travel time and run-up), a database of prescenarios has been created for the entire Indian Ocean. For operational early warnings, a database of prerun numerical simulations has been created that can be accessed at the time of an earthquake event to generate a forecast of the tsunami travel time and run-up estimates for different areas of the Indian Ocean coastline. The model scenarios provide information on 1800 forecast points that are generally towns, cities, and settlements, as well as the locations of the bottom pressure recorders (BPRs) and sea-level stations (Nayak and Kumar, 2008).

An important global tsunami model that has been able to provide a global picture of the December 2004 tsunami distribution and capture the "new physics" of tsunami propagation in the world oceans is the one developed by Kowalik and colleagues. Tsunamis are typically computed without the Coriolis force because their periods are much smaller than the inertial period. However, as propagation proceeds over long distances, the compounding effect of the Coriolis force may combine and significantly increase. On this assumption, the Kowalik model takes into account the Coriolis force. Kowalik and colleagues (2007a) computed the December 2004 global tsunami maximum amplitude distributions, one with and one without the Coriolis force, and the difference between the two (i.e., the residual) increased toward the polar regions, since the Coriolis effect increases poleward from the equator. The observed change was particularly significant along the South Pacific Oceanic Ridge. The increase in residual tsunami amplitude due to the Coriolis force was about 25%, thereby indicating that the Coriolis force plays a dominant role in the energy trapping along oceanic ridges. According to the model estimates computed by Kowalik and colleagues (2006), the larger tsunami amplitudes in the open ocean are located above the oceanic ridge, and the energy flux is directed along the ridge. The estimated period of the tsunami wave propagating over the ridge in the southwestern part of the Pacific Ocean was about 2 hours, which is much larger than the observed period at several locations during the December 2004 tsunami. As already discussed, the period of tsunami waves traveling over large distances can undergo an increase during propagation (Munk, 1947). However, the reason for such an unexpectedly long period for a free tsunami wave is interpreted to be the influence of the Coriolis force.

Other findings obtained with the application of the Kowalik model include local trapping of tsunami energy by oceanic ridges due to refraction and focusing of off-ridge energy toward the ridge, the possibility of a resonance interaction of the tsunami waves and ridge bathymetry, reflection, ducting along oceanic ridges, and complicated temporal and spatial patterns in tsunami propagation, among others.

The importance of tsunami modeling incorporating high-resolution bathymetric and topographic data and tsunami inundation mapping efforts based on more conservative "worst-case but realistic scenarios" has become increasingly evident in recent years. For example, incidents such as a relatively small tsunami that caused extensive damage in Crescent City harbor on November 15, 2006, served as a wake-up call that the tsunami threat is still not well understood for many coastal communities (Barberopoulou et al., 2009). These incidents attracted new attention to tsunami scientists and emergency managers who demanded a more systematic, consistent, and efficient response system. This implies that tsunami forecasting for coastal regions based on precomputed synthetics is still in its infancy, and, therefore, the forecasts need to be thoroughly evaluated based on consistency among methods and on a variety of quality control factors. Nevertheless, it holds the promise of being the most useful tool available for rapid and accurate decision making by TWCs and for possibly basing regionalized and multilevel warnings in the future. Such forecasting capabilities are expected to make it possible to provide adequate warning protection to areas at risk while limiting the adverse impacts of full evacuations only to coasts where it is really necessary (McCreery, 2005). However, real-time local or regional forecasting capabilities are most desirable and need to be pursued vigorously.

Chapter 9

The Role of IOC-UNESCO in Tsunami Early Warnings

The need for a reliable system for early tsunami detection and warnings was made painfully clear by the hundreds of lives lost to the disastrous tsunami in Hawaii on April 1, 1946, from an earthquake generated in the Aleutian Islands off the coast of the United States (Okal et al., 2002). As a result of this event, the Intergovernmental Oceanographic Commission (IOC) of the United Nations Educational, Scientific, and Cultural Organization (UNESCO) took the initiative to establish the Pacific Ocean Tsunami Warning and Mitigation System during the IOC Fourth Session of the IOC Assembly (November 3–12, 1965), through Resolution IV-6, to create the International Coordination Group for the Tsunami Warning System in the Pacific to detect tsunamis and provide prompt notification of tsunami threats to member states. The Pacific Ocean Tsunami Warning System (PTWS) is based in Ewa Beach, Oahu Island, Hawaii. The system contains many member countries and has been administered by the IOC of UNESCO in Paris since 1965. Because tsunamis are rare in the Atlantic and Indian Oceans (at least compared to the Pacific Ocean), until recently there have been no tsunami early warning systems for these two oceans.

In the present era, the December 2004 tsunami demonstrated that catastrophic events that were considered rare could be devastating for unprepared coastlines. Although the major victim of this tsunami was the Indian Ocean, the Atlantic and several marginal seas such as the Caribbean Sea, the Mediterranean Sea, and the East China Sea were not without risk. For example, while the northwest Indian Ocean regions are sensitive to potential tsunami risks from the Makran Subduction Zone (in the Northern Arabian Sea) that extends off the coasts of Iran and Pakistan, a devastating earthquake of estimated $M_w = 8.5/9.0$ destroyed Lisbon (Portugal) on November 1, 1755, killing over 60,000 people, and was felt throughout Western Europe. It generated a huge tsunami that reached the coastlines from Morocco to southwestern England with local run-up heights as high as 15 m in Cap St. Vincent (Portugal) and Cadiz (Spain). Large waves (3 m) were reported in Madeira Islands and as far as the West Indies, causing large-scale destruction (Roger et al., 2009). The tsunami waves generated by the 1755 Lisbon earthquake in the Atlantic Ocean affected even the coasts of the French West Indies in the Caribbean Sea (Zahibo et al., 2009).

Tsunamis: Detection, Monitoring, and Early-Warning Technologies. DOI: 10.1016/B978-0-12-385053-9.10009-2

The Gulf of Cadiz, located close to the eastern part of the Nubia-Eurasia plate boundary, has been the location of several tsunamis, such as the well-known event of November 1755. This area, which is part of the northeast Atlantic, the Mediterranean, and connected seas (NEAM) region, has no warning system in place even today.

Historical records show that the eastern Mediterraneann region encountered numerous large earthquakes and tsunamis. The eastern segment of the Hellenic arc trench is one of the most active in the Mediterranean region, producing large earthquakes and tsunamis. The island of Rhodes has been hit many times by large earthquakes and associated destructive tsunami waves in AD 148, 1303, 1481, 1609, 1741, and 1851. A local aseismic tsunami inundated part of the northwest coastal segment of the city of Rhodes on March 24, 2002 (Fokaefs et al., 2009). The Mediterranean coasts of Africa are also known to have experienced strong tsunami impacts in history, mainly related to far-field tectonic sources and in particular to different sections of the Hellenic arc and in the Cyprean arc thrust zones (Tinti et al., 2009a, b). In the Mediterranean, the cataclysmic Santorini tsunami struck Alexandria in 1638 BC, and smaller tsunamis hit in AD 363 and 1303. Tsunami simulation studies have shown that the town of Fethiye and Fethiye Bay on the southwest coast of Turkey are likely to be directly affected by tsunamis that can be triggered by the rupture of earthquakes in the eastern Mediterranean (Dilmen et al., 2009). Likewise, southern Italy as a whole, and in particular Apulia, Calabria, and Sicily, are some of the Mediterranean regions that are most exposed to tsunami attacks of both local and remote origin. Consequently, implementation of both national and regional tsunami warning systems (TWSs) in these regions is an urgent requirement.

After the devastating December 2004 tsunami, a systematic effort was initiated for the establishment of regional, national, and local early TWSs in Europe and Africa under the coordination of IOC. In 2005 IOC-UNESCO decided to implement a global TWS in the Indian Ocean, the Atlantic Ocean, and the NEAM regions. The European Union has also funded a major research program to study tsunami risk along European coastlines, with special emphasis on the Mediterranean coasts. Because of its low population density, no priority has been attached so far for a tsunami early warning system for the Arctic Ocean.

A true tsunami warning network should be an end-to-end system for providing protection from tsunami hazards to the people residing at the tsunami-prone oceanic rim regions. Accordingly, the architecture of the international TWS put in place in the Indian Ocean under the leadership of the IOC of UNESCO is composed of two different networks: the upstream detection network of instruments (i.e., seismographs, sea-level gauges, and deep-ocean pressure sensors) and the downstream network of national tsunami centers. The latter have the responsibility of delivering warnings to the people at risk, with at least one national center in each participating nation. According to the guidelines established by IOC-UNESCO, in a minimal configuration, the national tsunami center must have the operational capability of receiving warnings around the clock and of disseminating these warnings both to the responsible authorities and to the general public. National tsunami warning centers must also be capable of defining national preparedness procedures and of putting in place national education and awareness plans.

A tsunami warning system (TWS) is a network of seismographs, sea-level gauges, and high-speed communication devices intended to predict and warn against the approach of a tsunami wave train. The main objective of a TWS is to detect, locate, and determine the magnitude of earthquakes that have the potential for triggering tsunami waves and validating the warning based on real-time sea-level measurements at multiple locations. Seismic stations that are spread over many spatially distant geographical locations provide earthquake information from each of these locations. If the location and magnitude of an earthquake meet the criteria for triggering a tsunami, the tsunami warning center (TWC) issues a warning of an imminent tsunami hazard. Tsunami warnings, including predicted tsunami arrival times at selected coastal regions and related information bulletins, are disseminated to the appropriate emergency officials and the general public via a variety of communication channels, such as commercial radio, television, weather radio systems, marine radio systems, and so forth. Because tsunamis travel relatively slowly in shallow waters and in a predictable pattern, in contrast to the shock waves generated by the earthquakes, there would be some time to warn populated areas about the oncoming potential disaster. The following capabilities seem to be necessary for an operational TWS (Lockridge, 1989):

- Rapid and accurate detection of the location of an earthquake
- Determination of the actual existence of a tsunami (sea-level recording allows confirming or refuting the occurrence of a regional tsunami)
- Accurate calculation of the expected time of arrival of the tsunami at given coastal locations (the approach is based on computation of travel times and modeling of wave propagation)

This method would estimate the time required to warn of the tsunami, which requires a much longer time (in contrast to the high-speed seismic shock waves) to reach the coast from its point of generation.

Because almost all of the tsunamis that have occurred in the world prior to the December 2004 tsunami were generated and found in the Pacific Ocean, only two major Pacific Ocean Rim countries (United States and Japan) were initially involved in the establishment and maintenance of TWCs on an operational basis. Some efforts were made by Russia as well. In a nutshell, a TWS consists of many elements (Whitmore et al., 2009):

- Observational networks that record tsunamis and other associated phenomena

- Warning centers that process/interpret data and disseminate threat information
- Communication networks that transmit the threat information to coastal residents and emergency management organizations (EMOs)
- EMOs that respond to official warnings, as well as nature's warning signs, and prepare themselves and educate nearby populations to respond properly.

Chapter 10

Earthquake Monitoring for Early Tsunami Warnings

As discussed in Chapter 7, seismic waves (especially *P*-waves and *S*-waves) propagate with tremendous velocities. It is possible to determine the source location and magnitude of an earthquake from analysis of seismic waveform data from a spatially distributed (preferably global) network of land-based and seafloor-based seismometers that detect and measure the seismic waves. This is the logical basis behind the widespread use of seismometer networks for accurate determination of the time of generation, hypocenter, and magnitude of an earthquake, made in addition to communication of this information for early earthquake warning purposes. Further, it is possible to estimate tsunami generation to a certain extent from seismic wave analysis. Deployment of ocean bottom seismometers (OBSs) became necessary because with only land-based seismometers it is sometimes difficult to accurately determine the earthquake source location in the sea. OBS is a system specially designed for measuring seismic waves in the sea for effective early detection of earthquakes in the sea. *P*- and *S*-seismic waveforms are typically used for calculation and analysis purposes.

In cases of tsunamigenic events in the near zone (i.e., local events), the parameters of the earthquakes are estimated by seismic stations (SS). The initial tsunami warning is provided by the same SS. Estimation of tsunami threat is based on the magnitude-geographical criterion. Criteria for the warning notification are based on the magnitude and the location of the tsunamigenic earthquake. Different criteria have been used by different countries. The Pacific Ocean rim countries such as Japan, Chile, and Russia have historically been victims of several large- and medium-scale tsunamis because

Tsunamis: Detection, Monitoring, and Early-Warning Technologies. DOI: 10.1016/B978-0-12-385053-9.10010-9

of the strong seismicity of the zone covering the Kuril Islands and the Kamchatka Peninsula in the western Pacific.

10.1. SEISMIC NETWORK OF THE PACIFIC TSUNAMI WARNING CENTER

As recently as 1996, the Pacific Tsunami Warning Center (PTWC) relied on only a very limited set of seismic data to locate and determine the magnitude of distant earthquakes (McCreery, 2005). These data were transmitted from the U.S. Geological Survey's (USGS) National Earthquake Information Center (NEIC) to PTWC via a modem over a dedicated circuit. The dynamic range of the data was very limited because the system was based on a 12-bit digitizer. Because the data were often contaminated with frequent spikes, modern processing such as filtering or automatic arrival picking was not feasible. The data could, however, be used for event detection, manual arrival picking, and manual amplitude scaling for magnitude. These data were supplemented with time-series data from Hawaii SS, automatic first arrival picks from NEIC, and first arrival times transmitted to PTWC from a few cooperating international observatories. These data were usually adequate for computing shallow epicenters to within a degree, but typically they provided little depth control, since the closest stations were often too far away. Also, depth phases were difficult to recognize on the narrow-band records. Finally, computation of the surface wave magnitude, on which the warning criterion was based, was very slow for earthquakes in the southern or western Pacific.

Since 1997, PTWC began importing very high-quality data over the Internet from a growing number of international broadband SS having data available in near real time. This enabled PTWC to apply modern seismic analysis techniques to the tsunami warning problem. The PTWC now receives data from about 90 broadband vertical seismic sensors located around the Pacific, including stations in South America, Antarctica, New Zealand, Australia, southeast Asia, Japan, Russia, and some Pacific islands, as well as in Alaska and the continental United States (Figure 10.1). These data are typically digitized at 20 samples per second with a 24-bit digitizer. These high-quality seismic data with their wide dynamic range allow accurate timing of high-frequency P-wave arrivals, as well as precise magnitude measurements, thus providing the foundation for improved warning center performance. Their extensive geographical distribution permits earlier detection of an earthquake and more rapid and accurate hypocenter calculations. The broader-frequency band of the data and the larger number of traces permit recognition of depth phases for more accurate depth determinations. In addition, many lower-frequency seismic waveforms allow techniques for determining *moment* magnitude, which is a more accurate measure of size than the surface wave magnitude for the largest earthquakes with the most tsunamigenic potential. PTWC now routinely calculates M_{wp}, the moment magnitude based on the

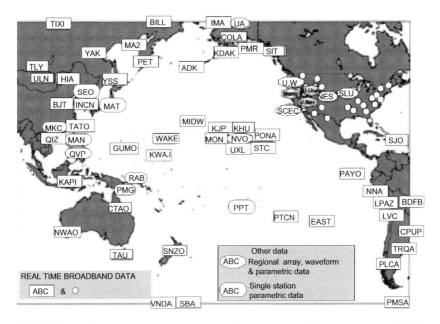

FIGURE 10.1 Seismic stations and networks providing waveform and parametric data to PTWC. *Source: McCreery, 2005; Natural Hazards; Reproduced with kind permission of Springer.*

first-arriving *P*-waves (Tsuboi et al., 1995), and mantle magnitude, M_m (Okal and Talandier, 1989), from surface waves that can be directly converted to moment magnitude. These computations are done for each of the available broadband seismic signals, and the final values are typically based on 30 to 50 independent measurements. As a result of having these capabilities, M_w was adopted in June 2003 as the magnitude to be used in bulletins and for warning criteria. The new data also make possible automatic teleseismic epicenter determinations. PTWC has now implemented the teleseismic *P*-wave picker and associator developed by WC/ATWC, and these two centers exchange their automatic hypocenters as they are produced in the minutes following an earthquake. The broadband data also facilitate techniques for the discrimination of the so-called "tsunami" or "slow" earthquakes that carry an especially high tsunamigenic potential (Kanamori, 1972). These events are usually recognized by unusually high ratios between low- and high-frequency seismic energy, and PTWC routinely computes M_w–M_s and Theta values (Newman and Okal, 1998) as discriminants to check for this possibility. Techniques are now available for rapid computation of the centroid moment tensor, slip distributions, and fault rupture dynamics. This type of source information is useful not only for quickly estimating tsunamigenic potential but also to constrain initial conditions of the water wave numerical models used for forecasting tsunami impacts.

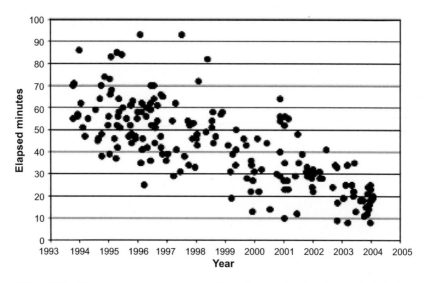

FIGURE 10.2   Elapsed minutes after a large Pacific earthquake after the issuance of a bulletin by PTWC containing a preliminary evaluation of the earthquake and its tsunamigenic potential. *Source: McCreery, 2005; Natural Hazards; Reproduced with kind permission of Springer.*

An important measure of how these seismic enhancements have helped improve PTWC performance is the elapsed time from the earthquake occurrence to the issuance of a bulletin. From 1994 through 1998, it took 30—90 minutes to issue a bulletin. After 1999, with improved technology, it took only 20—60 minutes. Since June 2003, when procedures were officially changed to use M_w instead of M_s for magnitude criteria, it takes 2—4 minutes or less (Figure 10.2). This improved response time will allow early receipt of warnings to areas at risk. In short, real-time access to very high-quality seismic data permit more rapid and comprehensive characterization of seismic sources that may trigger tsunamis. This has resulted in a significant reduction in elapsed time between the earthquake and the issuance of an initial bulletin for both distant and local events.

10.1.1. Seismic Network in Japan

Japan has a long history of relatively frequent regional earthquakes because of the presence of several seismically active regions off Japan and around the Japan Trench & Kurile Trench (Figure 10.3). All of the major Japanese islands have been struck by devastating tsunamis. A total of 68 destructive tsunamis have struck Japan between AD 684 and 1984, with thousands of lives lost and enormous property damage to its coastal communities. Therefore, it is important that Japan's tsunami warning systems be extremely vigilant and in continuous operation year round. Consequently, besides routine coastal sea-level measurements from a network of dense coastal sea-level stations and land-based

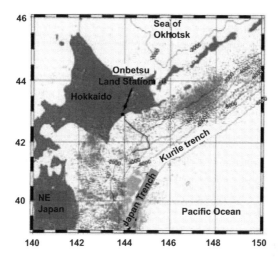

FIGURE 10.3   Seismically active regions off Japan and around the Japan and Kurile trenches. *Source: Hirata et al., 2002; IEEE J. Oceanic Engg; Reproduced with kind permission of IEEE.*

seismicity measurements from about 180 seismometers (Figure 10.4) with real-time data transmission through dedicated telephone lines, additional continuous seismic and tsunami measurements in real time from its offshore regions were considered to be more appropriate for Japan's tsunami warning needs. Because continuous measurements would need a continuous supply of electric power, the incorporation of cable-driven measuring devices was the most pragmatic option for Japan's long-term offshore measurements.

In Japan, two agencies are involved in tsunami-related research and operational activities: Japan Meteorological Agency (JMA) and JApan Marine Science and TEchnology Center (JAMSTEC). From the late 1960s to early 1980s, Japan introduced telemetry technology to transmit seismic waveform data from observatories to the headquarters and a processing computer. Earthquake detection capability was improved by monitoring collected seismic waveforms at one place, and the *P*- and *S*-picking precision was also improved substantially by the introduction of a digitizer. By these improvements, dissemination time was reduced to 12–13 minutes.

After the devastating 1983 mid–Japan Sea earthquake-generated tsunami, the JMA deployed a more sophisticated computer system for seismic waveform processing. A graphical man–machine interface was introduced for more accurate and quicker phase reading and hypocenter- and magnitude-calculations. Still, the empirical method was used for the tsunami amplitude estimation. Dissemination time was reduced to between 7 and 8 minutes. After the 1993 southwest off Hokkaido earthquake-generated tsunami that destructively struck the coast before tsunami dissemination, the JMA totally replaced the seismic network.

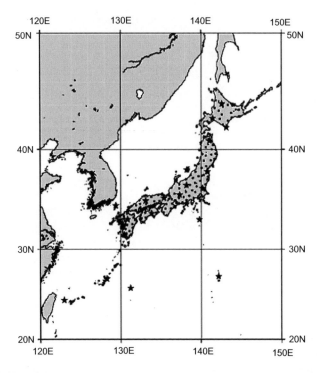

FIGURE 10.4 Seismic network of the Japan Meteorological Agency (JMA) used for tsunami forecasting. The solid circles and stars denote the locations of seismic stations (SS) operated by the JMA. The average spacing between the almost 180 stations is 50 to 60 km. The stars denote the location of stations where STS-2 velocity broadband seismometers have been installed in addition to the Japanese short-period velocity-type seismometers and accelerometers. *Source: Kamigaichi, 2009. Courtesy of Dr. Osamu Kamigaichi, JMA; Reproduced with kind permission from Springer Science+Business Media: Encyclopedia of Complexity and Systems Science, R. A. Meyers (Ed.), Tsunami forecasting and warning, 9592–9617.*

10.1.1.1. Seafloor Submarine Cable-Mounted Observatory of JMA

At JMA, the Seismological and Volcanological Department is responsible for the establishment and maintenance of land-based and ocean-bottom seismographs. Because most tsunamis in Japan had been generated by large earthquakes occurring in oceanic areas, and it is possible to estimate tsunami generation to a certain extent from seismic wave analysis, JMA has been watching seismic activity in and around the Japan arc. Seismic observation around the Tokai region was of particular interest in the recent past because of an expected large earthquake. In order to understand the seismic activity in detail, JMA installed OBSs, using marine cable at assumed focal regions of the Tokai earthquake in 1978. As a result, the capability of hypocenter determination in and off the Tokai region has improved.

Recent studies have led to a revision of the assumed focal region of the Tokai earthquake, which extends to the west of the previous one. It is also expected that the Tonankai earthquakes might occur during early parts of this century. However, Japan does not have enough knowledge of the seismic activity in the assumed focal region of the Tonankai earthquake.

To improve the capability of hypocenter determination in the revised assumed focal region of the Tokai and the Tonankai earthquakes and make the estimation of hypocenter and magnitude more accurate for the calculation of early earthquake warning, in 2004 JMA started a project to install new OBSs using marine optical cables and a repeater in and around the assumed focal regions of the Tokai and Tonankai earthquakes. Each device in the system is connected to the optical cable.

Deployment of OBSs became necessary because with only land-based seismometers it is sometimes difficult to accurately determine the earthquake source location in the sea. OBS is a system specially designed for measuring seismic waves in the sea for effective early detection of earthquakes. However, seafloor mounted pressure gauge (PG) systems (known as a tsunami gauges) are also considered an integral part of the OBS. JMA's OBS observation system consists of seismometers, tsunami meters, repeater devices (Figure 10.5), and a land system. The seismometer of the OBS system contains a short-period velocity-type sensor and an acceleration-type sensor (Figure 10.6). The records from the short-period sensor are mainly used for precise picking of the onset times of *P*- and *S*-phases that are required for an accurate hypocenter determination. Accelerometer records are used for calculating the magnitude of large earthquakes in cases where the ground motion amplitude exceeds the dynamic range of the short-period sensor (Kamigaichi, 2009). The OBS network operated by JMA is listed in Table 10.1. Location maps of OBS installation sites are illustrated in Figures 10.7 and 10.8). A schematic illustration of JMA's new OBS array incorporating optical cable is shown in Figure 10.9. A schematic illustration of the OBS array installation on the ocean floor is provided in Figure 10.10.

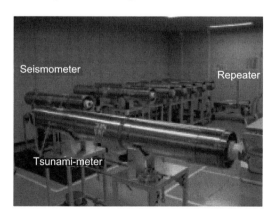

Seismometer

Repeater

Tsunami-meter

FIGURE 10.5   Seismometer, tsunami-meter, and repeater device in JMA's OBS system. *Source: Courtesy of Dr. Osamu Kamigaichi of JMA.*

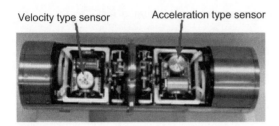

Velocity type sensor   Acceleration type sensor

FIGURE   10.6 Velocity and acceleration seismometer sensors in JMA's OBS system. *Source: Courtesy of Dr. Osamu Kamigaichi of JMA.*

Because the installation of the OBS is a very complicated and costly endeavor, the selection of the best cable path is an important aspect in the establishment of an OBS array. Because the cable and the sensors are very stiff and heavy, it is desirable to avoid steep slopes when laying the cable. It is best to lay the OBS array cable as far away from other marine cables as logistically feasible. If other cables (for example, submarine telegraph cables) are encountered, then it is recommended that the crossing angle of the OBS array be perpendicular to the existing cables. With JMA's new OBS array, data are carried by an optical cable, the cable is robustly protected from high water pressure, and any cable length that exceeds a specified limit is provided with an ocean-bottom optical amplifier to boost the attenuated signal.

The land-based system controls the underwater devices and receives the observed signals. The received signals are, in turn, transmitted in a real-time basis to JMA headquarters (at Tokyo) and Osaka District Meteorological Observatory. The calculations for early earthquake warnings are set in motion immediately after the detection of an earthquake by seismometer devices. Upon completion of calculations, the results are transmitted to both headquarters and Osaka immediately.

The first Japanese cable-mounted seafloor seismic observatory system was established in the Nankai trough off Cape Muroto, Shikoku Island, Japan, in March 1997 (Momma et al., 1997). The success of its first cable-based seafloor

TABLE 10.1 JMA Ocean-Bottom Seismometers (OBSs) and Tsunami Gauges (TGs) for the Early Detection of Earthquakes and Tsunamis

Area	Installation Year	Seismograph	Tsunami Gauge	Cable Length (km)
Off Tokai	1978	4	1	154
Off Boso	1986	4	3	126
Off Tokai (new)	2008	5	3	200

Source: Courtesy of Dr. Osamu Kamigaichi, Shouji Saitou, Ken Moriwaki, and Yasuyuki Yamada of JMA.

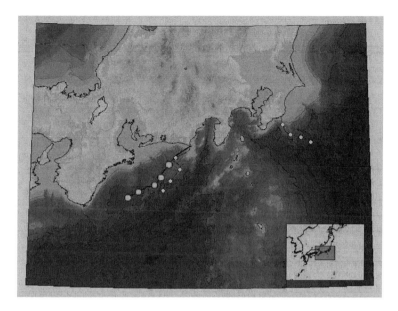

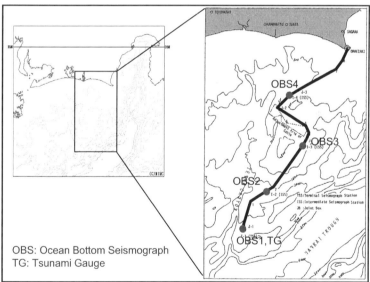

FIGURE 10.7   Ocean-bottom seismometers (OBSs) and tsunami gauges (TGs) established by JMA off Tokai for early detection of earthquakes and tsunamis. *Source: Courtesy of Dr. Osamu Kamigaichi of JMA.*

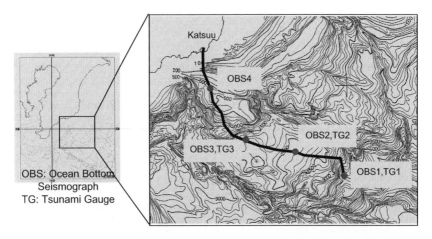

FIGURE 10.8   Ocean-bottom tsunami monitoring network off Boso Peninsula established by JMA. *Source: Courtesy of Dr. Osamu Kamigaichi of JMA.*

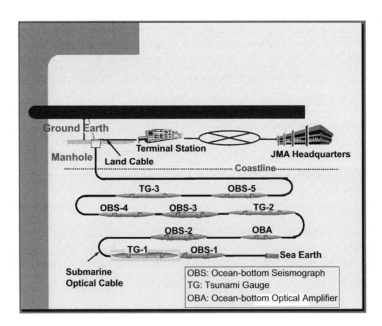

FIGURE 10.9   JMA's new OBS array incorporating optical cable. *Source: Courtesy of Dr. Osamu Kamigaichi of JMA.*

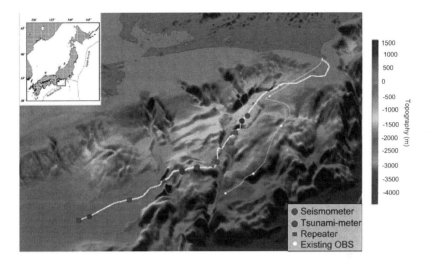

FIGURE 10.10 Cable route (thin white line represents existing route; thick white line represents the new route) of the ocean-bottom tsunami monitoring sensor network of Japan Meteorological Agency (JMA) using marine cable installed at the Sea of Enshu to the Sea of Kumano. *Source: Courtesy of Dr. Osamu Kamigaichi, Shouji Saitou, Ken Moriwaki, and Yasuyuki Yamada of JMA.*

observatories encouraged Japan to deploy more cable lines with seafloor pressure sensors as well as seismometers.

10.1.1.2. Seafloor Submarine Cable-Mounted Observatory of JAMSTEC

After the January 1995 Kobe earthquake, the schedules for expanding earthquake and tsunami observation networks on the seafloor have been accelerated. JAMSTEC developed an advanced real-time seafloor measurement system based on a conventional in-line monitoring system and multidisciplinary real-time seafloor observatory to deal with this requirement. This type of submarine cable-connected measurement system has been developed and installed since 1995 to bring a series of in-line geoscientific observation networks at typical seismogenic zones around Japan to completion. The first system was deployed off the Muroto area in 1997 (Momma et al., 1997; Fujiwara et al., 1998; Momma et al., 1998; Momma et al., 1999), and the second observation system was deployed in July 1999 on the southern Kuril subduction zone in the southeastern part of Hokkaido, off the Kushiro-Tokachi area (Kawaguchi et al., 2000). To deploy as many sensors as those existing on land, the second observation system was designed to extend the existing cable observation using adaptable and mobile seabed observatories (Kawaguchi et al., 2002).

Although all the issues concerning operational uses of the data obtained from seismometers and ocean-bottom pressure gauges (OBPGs) are looked after by JMA, other Japanese agencies also operate such offshore devices to record earthquakes and tsunamis offshore of Japan. For example, JAMSTEC developed permanent seafloor observatory systems incorporating submarine fiberoptic cable and deployed them at the continental slope southeast of Hokkaido and on the deep seafloor in the southern Kurile subduction zone during 1999 for real-time measurement of geophysical parameters, including seismicity and tsunamis (Hirata et al., 2002).

The system is made up of the following elements:

1. A backbone cable system
2. Three ocean-bottom broadband seismometers with hydrophones (OBSHs)
3. Two tsunami PGs
4. A cable-end multisensor station
5. Two expandable interfaces for adaptable observation systems (AOS)

The system (Figure 10.11) was designed to detect an $M \sim 2$ seismic event with a location error of a few km. This observatory is the second one in the cable-connected seismic observatory projects of JAMSTEC.

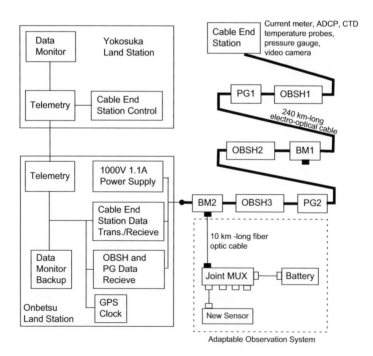

FIGURE 10.11 System block diagram of JAMSTEC's permanent seafloor observatory incorporating submarine fiberoptic cable. *Source: Hirata et al., 2002; IEEE J. Oceanic Engg; Reproduced with kind permission of IEEE.*

An optic/power submarine cable approximately 240 km long, a power/data management shore station, and a control land station constitute the backbone system. Real-time data from all the undersea sensors are transmitted through the main electro-optical cable to the land station.

The OBSH units (powered by 1.1 A at 40 Volts DC) incorporate a three-axis broadband acceleration sensor (JA-5III manufactured by Japan Aero Electronics Corporation), which has a flat output voltage sensitivity over a wide frequency band and is housed in a beryllium-copper alloy cylinder. The JA-5III has been used previously also for ocean-bottom seismicity measurements (Kanazawa and Hasegawa, 1997) because it is a small and tough sensor. Data from low- and high-gain channels for each orthogonal component are obtained. Gimbal units are not used in the OBSH because the inclination of each axis of the sensor can be calculated from the DC offset in the low-gain channels. The seismograms are digitized using 24-bit sigma-delta A/D converters. The digital signals are time-division multiplexed and then converted to optical signals that, in turn, are transmitted through the optical fiber to the *Onbetsu* land station. At the land station, the optical signals are converted back to the original digital signals.

Figure 10.12 illustrates the cable-end station attached to the seafloor end of the submarine optical cable of JAMSTEC's permanent seafloor observatory. The cable-end station, in addition to seismometers and ocean-bottom tsunami

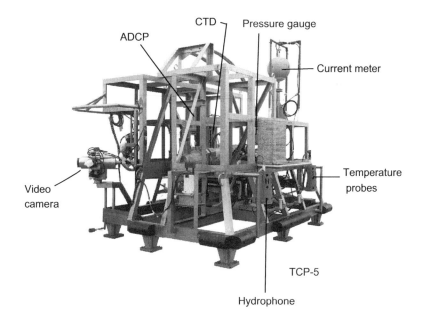

FIGURE 10.12 Cable-end multisensor station attached to the seafloor end of the submarine optical cable of JAMSTEC's permanent seafloor observatory. *Source: Hirata et al., 2002; IEEE J. Oceanic Engg; Reproduced with kind permission of IEEE.*

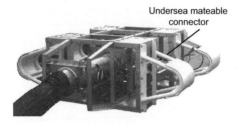

Undersea mateable
connector

FIGURE 10.13 Data-telemetry unit attached to the seafloor end of the submarine optical cable of the permanent seafloor observatory of JAMSTEC. *Source: Hirata et al., 2002; IEEE J. Oceanic Engg; Reproduced with kind permission of IEEE.*

pressure gauges, has several environmental measurement sensors, such as a current meter, an acoustic Doppler current profiler (ADCP), a CTD, temperature probes, a hydrophone, a video camera, and an additional quartz crystal pressure gauge manufactured by Paroscientific Company. All the sensors and the data-telemetry unit (Figure 10.13) are installed in an open titanium frame.

An AOS is a concept used for the future expansion of the deep seafloor observatory (Kawaguchi et al., 2002). The backbone submarine cable system has two expandable interfaces for the AOS, known as a branch multiplexer (B-MUX). The B-MUX branches one of the six main optical-fiber lines and enables the installation of a joint multiplexer (J-MUX) at the end of the branching optical-fiber line. The J-MUX equips a telemetry unit and a multiplexer, and it works as a hub for adaptable sensor packages, such as broadband seismometers. The B-MUX consists of a telemetry unit and a fiberoptic underwater mateable connector. The telemetry unit contains an optical repeater and physically divides an optical-fiber line from the backbone cable system to the AOS to facilitate two-way data communication between land and shore stations to the AOS. The telemetry unit additionally has a power circuit and functions to amplify the optical data from the other measurement equipment in the backbone cable system (such as the in-line OBSH, in-line PG, and multi-sensor station).

Three different kinds of cable armoring are used: double armored, single armored, and nonarmored. The double-armored cables are used at shallow depths near the shore, and the nonarmored cables are used for most of the ocean floor. The single-armored cable is used between these extremes. The main electro-optical cable has six-core single-mode fibers (Hitachi Cable Company, Ltd.). The structure of a simple submarine thin fiber cable is illustrated in Figure 10.14.

The *Onbetsu* land station contains terminal equipment such as a data transmission unit, DC power supply, data monitoring and control unit, telemetry, data displays, and data storage unit. This land station supplies 480–850 VDC at 1.1 A to the undersea units. The power for all electronic or electric devices is fed in series connection from the *Onbetsu* land station with a constant current of 1.1 A. Each device obtains its own power by deriving the necessary voltage with the use of several kinds of Zener diodes (ZDs). Incorporation of switching regulators minimizes electric power dissipation in all the power-feed

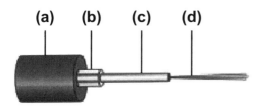

(a) (b) (c) (d)

FIGURE 10.14 Structure of a simple submarine thin-fiber cable used with the cable-mounted seismic and seafloor bottom pressure real-time data reporting system used in Japan. (a) HDPE outer jacket, (b) adhesive polyolefin inner jacket, (c) stainless steel pipe tension member, and (d) optical-fiber lines with water block compound. *Source: Kawaguchi et al., 2002; IEEE J. Oceanic Engg; Reproduced with kind permission of IEEE.*

circuits. Also, the power conductor resistance of the underwater cable is about 0.7 ohm/km, so the total resistance of the submarine electrical cable is about 168 ohms. The power dissipation in the cable is 203 W.

Prior to the deployment of the seafloor observatory, presite surveys such as side-scan sonar imaging, piston coring, and so forth are usually carried out to obtain sufficient information about the bathymetry and surface sediment conditions in the continental slope for cable route planning. Dive expeditions of unmanned submersibles are also undertaken to determine the exact positions of the observation units. Deployments of the cable and seafloor probes/observatories are carried out using cable-laying ships. Armored submarine cables are usually attached with steel-made guard tubes and dropped into gutters dug on hard-bottom regions with a diver-operated excavator to shelter the cables from chafing on the hard rocks (Hirata et al., 2002). The burial depth usually used in Japan is 0.6−1.0 m below the seafloor. However, if the selected region is a vigorous fishing area, the burial depth would need to be larger than usual. The submarine cable route and positions of the seafloor observatory of JAMSTEC are shown in Figure 10.15.

All of the data, except video camera images, are also sent through a 256-kbps digital data link with the TCP/IP protocol to the *Yokosuka* land station at the headquarter of JAMSTEC. The WIN format, which is a standard seismic data format in Japan, is used for transmission of seismic, hydrophone, and tsunami data between the *Onbetsu* and *Yokosuka* land stations. Real-time OBSH and tsunami PG data are transmitted from the *Onbetsu* land station to the Sapporo Meteorological Observatory of JMA to routinely monitor seismicity and tsunamis.

All of the data from the undersea sensors, power feed voltage and current, and the status of equipment in the *Onbetsu* land station are monitored with a supervisory system at the *Onbetsu* and *Yokosuka* land stations. When arrival of any particular dataset from an undersea sensor stops, an abnormal change in voltage occurs, or any package failure such as a blown fuse in the *Onbetsu* land station occurs, a centralized, visible alarm with a buzzer is issued at the land stations. This scheme allows the trouble to be known immediately, though only roughly. However, it allows some countermeasures to be initiated with a remote

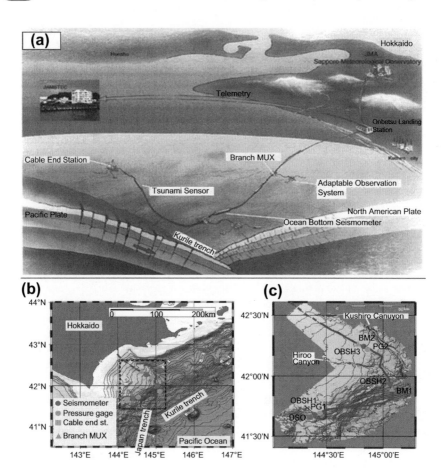

FIGURE 10.15 (a) Submarine cable route and positions of the permanent seafloor observatory of JAMSTEC off Hokkaido, Japan, in the Pacific Ocean. (b) The solid line indicates a part of the cable laying on seafloor surface, and the dashed line indicates the buried part of the cable. (c) A more precise map of a deeper part in the bottom left map. Green and yellow circles indicate the OBSHs and the PGs, respectively. Triangles are the branch units. Squares are the cable-end stations. *Source: Hirata et al., 2002; IEEE J. Oceanic Engg; Reproduced with kind permission of IEEE.*

control system in the *Yokosuka* land station. Subsequently, the trouble can be fixed either in the *Onbetsu* land station or on the ocean bottom with the help of submersibles.

10.1.1.3. Tsunami Measurements by the Seafloor Observatory of JAMSTEC

The submarine cabled seafloor observatory of JAMSTEC routinely provides real-time measurements that are relevant to earthquake and tsunami monitoring

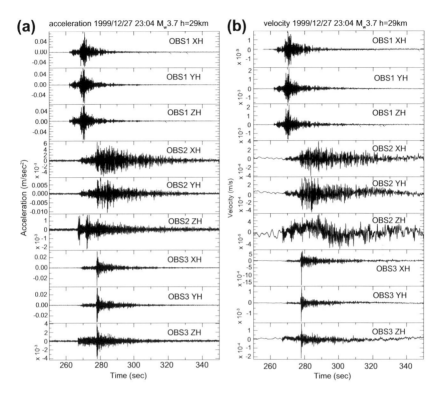

FIGURE 10.16 (a) An example of observed acceleration seismograms from a nearby earthquake on December 27, 1999, recorded by the permanent seafloor observatory of JAMSTEC off Hokkaido, Japan, in the Pacific Ocean. (b) Velocity seismograms converted from those in (a). No filter was used in the conversion. *Source: Hirata et al., 2002; IEEE J. Oceanic Engg; Reproduced with kind permission of IEEE.*

applications. Figure 10.16 shows examples of observed seismograms from a nearby earthquake with $M_w = 3.7$ that occurred on December 27, 1999, recorded by the permanent seafloor observatory of JAMSTEC off Hokkaido, Japan, in the Pacific Ocean.

10.1.1.4. Earthquake Monitoring by JMA

As indicated earlier, although JMA has its own coastal and deep-sea monitoring systems, it is collecting data from cable-type OBSs and pressure sensor systems operated by several other organizations as well. Table 10.2 provides the specifications of the OBSs and tsunami gauges (TGs) established by various agencies in Japan for early detection of earthquakes and tsunamis, as well as for research applications. JMA's man—machine interface system to facilitate real-time viewing of a just-arrived seismic event and one in progress that facilitates preliminary seismic data processing is shown in

TABLE 10.2 Specifications of Ocean-Bottom Seismometers (OBSs) and Tsunami Gauges (TGs)

Installation Sea Area	Installation Organization	Ocean-Bottom Seismograph	Tsunami Meter	Transmission Line and Transmission Method	Installation Year
Off Toukai. Maximum depth of route: 2.2 km Cable length: 154 km	JMA	3 observation points Composition of observation points: • Displacement: 3 components. Frequency band: 2 ~ 20 Hz. Range: 0.02 ~ 400 μm • Velocity meter: Upper and lower component Frequency band: 2 ~ 20 Hz. Range: 0.02 ~ 20 mkine	1 quartz pressure gauge Depth range: 0–4000 m Wave height resolution: 1 mm	Submarine co axial cable Transmitted by an FM-FDM with S/N of 72 dB in the sea DC constant current.	1979
Off Bousou. Maximum depth of route: ~4 km Cable length: 126 km	JMA	3 observation points Composition of observation points: • Displacement: 3 components. Frequency band: 2 ~ 20 Hz. Range: 0.05 ~ 400 μm • Velocity meter: Upper and lower component Frequency band: 2 ~ 20 Hz. Range: 20 ~ 8000 mkine	2 observation points Composition of observation points: • Quartz pressure gauge. Depth range: 0–4000 m. Wave height resolution: 1 mm • Crystal thermometer. Resolution: 0.001°C.	Submarine coaxial cable Transmitted by an FM-FDM with S/N of 72 dB in the sea DC constant current	1985

Location	Institution	Composition of observation points	Transmission method	Year
Off east of Izu Maximum depth of route: 1.35 km Cable length: 33 km	ERI	3 observation points: Composition of observation points: • Servo accelerometer: 3 components Frequency band: 0.05 ~ 200 Hz. Rane: 150 μgal ~ 1100 gal	Submarine optical cable • Optical digital method DC constant current	1994
Sagami bay Maximum depth of route: 2.34 km Cable length: 125 km	NIED	6 observation points: Composition of observation points: • Velocity meter: 3 components Frequency band: 1 ~ 30 Hz Range: 2.5 μkine ~ 83 mkine • Accelerometer: 3 components Frequency band: 0.05 ~ 30 Hz Range: 15 mgal ~ 500 gal; 3 observation points Composition of observation points: • Quartz pressure gauge. Depth range: 140 m ~ 7700 m Wave height resolution: 1 mm • Crystal thermometer. Resolution: 0.001°C	Submarine optical cable • Optical digital method DC constant current	1996
Off Sanriku. Maximum depth of route: 2.7 km Cable length: 126 km	ERI	3 observation points: Composition of observation points: • Servo accelerometer: 3 components Frequency band: 0.05 ~ 200 Hz Range: 150 μgal ~ 3000 gal; 2 observation points Composition of observation points: • Quartz pressure gauge Depth range: 300 m ~ 4000 m Wave height resolution: 0.5 mm • Crystal thermometer. Resolution: 0.01°C	Submarine optical cable • Optical digital method DC constant current	1996

(Continued)

TABLE 10.2 Specifications of Ocean-Bottom Seismometers (OBSs) and Tsunami Gauges (TGs)—cont'd

Installation Sea Area	Installation Organization	Ocean-Bottom Seismograph	Tsunami Meter	Transmission Line and Transmission Method	Installation Year
Off Muroto Maximum depth of route: 3.46 km Cable length: 130 km	JAMSTEC	2 observation points: Composition of observation points: • Servo accelerometer: 3 components. Frequency band: 0.05 ~ 200 Hz Range: 50 µgal ~ 3000 gal	2 observation points Composition of observation points: • Quartz pressure gauge Depth range: 300 m ~ 4000 Wave heigh resolution: 1 mm • Crystal thermometer. Resolution: 0.01°C	Submarine optical cable • WDM • Optical analog method DC constant current	1999
Off Toukai. Maximum depth of route: 2.06 km Cable length: 200 km	JMA	5 observation points: Composition of observation points: • Velocity: 3 components. Frequency band: 1 ~ 30 Hz Range: 0.1 µkine ~ 55 mkine • Accelerometer Frequency band: 0.05 ~ 30 Hz Range: 15 mgal ~ 500 gal	3 observation points Composition of observation points: • Quartz pressure gauge Depth range: 140 m ~ 7700 m Wave height resolution: 1 mm • Crystal thermometer Resolution: 0.01°C	Submarine optical cable • Optical digital method • Ocean bottom optical amplifier DC constant current	2008

Source: Kanazawa, 2000; *Earthquake Journal, Reproduced with kind permission from the Association for the Development of Earthquake Prediction (ADEP).*

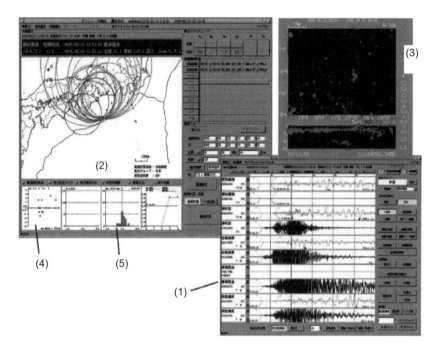

FIGURE 10.17 JMA's man—machine interface screen image in the seismic data processing system. The numbers 1—5 are (1) seismic waveforms to facilitate phase pickings and maximum amplitude readings; (2) a hypocenter plot map in which the circles denote the calculated epicentral distance from each seismic station; (3) a background seismicity map; (4) travel time residuals as a function of epicentral distance; and (5) histograms of station magnitudes. *Source: Kamigaichi, 2009. Courtesy of Dr. Osamu Kamigaichi, JMA; Reproduced with kind permission from Springer Science+Business Media: Encyclopedia of Complexity and Systems Science, R. A. Meyers (Ed.), Tsunami forecasting and warning, 9592—9617.*

Figure 10.17. The seismic monitoring system provides a graphical illustration of the following (Kamigaichi, 2009):

1. Seismic waveforms to facilitate phase pickings and maximum amplitude readings
2. Hypocenter plot map in which the circles denote the calculated epicentral distance from each SS (if the circles intersect densely at one point and if the used seismic stations ensure a wide azimuthal coverage around the source, then the calculation is judged as accurate)
3. Background seismicity map
4. Travel time residuals as a function of epicentral distance
5. Histograms of station magnitudes

In addition to the data from the domestic seismic network, JMA utilizes Internet-accessible seismic waveform data from global networks for hypocenter and magnitude calculations. JMA makes good use of bulletins from the

PTWC as well. JMA has now established a fairly large tsunami simulation database to cover even distant events in which Coriolis force effects for long-distance propagation have been accounted for. Besides the sea-level data from its domestic network, JMA monitors data collected by geostationary meteorological satellites and circulated through GTS (global telecommunication network) operated by WMO (World Meteorological Organization) of the United Nations.

Since 1999, the JMA has been providing a tsunami forecast service, called a "quantitative tsunami forecast" for both local and distant tsunamis based on a precomputed tsunami database constructed by numerically simulated tsunamis from a great number of assumed earthquake faults (Tatehata, 1997). Once earthquake parameters such as location and magnitude are estimated from seismological observations, appropriate tsunami forecast solutions are extracted from the database so an initial tsunami forecast is issued. There is a planned deployment of a deep-ocean cabled observatory with 20 OBPGs off the Peninsula of Kii in 2009 by JAMSTEC. Therefore, a much denser combined network of deep offshore stations will be constructed along the Nankai trough. It may then be possible to construct an experimental tsunami forecast system based on real-time offshore observations for earthquakes along the Nankai trough in the future.

10.1.1.5. Earthquake Monitoring in Australia

In Australia, the seismic detection and monitoring hub based at Geoscience Australia (GA) in Canberra automatically issues seismic solutions for potentially tsunamigenic earthquakes to the tsunami warning and sea-level verification hub in Melborne. After further assessment, the GA issues a manually assessed seismic solution. Electronic notification is relayed between the two organizations via a dedicated fiberoptic cable. Receipt of the manual seismic assessment automatically triggers tsunami travel time software at the bureau. Predicted tsunami amplitudes from deep-ocean model tsunami scenarios are then used to determine whether a *nil threat bulletin* or a *potential threat bulletin* will be issued through the bureau's dedicated dissemination systems

The recent South Pacific tsunami that struck Samoa reached shore 15 minutes after the generating earthquake. The warning message took 16 minutes, so obviously the entire detection and communication system needs to be better understood and applied. However, if the message is not understood by authorities and communicated to a public that already understands the nature of the threat and how to respond, detection, monitoring, and reporting become almost academic exercises.

The GA accesses real-time seismic data from a network of national and international SS (Figure 10.18a). The Australian National Seismic Network (ANSN), which was initially developed to detect local earthquakes in Australia, is proposed to be upgraded and expanded using very broadband seismometers to enhance the detection of tsunamigenic earthquakes (Figure 10.18b). The

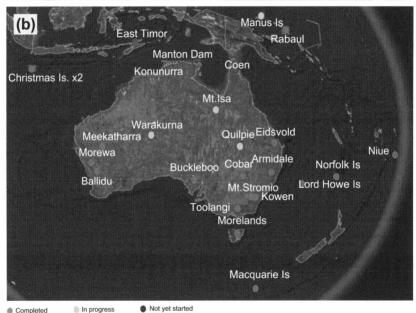

FIGURE 10.18 (a) Locations of national and international SS, which are monitored in real time by Geoscience Australia for earthquakes that may generate a tsunami. (b) Proposed to be upgraded and expanded using very broadband seismometers. *Source: National Report of Australia to IOC-UNESCO, February 2009; Courtesy of IOC-UNESCO; Reproduced with kind permission of IOC-UNESCO.*

enhanced network has been designed so potentially tsunamigenic earthquakes in the Indian and southwest Pacific Oceans can be monitored. The suite of software and tools used to detect potentially tsunamigenic earthquakes has been substantially enhanced and refined. Further enhancements are planned, including implementation of a new technique for rapid estimation of earthquake focal mechanism and magnitude, and introducing array processing.

Open Ocean Tsunami Detection

Warning systems and evacuation plans are crucial to the effective mitigation of the tsunami hazard. However, a false tsunami alarm can be costly in terms of unnecessary evacuation, and it is also a matter of lost credibility to the scientists and the tsunami warning agencies. Permanent monitoring of earthquakes, tsunami waves, and associated hazards in such situations becomes an important socioeconomic problem. Regular geophysical monitoring and a modern early warning scheme are therefore considered vital.

Although the initial early tsunami warnings (at least for earthquake-generated tsunamis) are based primarily on seismic parameters, verification of the actual existence of a tsunami is required by means of one or more different kinds of open ocean, coastal, and space-based instrumentation so a full alert can be made. In other cases, when a seismic signal is absent (e.g., due to a landslide without a preceding earthquake), then the sea-level instrumentation just mentioned can become the primary warning mechanism, and in this case, the tsunami signals have to be separable from the spectra of ocean signals in the instrumental data. Until recently it was necessary to wait until a potential tsunami reached the nearest sea-level station to confirm or deny its existence and evaluate its character. In the past, during a tsunami event, the Pacific Tsunami Warning Center (PTWC) had to rely on Telex messages from "tide observers" around the Pacific coast. The Local Automatic Remote Collector (LARC) stations were dial-up types only. In subsequent technology developments, the PTWC sea-level stations started sending their data continuously and in real time. Since the 1980s, PTWC started receiving sea-level data via

Tsunamis: Detection, Monitoring, and Early-Warning Technologies. DOI: 10.1016/B978-0-12-385053-9.10011-0
149

satellite from stations around the Pacific Ocean (Figure 11.1). National Ocean Service (NOS) sea-level stations send their data via satellite in hourly transmissions but will send more frequent data if they detect a tsunami. The PTWC uses about 100 sea-level stations that are operated by PTWC and various other organizations in the United States, Japan, Russia, Chile, and Australia, often for a variety of purposes other than just tsunami detection and evaluation.

One of the serious problems in issuing early tsunami warnings based on numerical model results is that the rupture parameters of the tsunamigenic earthquakes are difficult to obtain in real time, which is needed as input of the computer models. In the absence of this information, the next best piece of information is the deep-ocean signature of the tsunami. Several attempts have been made in the past (Murty, 1977) to deduce the deepwater signature from coastal tide gauge records. These attempts were largely unsuccessful mainly because coastal sea-level data have significant shortcomings when applied to the problem of tsunami forecasting. They are typically located in the shallow, protected waters of harbors and bays to provide security and a relatively benign ocean environment for instrument longevity. In a harbor/bay environment, tsunami waves arriving from deep waters are highly modified in nonlinear ways as they shoal and interact with the shoreline. The nonlinearity and the

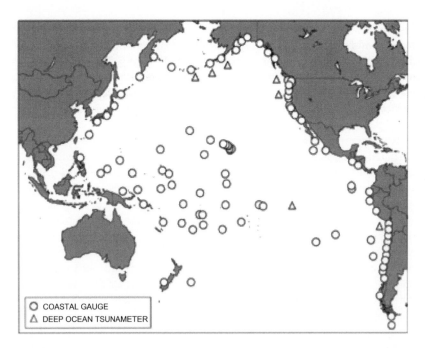

FIGURE 11.1 Sea-level stations used by the Pacific Tsunami Warning Center (PTWC) to detect and evaluate tsunami waves propagating across the Pacific. *Source: McCreery, 2005; Natural Hazards; Reproduced with kind permission of Springer.*

contamination of coastal tide gauge records by local resonances in inlets, bays, and gulfs severely limit the predictive usefulness of such records. The inverse problem of obtaining deepwater signatures from coastal records is a mathematically ill-posed problem and, therefore, has no unique solution (Murty et al., 2007b).

Another issue that is specific to the Pacific Ocean is the limited number of islands, which means that vast portions of the northern and eastern Pacific are devoid of sea-level stations. Therefore, tsunamis from some of the most dangerous tsunamigenic zones stretching from northern Japan to Kamchatka to the Aleutian Islands and even down to Peru and Chile travel a long distance without being detected until they reach some strategically located coastal sea-level stations seaward of the source. Some methods have successfully been used for tsunami detection from the open ocean. Some scientific principles, which have theoretically sound bases and evidence on the basis of practical observations, are expected to be realized into promising technologies in the coming decades. These technologies are addressed in the below sections.

11.1. DETECTION USING SEAFLOOR PRESSURE MEASUREMENT

The practical difficulties in obtaining the rupture parameters of tsunamigenic earthquakes in real time and the drawbacks in the use of coastal sea-level data for early tsunami warning applications led to the development (and deployment in the Pacific Ocean) of a device known as deep-ocean assessment and reporting of tsunamis (DART) buoys network system by Pacific Marine Environmental Laboratory (PMEL) of National Oceanic and Atmospheric Administration (NOAA) (Bernard, 2005). The DART system is capable of being moored at offshore locations for detection, storage, and real-time reporting via satellite. Indeed, an early and dependable assessment of a tsunami threat requires detection of the tsunami wavetrain in the open ocean away from the shore.

Seafloor-mounted pressure stations have the advantage that they can be deployed near potential tsunami sources and thereby report direct tsunami measurements, often hours before the tsunamis can propagate to major coastal communities. Observing tsunamis in the open ocean also has the advantage of being able to see the tsunami before it encounters the complicating, often fine-scale, effects of the continental shelf and the coastal region, as well as resonance effects in bays and harbors. This permits a more straightforward interpretation of the tsunami waves and their relationship to their sources (Mofjeld, 2009).

Advancements in acoustic modem technology allowed the transmission of seafloor measurements to a floating buoy. Based on the latest developments in this new field, the NOAA deployed seafloor tsunami detection sensors with the support of deepwater moorings in the vicinity of seafloor spreading centers and this allows transmission of event detection data by acoustic modems to a surface buoy installed module that can be interfaced to satellite data transmission modems and subsequent retransmission to land-based laboratories (Anonymous, 1999). These developments in technology are expected to go a long way in the early forecast of tsunami events, thereby reducing loss of life and property in coastal communities. The DART system (Figure 11.2) is an open ocean moored buoy system consisting of a seafloor-mounted pressure recorder claimed to be capable of detecting tsunamis as small as 1 cm and a surface buoy for real-time communication via satellite. The bottom pressure recorder (BPR) in the DART system uses a Digiquartz pressure transducer, manufactured by Paroscientific, Inc., which provides high accuracy and excellent resolution. The seafloor-mounted tsunami instrument (tsunameter) consists of a Digiquartz pressure

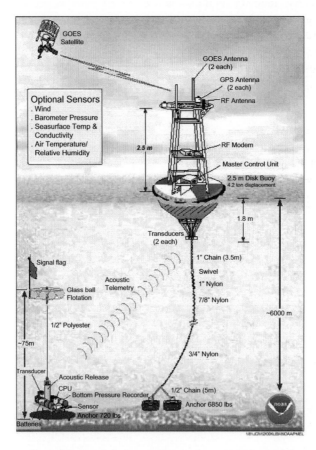

FIGURE 11.2 Open ocean moored buoy system, known as DART, consisting of a seafloor bottom pressure recorder capable of detecting tsunamis as small as 1 cm and a surface buoy for real-time communication via satellite. *Source: Courtesy of PMEL/NOAA.*

transducer (which measures both pressure and temperature), a microprocessor, a communication system, a power supply using batteries, and a pressure housing to protect the components. The seafloor-mounted unit also has an anchor-platform, acoustic release, and flotation to facilitate its recovery along with the data it has recorded internally. Pressure readings from the seafloor-mounted unit are acoustically transmitted to the surface buoy.

There are multiple versions of the DART system. In the initial version, the data used to be relayed via a GOES (Geostationary Operational Environmental Satellite) link to ground stations, which demodulated the signals for immediate dissemination to NOAA's tsunami warning centers. However, the DART II systems use the global Iridium satellite system, which allows for efficient two-way communication. The present battery capacity of the DART II system is sufficient to allow servicing every 24 months. Detailed descriptions of the DART II system and its depth limits (minimum deployment depth: 1000 m; maximum deployment depth: 6000 m) are provided by Mofjeld (2009). Normally, hourly transmissions provide four 15-second average values of bottom pressure separated by 15-minute intervals, but the full sequence of 15-second averages is recorded internally. Transmission of 15- and 60-second data (event mode) occurs when an onboard tsunami detection algorithm is triggered by a difference between the measured value and a predicted value that is greater than a threshold, typically 3 cm (Gower, 2006). Real-time tsunami data provided by the DART system by event mode triggering mechanism is shown in Figure 11.3. Data from these tsunami meters, free of the coastal effects, provide accurate forecasts of tsunamis by assimilating real-time data into nested numerical models (Titov et al., 2005b).

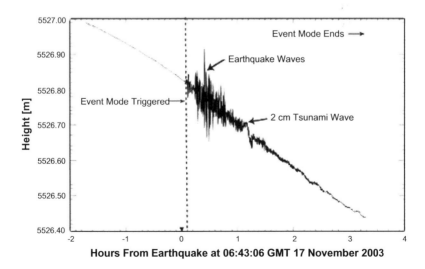

FIGURE 11.3 Real-time tsunami data provided by the DART system by an event mode triggering mechanism. *Source: Courtesy of PMEL/NOAA.*

The tsunami buoy systems are aimed at quickly confirming the existence of potentially destructive tsunamis and reducing the incidence of false alarms. It is expected that a series of tsunami alert buoys that float in the ocean might provide a much-improved and reliable early warning of tsunamis.

The tsunami alert buoys that form part of the DART program are designed to provide as much warning as possible (Anonymous, 1999). Formidable logistical challenges have long discouraged implementation of comprehensive programs aimed at effectively protecting the coastal communities from dangerous tsunamis. The tsunami buoy systems are expected to herald a new beginning in our efforts toward achieving a reliable scheme for tsunami warnings. Together with deployment of open ocean instrumentation and the effective use of computer models that simulate tsunami inundation, it would be possible to arrive at an estimate of the extent of coastal flooding. Using worst-case inundation scenarios, these computer models would help map and demarcate probable escape routes and evacuation zones so coastal communities can be evacuated quickly in the event of a tsunami threat. These would then be an integral aspect of planning and preparedness against tsunami hazards.

Prior to the December 2004 Indian Ocean tsunami, a limited DART network existed in the eastern Pacific Ocean. Although the December 2004 great Sumatra earthquake and the resulting tsunami were very distant from the northeast Pacific Ocean, the U.S. DART array in the northeast Pacific successfully demonstrated high sensitivity and provided useful data for understanding the propagation of the tsunami in this remote region (Gower and Gonzalez, 2006). As a result of the 2004 Indian Ocean tsunami, the network of offshore tsunami stations is presently under rapid expansion. The United States has completed a major expansion, with a total of 39 DART systems deployed so far in the Pacific and Atlantic regions.

The utility and benefits of DART systems and parallel enhancements of numerical tsunami simulation models have started trickling in. The successful real-time forecast of the Peruvian tsunami of August 2007 for U.S. coastlines is a glaring example (Wei et al., 2008). On August 15, 2007, an offshore earthquake of magnitude 8.0 severely damaged central Peru and generated a tsunami. The first real-time tsunami data available came from a deep-ocean tsunami detection DART buoy within 1 hour of tsunami generation. These tsunami data were used to produce initial experimental forecasts within 2 hours of tsunami generation. The forecast scenario of tsunami energy projection in the Pacific for this earthquake-generated tsunami is shown in Wei et al. (2008). Comparison with real-time sea-level data showed very accurate forecasts.

Forecasts were made based on the methodology of Titov and colleagues (2001), which uses precomputed tsunami propagation models that match deep-ocean tsunami measurements to generate offshore tsunami characteristics. The offshore scenario is then used as the initial and boundary conditions for the high-resolution Standby Inundation Models (SIMs). These forecast models are an optimized version of the method of splitting tsunami (MOST) model (Titov and

Gonzalez, 1997b), which has been extensively validated and verified (Synolakis et al., 2007). SIMs for specific communities have been carefully developed, tested, and validated by known historical data and designed to provide at least 4 hours of coastal tsunami simulation in less than 10 minutes of real-time computation. The key component of the forecast system is the data assimilation method that inverts the real-time deep-ocean tsunami measurements from DART buoys to constrain the tsunami source, which is then used to provide the propagation forecast (precomputed) and the coastal inundation forecast (real-time). The application of this approach to the November 17, 2003, Rat Island tsunami paved the way for developing a more comprehensive and time-efficient tsunami forecast methodology by Wei and colleagues (2008). Precomputed models that incorporate real-time data from the DART network have several hundreds of scenarios to cover Pacific tsunami sources. The time line of the experimental real-time forecast with respect to the PTWC bulletin series during the August 15, 2007, Peruvian tsunami is shown in Wei et al. (2008). Twelve minutes after the earthquake, PTWC disseminated its first information bulletin reporting the earthquake location with a preliminary magnitude $M_w = 7.5$. Potential tsunami threats were advised but limited to coasts near the epicenter. Twenty-six minutes later, PTWC upgraded the earthquake magnitude to $M_w = 7.9$ and issued a regional tsunami warning to the entire Pacific coast of South America, as well as a tsunami watch to the Pacific coast of Central America. The success of this and similar real-time tsunami forecasts suggests that the inversion using real-time data from the DART network produces robust and meaningful results.

11.2. DETECTION USING ORBITING SATELLITE ALTIMETERS

While tsunamis are routinely recorded by sea-level gauge stations located at continent or island shorelines, recordings from deep oceans have been very limited (e.g., using ocean-bottom pressure gauges that recorded the Petatlan, Mexico, tsunami on March 14, 1979, during deployment at the mouth of the Gulf of California (Filloux, 1982) and similar devices later deployed in the Gulf of Alaska, which detected the tsunamis generated by the intraplate earthquakes of November 30, 1987, and March 6, 1988 (Gonzalez et al., 1991)) until the deployments of cable-mounted ocean-bottom pressure gauges by Japan's JMA and JAMSTEC and the United States' PMEL/NOAA's DART buoys in recent years. Such measurements suffer from depth limitations. Tsunami wave generation occasionally involves poorly known parameters (such as the precise geometry of the rupture) or structures (such as sedimentary layers). While recognizing the practical utility of DART systems, it would be of great interest to obtain additional independent constraints in the form of direct measurement of the amplitude of a tsunami wave on the high seas because it would then be feasible to separate the true tsunami signature from its coastal interaction (i.e., from earthquake source to high-seas amplitude and from high-seas amplitude to run-up amplitude or sea-level gauge record). In the absence of high-seas measurements, it would be possible to study only the product of the two effects (Okal et al., 1999). Due to a combination of relatively small tsunami wave amplitudes (estimated at less than 1 m, even for exceptionally large events) and very large wavelengths (≥ 100 km), any direct observation of the deformation of the surface of the deep ocean was not practicable for a long time. A number of attempts have been made to obtain such measurements immediately after the dawn of satellite altimetry era, but many of them proved to be unsuccessful for various reasons (Callahan and Daffer, 1994).

Okal and colleagues (1999) made the first direct observation of the deformation of the surface of the ocean upon passage of a tsunami wave on the high seas, far from the influence of the shorelines and the continental shelves using satellite altimetry. They sought to detect the tsunami signal where it was expected to be the largest—that is, when the satellite track would cross the initial wave-front, hopefully at a steep azimuthal angle and in the first few hours following its generation, before geometrical spreading on the spherical Earth and dispersive propagation significantly reduce the amplitude of the wave. In realizing this, they used satellite altimetry data from the ERS-1 and TOPEX/POSEIDON (T/P) programs that was complemented by spectrogram techniques and synthetic maregrams to examine the case of seven tsunamigenic earthquakes.

The satellite altimetry data series consists of measurements taken at regular time intervals aboard the satellite (occasionally corrected by linear interpolation when a few points were missing). Since the satellite orbits at a ground velocity of approximately 6–7 km/s, each point in the series is displaced from the previous one both in time ($\Delta t = 0.98$ s for ERS-1; 1.17 s for T/P) and space on the ocean surface ($\Delta d = 7.06$ km for ERS-1; 6.77 km for T/P). Because of the large disparity in surface velocity between the spacecraft (7.2 km/s for ERS-1; 5.8 km/s for T/P) and the expected tsunami wave (typically 210 m/s for an ocean depth of 4500 m), any data series represents more a snapshot of the surface of the ocean at a given time than a maregram (i.e., time-series) at a given point. Thus, the satellite altimeter time-series data approaches more a "space-series" than a "time-series." Accordingly, the "frequency" in the Fourier space resulting from the application of a standard transform to the data series is hybrid in nature but approaches a spatial frequency sampling the wave number space, which is conveniently characterized with units of inverse distance (km^{-1}). Spectral amplitudes will then have units of squared length (m^2). It must be noted that the resulting wave numbers k_i are measured along the satellite track, which is generally oriented at an angle to the local tsunami ray and thus may not be representative of the true wave numbers k of the tsunami field at the corresponding space-time combination.

For analyzing the satellite altimetry time-series data, Okal and colleagues (1999) used the standard spectrogram technique, originally introduced in geophysics by Landisman and colleagues (1969) and consisting of Fourier-transforming a moving window in the original data set. This identifies the exact portion(s) of the data series contributing energy at a given k_i, especially in relation to the causality of any window with respect to the expected timing of the tsunami wave field. For all spectrograms created in their analysis, the length of the moving window was 2^6 points (452 km for ERS-1 and 433 km for T/P), and the interval between windows was 7 samples (49.4 km for ERS-1 and 47.4 km for T/P). The spectrograms were then interpolated and plotted using 50-km pixels.

While Okal and colleagues (1999) succeeded in clearly detecting the tsunami signal in the case of three events, they failed to detect the tsunamis of five other large events, including the 1996 Biak and 1996 Peru earthquakes, primarily because of unfavorable source directivity in the geometry of existing satellite tracks and the strong and incoherent noise produced by large current systems, such as the Kuroshio in the Northwest Pacific.

As just noted, satellite altimeters have detected a limited number of mid-ocean tsunami waves in the past. The December 2004 tsunami changed this picture dramatically. This event was the strongest to have occurred since satellite altimetry started in the 1970s. The altimeters now in orbit can measure sea surface height (SSH) with sufficient resolution (a few cm for the average of a 5-km circle of ocean) but can measure only along their tracks and cannot provide a full picture of an event. Also, they are unlikely to be optimally placed for early detection near the source earthquake.

Following the T/P mission that provided sea-level data since 1992, the Jason 1 altimeter was launched in 2001 in a joint program between NASA and the French space agency Centre National d'Etudes Spatiale (CNES). By chance, however, Jason 1 was in the right place at the right time.

When the earthquake occurred, the Jason 1 altimetry satellite was about 1500 km south of Sri Lanka, heading northeast toward the Bay of Bengal. The satellite passed over the December 2004 tsunami wave front at about $-5°$ latitude (Gower, 2005). At this point, the sea surface started to rise above the height measured in previous cycles to a maximum of about 70 cm above the previous average. The subsequent 30- to 40-cm drop below the average of the previous cycles shows a total trough-to-crest wave height of about 1 m (Gower, 2005; Godin et al., 2009). The measurements show an initial dominant wave-length of about 500 km, followed by significantly greater height variations in the Bay of Bengal compared with those observed in earlier cycles before the event. High-pass filtered SSHs from pass 129 of Jason-1 on cycle 109 during the passage of the tsunami in the Indian Ocean (Figure 11.4) indicates a clearly evident area of enhanced wave energy between latitudes 3° S and 8° N on cycle 109 of the pass over the tsunami. Figure 11.5 shows the spatial extent and height of the devastating December 2004 tsunami as measured by Earth-orbiting radar satellites two hours after the event.

Detection of the tsunami by satellites is quite encouraging, even though at present the satellite altimetry coverage of the global oceans for real-time detection of tsunamis is not adequate. Just by sheer coincidence, a satellite

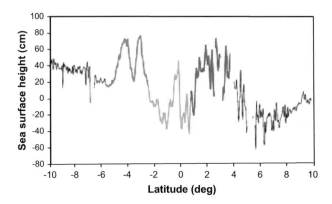

FIGURE 11.4   December 2004 North Sumatra tsunami manifestations in the open sea indicated in the sea surface height (SSH) data from Jason-1 satellite's ascending pass 129 for cycle 109. Breaks in the graph reflect gaps in the available SSH data. The track of pass 129 ascends from southwest to northeast into the Bay of Bengal, east of India, 2 hours after the tsunami earthquake. *Source: Godin et al., 2009; Natural Hazards and Earth System Sciences; Copernicus Publications.*

happened to be at the right time and place to detect this tsunami, and we cannot expect such a coincidence to happen during every tsunami event. However, with adequate satellite altimetry coverage of the global oceans in the future, it is reasonable to expect real-time remote detection of the tsunamis.

11.3. DETECTION USING OPTICAL DEVICES IN SATELLITES AND AIRCRAFTS

A number of eyewitnesses located at different points along the shore of the island of Oahu, Hawaii, observed, in a cloud-free environment, a darkened strip on the ocean surface speedily propagating from the horizon toward the shore, with arrival time of the strip, its velocity, and shape coinciding with that of a front of a small tsunami generated by the October 4, 1994, earthquake at Shikotan (Kuril Islands), Japan. The eyewitness observations are supported by a videorecording made on Oahu's northeastern shoreline. Surprisingly, this tsunami was small (the tsunami wave height was estimated to be only a few cm in water depths of about 2 km). Professor Daniel Walker of the University of Hawaii in Honolulu took keen interest in this observation, which he termed the "tsunami shadowing" effect, and probably foresaw in this phenomenon a unique "discovery" that could have the potential to be translated into a technology of the future for detection of tsunamis in the open ocean. The shadow was actually a uniformly wide band (of 1−2 km width) moving parallel to the horizon toward the observer. The immediate physical cause of a tsunami shadow is a change in the characteristic slope of the sea surface that affects surface reflection of the visible light radiated by the sun and scattered in the

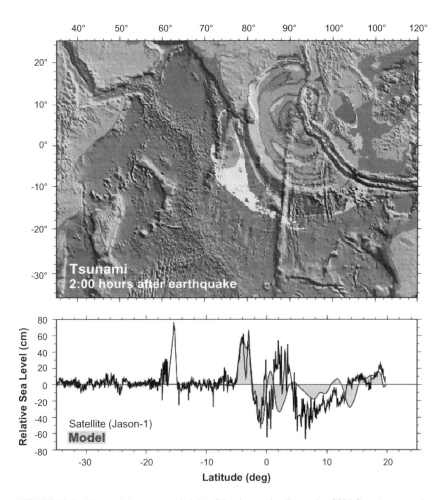

FIGURE 11.5 The spatial extent and height of the devastating December 2004 Sumatra tsunami as measured by Earth-orbiting radar satellites two hours after the event. *Source: Courtesy of PMEL/NOAA.*

atmosphere. However, away from the shore, slopes of the mean ocean surface due to a tsunami are extremely small—typically $(1 \text{ to } 6) \times 10^{-6}$. These slopes can neither account for the observation of the tsunami shadow with the naked eye nor can they be directly measured by other means. The scientific knowledge available at the time on the physical processes at the ocean surface indicated that long surface gravity waves in the ocean modulate short gravity and gravity-capillary waves and alter the ocean surface roughness through air–sea interaction (Hara and Plant, 1994; Troitskaya, 1994). In the light of this, Walker (1996) reasoned that the tsunami shadow observed in Hawaii for the October 4, 1994, tsunami might have been produced by an increase in ocean surface roughness that lowered the amount of skylight reflected toward the observer from the ocean's surface along the crest of the tsunami. Formation of the "shadows" as areas with a different root-mean-square (RMS) surface slope has been tentatively explained theoretically as a result of air—sea interaction. Based on observations and backed by sound theoretical basis, Walker reasoned that it might be possible for low-flying aircraft or appropriately located satellites to track the tsunami out in the deep ocean and provide estimates of its destructive potential well in advance of its arrival in populated coastal areas. He believed this could be a potentially useful remote-detection tool for early tsunami warnings that would avoid false alarms.

An earlier observation of a tsunami shadow in the open ocean is described by Dudley and Lee (1998). Two years later, Walker published his observations incorporating the information extracted from videorecordings and even the real-time radio broadcast captured in the audiorecord on the forecast of the possible arrival of an impending tsunami at this island. The feasibility of the remote detection of tsunami shadows in the open sea is strengthened by the eyewitness accounts of surface manifestation of a destructive tsunami from the April 1, 1946, Aleutian Islands earthquake reported by the pilot of an aircraft cruising about 2300 miles from the site of the earthquake (Dudley and Lee, 1998).

As with any curious mind, Professor Godin (University of Colorado and Environmental Technology Laboratory, U.S.) was intrigued that a tsunami in deep water could produce a variation in ocean surface roughness sufficient to bring about a noticeable change in sea color. Based on the available information, Godin (2004) suggested that although near-surface currents such as those associated with internal gravity waves can influence surface waves and modulate sea surface roughness (Hughes, 1978; Gotwols et al., 1988; Kropfli et al., 1999), such a hydrodynamic modulation of surface roughness is negligible for open-sea tsunamis. This is primarily because a tsunami-induced surface current in the open sea is very small. In the open sea (i.e., water depth $D > 50$ m, in which the linear long-wave model is applicable), the horizontal surface current amplitude u_o generated by a long gravity wave of surface displacement amplitude a is expressed as $u_o = a\sqrt{(g/D)}$, where g is the acceleration due to gravity. The amplitude of the vertical velocity of the sea surface is ηu_o, where $\eta = kD$ (Brekhovskikh and Goncharov, 1985), and $k = \omega/c$ for a monochromatic wave of frequency $\omega = 2\pi/T$. The open-sea tsunami wave period T is usually in the range of 10—60 min. As a consequence of small tsunami amplitude in the open sea, the value of u_o is generally less than 1 cm/s. Thus, a tsunami-induced surface current cannot explain the observed tsunami shadow effect.

Since the origin of surface manifestations of tsunamis remained unexplained, Godin (2004) evaluated the hypothesis that the tsunami shadows and underlying modulation of the surface roughness are caused by tsunami-induced changes in the mean wind profile very close to the ocean surface. He showed that despite negligible surface slopes and rather small surface current velocities associated with tsunamis on the high seas, these waves can indeed produce observable surface signatures as a result of the influence of tsunami-induced wind velocity perturbations on ocean surface roughness. Godin demonstrated that in the lowest few centimeters of the atmosphere above the ocean surface, wind velocity perturbations can exceed the velocity of tsunami-induced currents by a factor of 10 to 60 for representative values of tsunami and wind parameters. It has been found that the strong perturbations of the mean wind velocity close to the sea surface are a direct consequence of the coherent vertical motion of large expanses of the ocean surface in association with the tsunami wave propagation. Thus, "wind enhancement" of tsunami-induced perturbations is presently considered to be the key to explaining tsunami shadows. Enhancement of sea surface roughness through the generation of short gravity-capillary waves under the influence of a tsunami becomes feasible because of the fact that tsunami-induced perturbations of wind velocity are most prominent in the lower part of the atmospheric boundary layer. According to Godin's theoretical analysis, tsunami shadows are parallel to tsunami fronts and move with the velocity of tsunami propagation. The "shadows" are centered on the midpoint between the tsunami crest and the trough. The separation between successive "shadows" is close to the distance between the crest and the trough and is roughly half of a tsunami wave length. Correspondingly, separation of two consecutive tsunami shadows in time is close to half of a tsunami period. According to Godin, the contrast of a tsunami shadow image varies slowly along a tsunami wave front due to variations in the tsunami's amplitude and background wind velocity. The extent of the "shadow" across the wave front is determined by variations in the tsunami-induced vertical velocity of the ocean surface and equals a fraction of the tsunami wave length. These theories about the spatial and temporal structures of tsunami shadows are consistent with observations reported by Walker (1996) and Dudley and Lee (1998).

For potentially destructive tsunamis, expected changes in surface roughness are such that the tsunami shadows should be observable also on the high seas with airborne sensors. The spatial structures of the tsunami shadows—namely, their length of thousands of kilometers along the tsunami wave front and their width of a few tens of kilometers across the wave front—are conducive to tsunami detection from space. According to Godin (2004), of particular significance in their application to tsunami warnings is that the tsunami shadows propagate at a known and very distinct speed, which allows for their unambiguous differentiation from other features on the ocean surface.

11.4. DETECTION USING ORBITING MICROWAVE RADAR AND RADIOMETERS

Observations of tsunamis away from shore are critically important for improving early warning systems and understanding tsunami generation and propagation. Currently, tsunami observations in deep water rely on measurements of variations in the sea bottom pressure (through seafloor-mounted precision pressure sensor capsules) or SSH (through satellite altimeter). Tsunamis are difficult to detect and measure in the open ocean because the tsunami wave amplitude there is much smaller than it is closer to shore. Godin and colleagues (2009) demonstrated that this difficulty in reliably detecting tsunamis away from shore could probably be circumvented by measuring ocean surface roughness. The first detailed measurements of a tsunami's effect on SSH and radar backscattering strength in the open ocean were obtained from satellite altimeters during the 2004 Indian Ocean tsunami that originated in Sumatra. Through statistical analysis of satellite altimeter observations Godin and colleagues showed that the Indian Ocean tsunami effected distinct, detectable changes in sea surface roughness. The magnitude and spatial structure of the observed variations in radar backscattering strength were found to be consistent with hydrodynamic models predicting variations in the near-surface wind across the tsunami wave front. As noted in the previous section, tsunami-induced variations in sea surface roughness away from the shore were first observed in visible light (Walker, 1996; Dudley and Lee, 1998) as tsunami shadows—that is, extended darker strips on the ocean surface along a tsunami wave front. Tsunami-induced changes in sea surface roughness can be potentially used for early tsunami detection by orbiting microwave radars and radiometers as well. Of all the tsunami manifestations in the deep ocean, variations in ocean surface roughness are most relevant to tsunami detection from space provided that these can be detected by orbiting active (scatterometers) and passive (radiometers) scanning microwave sensors, which have broad surface coverage across the satellite ground track.

The December 2004 tsunami is the first for which detailed concurrent measurements of the SSH and radar backscattering strength at its nadir (σ_0) in the deep ocean are available. These measurements were made with microwave radars on board the Jason-1, Topex/Poseidon, Envisat, and Geosat Follow-On (GFO) altimetric satellites (Smith et al., 2005; Ablain et al., 2006; Gower, 2007). The SSH measurements were used by a number of investigators to study the properties of the Sumatra tsunami, its propagation, and its scattering from the coastline, as well as to improve characterization of the seismic source of the tsunami and to verify numerical tsunami models. Both Okal and colleagues (1999) and Zaichenko and colleagues (2005) discussed the detection of earlier, weaker tsunamis in less extensive SSH records. Understanding the physical processes on the leading front of a tsunami is of particular importance for tsunami early detection and warning. Godin and colleagues (2009) concurrently employed σ_0 data, which are a measure of the ocean surface roughness (Fu and Cazenave, 2001), and SSH data obtained by Jason 1 on December 26, 2004, to determine the tsunami's effect on surface roughness.

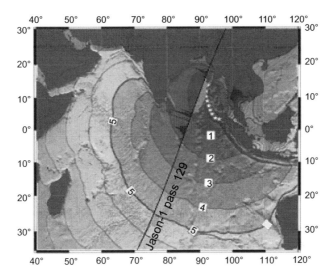

FIGURE 11.6 The Jason-1 satellite altimeter ground track and C-band σ_o data for pass 129 encountering the Sumatra-Andaman tsunami superimposed on contours of the tsunami leading wave front at hourly intervals after the earthquake. The σ_o data are shown for the portion of the ground track where the tsunami wave had arrived. White stars show the location of the tsunami wave sources. Tsunami wave front graphic is courtesy of the National Geophysical Data Center/ NOAA. *Source: Godin et al., 2009; Natural Hazards and Earth System Sciences; Copernicus Publications.*

Radar backscattering strengths measured in the Ku and C microwave frequency bands in the vicinity of the tsunami's leading front showed pronounced—up to ± 1 dB—variations that are not present in measurements along the same track on the last passage before and the first passage after the tsunami. Through statistical analysis of multiple years of satellite altimeter observations under various atmospheric conditions when no tsunami was present, they demonstrated (Figure 11.6) that the Sumatra tsunami induced distinctive variations in ocean surface roughnesses, which are detectable with microwave sensors already in orbit. The Jason-1 C-band σ_o data for pass 129 encountering the Sumatra-Andaman tsunami is shown in Figure 11.6.

While the existence of significant tsunami-induced variations in sea surface roughness has been established unambiguously, the critically important practical issues of optimal retrieval of a tsunami signal from various measures of sea surface roughness and of potential application of these measures to tsunami detection and characterization are subjects of future research. However, unlike the SSH, which is measured at nadir points along the satellite-borne altimeter's ground track, tsunami-induced variations in sea surface roughness can be potentially measured over wider swaths with spaceborne and airborne side-looking radar and scanning microwave radiometers. Godin and colleagues (2009) suggest that spatial averaging of radar backscattering strength or brightness temperature along hypothetical tsunami wave fronts could be used with these kinds of sensors to distinguish any tsunami signal from noise due to other sources of the roughness change. The much broader surface coverage of these sensors suggests that they are more promising for early tsunami detection than satellite altimeters and may be an important element in a future global system for tsunami detection and warning. This optimistic view arises primarily from the availability of a number of active and passive microwave sensors on several satellites in orbit, whereas operational satellite altimeters are limited. It is hoped that by complementing traditional seismic data and point measurements as provided by the DART buoys network and cable-mounted ocean bottom sensors; wide area satellite observations of tsunami manifestations can potentially improve the accuracy and timeliness of tsunami forecasts, increase the lead time of tsunami warnings, decrease the probability of false alarms, and help to avoid unnecessary evacuations.

Chapter 12

Land-Based Measurements of Inundation to Confirm Tsunamigenesis

In some locations, the time interval between the occurrence of an earthquake and the arrival of a tsunami on the nearest coastline is very short. Most fatalities occur in the near field of tsunamigenic earthquakes, and these will not be minimized substantially by current popular new technologies. In such situations, there is no time to wait for a confirmation of the existence of the tsunami, and therefore the optimal practical approach to the problem is to simply issue a tsunami warning based only on the occurrence of the earthquake. This could lead to false alarms in some instances, which cannot be easily avoided. Lives can be saved with rapid near-field detection and education. Land-based measurements of inundation to confirm actual tsunamigenesis after an earthquake and thus providing earlier warnings using cell phone or satellite transmission of flood detection on land near the source of the earthquake would prevent such false alarms.

Taking this requirement into account, the U.S. TWS has incorporated some newly developed remotely reporting real-time tsunami inundation detection systems (also known as tsunami run-up detectors) on the areas along the coast of Hawaii most likely to be affected by a tsunami (Figure 12.1). The run-up detectors will trigger and send a message to PTWC within seconds of their sensor being flooded by an advancing tsunami, positively indicating there is water on land where the detector is located. The sensors are about 2 to 4.5 m above mean sea level and are located some 18 to 120 m from the shoreline, where in the past significant run-ups have been observed. These detectors are outside the normal surface run-up and are insensitive to rain or moisture, except for serious flooding (McCreery, 2005). Based on home security alarm technology and cell phone communications, the detectors are relatively inexpensive (approximately, U.S. $1000 each), easy to install, and much easier to maintain than a normal sea-level station. At each station, any flooding over 1 foot (0.3 m) deep is detected and immediately communicated to warning centers via cell phone. One advantage of this type of tsunami detector system is that since the sensor is not in the sea, there are more options for positioning them along the

Tsunamis: Detection, Monitoring, and Early-Warning Technologies. DOI: 10.1016/B978-0-12-385053-9.10012-2

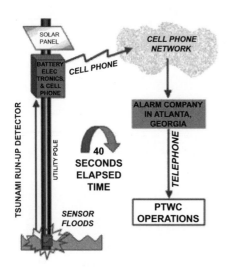

FIGURE 12.1 Tsunami run-up detectors used by PTWC, based on home security alarm technology. *Source: McCreery, 2005; Natural Hazards; Reproduced with kind permission of Springer.*

shoreline. Because this new type of tsunami run-up gauge observes inundation directly, the system is able to warn against local landslide-generated tsunamis that are not associated with earthquakes, as well as those that are. Also, this system reduces the chances of a false local warning. There is a history of both earthquake-generated and local landslide-generated tsunamis occurring around some islands in the Pacific. Because cellular communication systems are widely in place today, this type of system provides a very inexpensive supplement to tsunami warning systems (Walker and Cessaro, 2002). These new run-up detectors are expected to help PTWC to more quickly confirm any significant local tsunamis and eventually lead to a procedure whereby a warning is not issued unless an actual tsunami is confirmed (McCreery, 2005). Satellite-linked land-based water sensors are now in operation in Hawaii. The disadvantage of the land-based detector, however, is that it cannot provide a precise value of the tsunami amplitude, such as the one recorded by a sea-level gauge.

Chapter 13

The Technology of End-to-End Communication: Sending the Message

A timely, 100% accurate and precise warning is of no use as far as protecting citizens unless the information reaches them in time and unless they know how to respond to the emergency. Early warning is thus as much an issue of "soft" organizational technology, communication, and community-based systems as it is of "hard" science and technology, numerical modeling, and instrumental networks. Tsunami warning bulletins provided by warning centers and the flow of tsunami warnings from warning centers to the local communities are critical steps in the warning process. Public knowledge of natural phenomena that can be felt or witnessed (i.e., strong earthquake shaking, unusual water conditions such as rapid drawdown or a sudden rise in the water level of the ocean, tsunami wavetrains), in combination with robust, redundant, and widespread man-made warning systems will ensure that all residents and tourists in the inundation zone are warned in an effective and timely manner.

Man-made warning systems, such as emergency alert system (EAS), telephones, weather radios, and sirens, are most effective for warning about tsunamis from distant sources. Man-made systems are also useful for "all clear" notifications for distant tsunamis and, if the equipment survives the earthquake, for local tsunamis as well. If "all clear" notifications are made rapidly enough, community disruption by false alarms would be reduced. Repeated false evacuations are costly not only in terms of lost revenue but in lost credibility. EAS stations include radio, television, and cable networks. Although telephone alerts can be activated by the emergency operation centers or local public safety dispatch centers, telephone lines could become overloaded with calls from people wanting to verify what they heard of if shaking is indeed an earthquake. While weather radio is a useful device for receiving tsunami warning notifications, tourists visiting isolated beaches or other low-lying areas do not normally have a weather radio. Also, not all areas are within range of a transmitter because of the rugged nature of some coastlines.

Sirens are especially important because they can be heard from great distances. They are usually electromechanical or fully electronic devices (Figure 13.1) that can be triggered manually or automatically. Both of these

Tsunamis: Detection, Monitoring, and Early-Warning Technologies. DOI: 10.1016/B978-0-12-385053-9.10013-4

FIGURE 13.1 (a) Siren in Hawaii that is very useful in crowded areas, where access to warning devices is limited. (b) An All Hazards Broadcast System (AHBS) established at Mayaguez, Puerto Rico. *Sources: (a) Darienzo et al., 2005; Natural Hazards; Reproduced with kind permission of Springer. (b) von Hillebrandt-Andrade et al., 2008.*

devices project standard siren sounds, but electronic sirens can also broadcast public address announcements (Darienzo et al., 2005). Public address announcements are usually concise and, ideally, are prerecorded to avoid potential problems with unintelligible messages from a stressed system operator.

It is desirable that sirens project only one tone that sounds over a distinct period of time. This standardization reduces confusion for residents and transient populations along the coast. The sirens normally would prompt people to turn on their radio or television for more information. Relative to other types of man-made warning systems, sirens have the advantage of reaching all populations in the coverage area. Sirens are most useful in areas such as crowded beaches where access to warning devices like radio and television is limited.

It is an added benefit if public announcement siren systems are supplemented with EAS. Tsunami warnings through public announcement siren systems are an integral part of the U.S. tsunami warning system.

Tsunami warnings in French Polynesia are also done with sirens. More than 93 fully operational sirens have already been installed in French Polynesia. When the program is fully implemented, the number is expected to grow to 131. These electronic devices, which are powered from solar panels and equipped with Inmarsat mini-C satellite detectors, will be operated by high-commissariat departments in Tahiti.

IOC-UNESCO Tsunami Early Warning Systems

A tsunami is one of the most dangerous and destructive natural phenomena and one of the most formidable of all natural hazards. Before the December 26, 2004, Indian Ocean tsunami episode, the region most frequently affected was the Pacific Ocean because many severe earthquakes occurred along its margins. The 1960 Chile earthquake, the largest in recorded history, generated a terrible tsunami that struck the Chilean coast in just 15–20 minutes, Hawaii in about 15 hours, and then Japan about 22 hours later, causing damage and taking many lives in each locale. At that time, no real or near-real-time sea-level data to confirm a potential tsunami and no warning system were available to alert countries outside of Chile that a tsunami was on its way. This Pacific-wide destructive event resulted in the beginning of a tsunami warning system (TWS)

Tsunamis: Detection, Monitoring, and Early-Warning Technologies. DOI: 10.1016/B978-0-12-385053-9.10014-6

in 1965, now known as Pacific Tsunami Warning System (PTWS). We now know that no oceanic region in the world is exempt from the impact of tsunamis. Major earthquakes have produced destructive wave effects at trans-oceanic distances, and their repercussions have been felt in other oceanic regions as well. As noted earlier, transoceanic tsunamis are usually generated as a result of seismotectonic motions of the ocean bottom in the seismic source zone. Such tsunami waves propagate far from the source and can cause damage even in regions where the earthquake was not manifested. The unexpectedness of tsunamis is an additional risk factor. Taking into account the specific char-acteristics of a tsunami's destructive behavior, this natural hazard can be considered one of the most inevitable of all natural phenomena.

The most effective protection against a tsunami hazard is timely evacua-tion of people to safe regions and withdrawal of ships to the open sea. Obviously, advanced notification concerning the approaching wave is important in this case. A timely operative forecast of tsunami is, indeed, the most important aspect of this problem. In the case of far-field tsunamis, the interval between the moment of wave generation in the ocean and the arrival of the tsunami on the coast is usually sufficient to evacuate people from the possible flooding zone. However, the absence of an early warning system would render timely evacuation an impractical task. Thus, tsunami warning has worldwide importance.

The effectiveness of a tsunami warning, in terms of hazard reduction, depends greatly on the availability of accurate tsunami inundation maps. As the chair of the IUGG Tsunami Commission in 1989, Dr. Eddie Bernard gave an opening address for the International Tsunami Symposium (ITS-89), suggest-ing that the tsunami scientists could make a valuable contribution to the International Decade for Natural Hazard Reduction (1990–2000) by creating regional tsunami inundation maps. Tsunami scientists have far exceeded this initial scientific goal of inundation maps by pioneering new ideas of tsunami hazard reduction through a community-based mitigation program concept, including appropriate mitigation strategies to minimize the impact of future tsunamis and the implementation of effective warning and education programs to alert at-risk communities. Although much has been accomplished, numerous lives have been lost to tsunamis in the past 20 years. Society has great expectations that tsunami science and technology will save lives in the future. The biggest challenge tsunami scientists and technologists face is creating better, faster, and cheaper warning systems (Bernard, 2009).

Before the December 2004 tsunami, only a few nations (The United States, Japan, and, to a certain extent, Russia) had operational TWSs, and these were confined primarily to the Pacific Ocean. However, under the initiative of IOC-UNESCO, this tragic episode happened to be a major driving force in accel-erating the implementation of a network of tsunami monitoring and forecasting system within and beyond the Pacific Ocean. As a result, since the December 2004 tsunami, the number of tsunameters deployed throughout the world's oceans has increased significantly, especially in the Indian Ocean region. The main international TWS initiatives under various IOC regional activities include Pacific-, NEAM-, Caribbean-, and Indian-Ocean TWS. Several national TWS systems have slowly emerged or are in the process of being established under the umbrella of any of these international initiatives.

14.1. THE UNITED STATES TSUNAMI WARNING SYSTEM

The National Weather Service of the National Oceanic and Atmospheric Administration (NOAA) operates two tsunami warning centers (TWCs) in the United States: the Alaska Tsunami Warning Center (ATWC) and the Pacific Tsunami Warning Center (PTWC). The ATWC is located in Palmer, Alaska. This center serves as the regional TWC for Alaska, British Columbia, Washington, Oregon, and California. The PTWC is located at Ewa Beach in Hawaii. The center serves as the regional TWC for Hawaii and as a national/international warning center for tsunamis that pose a Pacific-wide threat. The PTWS, which consists of 26 participating international member states, monitors the seismological and the sea-level stations throughout the Pacific Basin. The system evaluates potentially tsunamigenic earthquakes and disseminates tsunami warning information.

PTWC is the operational center of the Pacific TWS. Headquartered at Honolulu, Hawaii, and covering the Pacific, the PTWC provides tsunami warning information to national authorities in the Pacific Basin. The island of Hawaii is both volcanically and seismically active, and in its 200-year historical record it has suffered two major local tsunamis (with maximum run-ups of ~15 m) and a handful of smaller local tsunamis. PTWC has the responsibility to provide warnings for such local and regional tsunamis. While PTWC may not be able to warn those nearest the epicentral region, its goal is to provide a warning to population centers as close as possible. The PTWC consists of a network of seismic stations (SS) (to detect earthquakes) and sea-level stations (to determine whether a tsunami is associated with the disturbance), all connected by a rapid communication link (McCreery, 2005).

TWCs of National Oceanic and Atmospheric Administration (NOAA) are tasked with issuing tsunami warnings for the United States and other nations around the Pacific. Tsunami warnings allow for immediate action by local authorities to mitigate potentially deadly wave inundation at a coastal community. The more timely and precise the warnings, the faster local emergency managers can take action, and the more lives and property that will be saved. Advances in tsunami measurement and numerical modeling technologies can be integrated to create an effective tsunami forecasting system. The Deep-ocean Assessment and Reporting of Tsunamis (DART) technology can detect tsunami waves of less than 1 cm in the open ocean (at depths up to 6 km) and reliably provide data to NOAA's TWCs in real time.

In general, the procedures used by PTWC to provide tsunami warnings are as follows (McCreery, 2005):

- Hardware and computer programs continually monitor seismic waveform data streams and alert watch-standers whenever large and widespread signals are detected from a significant earthquake.
- Watch-standers then locate the earthquake and determine its magnitude using a combination of automatic and interactive procedures. If the earthquake is shallow and located under or very close to the sea, and if its magnitude exceeds a predetermined threshold, a warning is issued based on the possibility that a destructive tsunami was generated.
- As sea-level data are received from the nearest gauges, the tsunami can be confirmed if it exists and is measured. These measurements are then evaluated in the context of any historical events from the region, any applicable numerical simulations, and other predictive tools based on the earthquake and sea-level measurements.
- Based on this evaluation, the warning is continued, upgraded to cover a larger area, or canceled. These procedures apply to cases of both destructive teletsunamis (i.e., transoceanic tsunamis) and local or regional tsunamis.

In recent years, the U.S. TWS has witnessed dramatic improvement in each element of the warning system (Whitmore et al., 2009). The density and quality of data provided by observational networks (seismic, deep-ocean sea-level, and coastal sea-level) have increased in addition to the enhanced timeliness of data transmission. U.S. warning centers have improved data processing and tsunami wave height/travel time forecasting capabilities and have implemented around-the-clock operations. Modern communication networks have greatly reduced the latency for receipt of timely information by affected populations, as well as produced a richer and more diverse set of message products. These networks include satellite communication networks, cellular networks, emergency alert systems, and Web-based message products. Tsunami response capabilities continue to improve thanks to programs such as the National Tsunami Hazard Mitigation Program and the National Weather Service Tsunami Ready Program.

NOAA's tsunami forecast methodology combines real-time deep-ocean measurements with model estimates to produce real-time tsunami forecasts for coastal communities. This forecast procedure (also known as a short-term inundation forecast, or SIFT) is implemented at NOAA's TWCs and is now being tested in warning operations. Tsunami forecasting provides site- and event-specific information about tsunamis before the first wave arrives at a threatened community. This scheme provides estimates of all critical tsunami parameters: amplitudes, inundation distances, and current velocities. The key observation data that ensure tsunami forecast accuracy are supplied by deep-ocean tsunami data. As mentioned earlier, the DART technology can detect tsunami waves of less than 1 cm in the open ocean (at depths up to 6 km) and provide data to the forecast system in real time during tsunami propagation. This accuracy allows fast, robust estimates of a tsunami source magnitude, even with a limited number of initial observations. The DART-estimated source magnitude is obtained separately from seismic observation and provides independent estimates of tsunami magnitude. A number of test forecasts for tsunami events during 2005–2009 were analyzed to estimate the accuracy of the DART-based predictions (Titov, 2009).

Despite the considerable success achieved from the DART systems, it must be noted that DARTs and the coastal sea-level stations perform different functions within the context of tsunami warnings and that both of them have their strengths and weaknesses. Therefore, these technologies must be used together to fulfill the mission of the PTWS (Weinstein, 2009).

14.2. THE JAPAN TSUNAMI WARNING SYSTEM

Recurrent major earthquakes associated with the circum-Pacific volcanic belt and subduction zones often generate powerful tsunamis that cause destruction both locally and on distant shores. High seismicity characterizes the Kurile subduction zone, where the Pacific plate is subducted beneath the Kurile Islands and Hokkaido, Japan. The Japanese islands are located in a tectonically active region on the edge of the Eurasian and North American Plates, where the Pacific and Philippines Sea Plates are subducted. Seismogenic zones are located near trenches or plate boundaries, which are, in many cases, under-water. Therefore, it is important to study the generation mechanisms of earthquakes around Japan in these seismic and geophysical underwater areas. This country has been particularly vulnerable to treacherous tsunami events, and therefore Japanese scientists place a lot of emphasis on tsunami research. This may explain why the phenomenon was given a Japanese term (from *tsu,* meaning "harbor," and *nami,* meaning "wave") (Wilson, 1964; Filloux, 1982). Interplate seismicity beneath the landward slope of the Japan Trench, off the Sanriku region, is active, and the northeastern part of Japan has experienced several large ($M_W > 7$) earthquakes (Tsuru et al., 2000). These large interplate earthquakes were sometimes accompanied by large tsunamis that caused serious damages to the coastal region (Hino et al., 2001). Studies of earth-quakes and tsunamis are considered to be important from the viewpoint of natural hazard mitigation. Real-time data telemetry from nearby offshore seafloor stations is essential in such studies because early detection of earth-quakes and tsunamis is needed to design and apply appropriate mitigation programs. Although high-efficiency seismic observation requires a high-density seismic measurement network, there is considerable technological difficulty in deploying as many sensors on the seafloor as on land. The existing observation network is still very small and insufficient for an ideal observation network covering Japan (Kawaguchi et al., 2002).

14.2.1. Historical Review of Japan's Tsunami Warning System

The Japan Meteorological Agency (JMA) is responsible for issuing tsunami warnings in Japan. JMA continuously monitors seismic activities in and around Japan. When an earthquake occurs, JMA immediately determines the hypocenter and magnitude of the quake. If the earthquake occurs in an ocean area with tsunamigenic potential, JMA conducts the tsunami forecast operation using the database containing the tsunami's amplitude and travel time calculated by numerical simulation. JMA issues warnings and advisories for 66 coastal regions that involve all of the coastal areas of the country. Such warning and advisory messages contain information such as the expected maximum tsunami height and the arrival time. JMA provides tsunami warnings and advisories for the national and local authorities and broadcasting stations for disaster prevention and mitigation. Governors of municipalities are responsible for giving instructions to the residents to evacuate from tsunami hazardous areas.

Real-time sea-level data are gathered in the JMA to monitor tsunami arrivals at the coasts. When confirming a tsunami arrival, JMA announces tsunami observations. Tsunami warnings and advisories are canceled when JMA concludes that the danger is over—that is, when the tsunami attenuates and the observed heights become lower (Japan's National Report to IOC-UNESCO, January 2009).

JMA started tsunami warning service in 1952 (Kamigaichi, 2009). Since then, JMA had been making serious efforts to integrate its TWS system. Initially, JMA had 46 seismometers at meteorological observatories. When earthquakes occurred, seismograms were read at observatories, and their results (P- and S-arrival times and maximum amplitudes) were transmitted to the headquarters in a telegram format. At the headquarters, hypocenter and magnitude were determined manually, and the tsunami warning grade was determined with an empirical chart based on the relationship among earthquake magnitude, epicentral distance to a coast of the past events, and tsunami amplitude. It used to take about 15 to 20 minutes to disseminate a tsunami warning. Japan has a network of dense coastal sea-level stations for routine coastal sea-level measurements. There are about 160 of these stations, including those operated by JMA, Japan Coast Guard, Ports and Harbors Bureau, and Geographical Survey Institute (Figure 14.1), with data sampling and transmission at 1-second intervals, in addition to a network of several offshore sea-level gauges and a few hydrophones. Some example of observed hydrophone records from three stations are shown in Figure 14.2. Japan's coastal and offshore sea-level gauge network (Figure 14.3) provides the requisite data (Figure 14.4) to verify the forecast. In recent years, differential GPS-based sea-level gauges on floating offshore platforms have been used in Japan's sea-level gauge network. JMA's man-machine interface system (Figure 14.5) facilitates real-time viewing of a tsunami from pressure gauges.

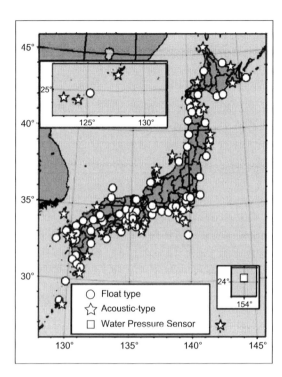

FIGURE 14.1 Tidal stations network used for tsunami monitoring by the Japan Meteorological Agency. More than 100 tidal stations are monitored in real time. Symbols denote instrumentation type. The circle, the star, and the square denote float type, acoustic type, and pressure sensor type, respectively. *Source: Kamigaichi, 2009, Courtesy of Dr. Osamu Kamigaichi of JMA; Encyclopedia of Complexity and Systems Science, R. A. Meyers (Ed.), Tsunami forecasting and warning, 9592-9617; Reproduced with kind permission from Springer Science+ Business Media.*

Japan's tsunami monitoring system is used to decide when the warning or advisory can be canceled, revise the grade of the warning or advisory in cases where the estimated and observed tsunami amplitudes differ significantly, and provide the public with an up-to-date status of tsunami observations (Kamigaichi, 2009).

Ocean bottom pressure gauges (OBPGs) can detect tsunamis with amplitudes of several mm to a few cm. This extremely precise pressure gauge (manufactured by NEC Corporation) uses a quartz pressure sensor (HP 2813 E, made by Hewlett-Packard Corporation) for seafloor pressure measurement. During an earthquake, an offshore OBPG usually records the tsunami and the preceding pressure fluctuations with frequencies higher than the period of the tsunami (Okada, 1995). The pressure fluctuations are primarily attributed to seismic Rayleigh waves travelling through oceanic lithosphere from a distant earthquake (Fillioux, 1982) or possibly to seismic body waves from a regional earthquake. In some cases, it masks tsunami signals for nearby earthquakes (Okada, 1995). Maximum amplitudes of the pressure signals observed with near-field OBPGs during the Tokachi-oki earthquake ($M_W = 8.0$) of September 25, 2003, were a few hundreds of kPa, while the maximum tsunami amplitude was estimated at only a few kPa. Thus, the seismic signals were much larger than the tsunami signal. Moreover, the tsunami and pressure signals completely

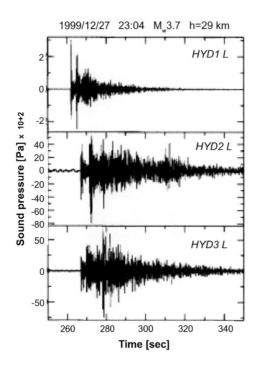

FIGURE 14.2 An example of observed hydrophone records from three HYDs, recorded by the permanent seafloor observatory of JAMSTEC off Hokkaido, Japan, in the Pacific Ocean. *Source: Hirata et al., 2002; IEEE J. Oceanic Engg; Reproduced with kind permission of IEEE.*

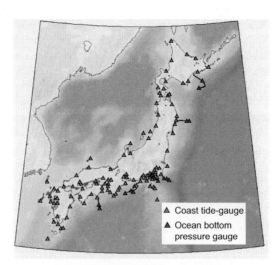

FIGURE 14.3 Japan's coastal and offshore sea-level gauge network. *Source: Courtesy of Dr. Osamu Kamigaichi of JMA.*

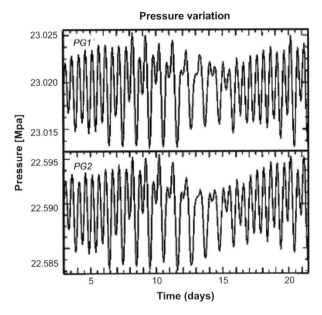

FIGURE 14.4 An example of observed seafloor pressure data from two pressure gauges, recorded by the permanent seafloor observatory of JAMSTEC off Hokkaido, Japan, in the Pacific Ocean. *Source: Hirata et al., 2002; IEEE J. Oceanic Engg; Reproduced with kind permission of IEEE.*

overlapped (Hirata, 2009). Therefore, the extraction of the tsunami signal from OBPG records observed during nearby large earthquakes is essentially important for a possible near-field tsunami forecast in the future. Presently, however, making a complete decomposition of an OBPG record into tsunami and other pressure signals for near-field tsunamis does not seem to be an easy task (Okada, 1995; Hino et al., 2001; Hirata et al., 2003). The observed pressure signals include low-frequency hydroacoustic (elastic) waves reverberating between the sea surface and the ocean bottom through water (Kajiura, 1970) and are closely related to tsunami generation or ocean-bottom motion due to an earthquake. Hydroacoustic waves, however, remain mostly in the source region such that they do not affect tsunami propagation and onshore run-up.

An interesting finding from the 2003 Tokachi-oki earthquake was that the OBPGs within or close to the source region had experienced a sudden temperature change of an order of 0.1°C per 10 minutes, probably due to a change in bottom currents (Hirata et al., 2003). Such sudden temperature changes cause artificial pressure changes that distort tsunami waveforms due to the transient thermal response of OBPGs (Takahashi, 1983; Hirata and Baba, 2006). Thus, a possible near-field forecast based on records monitored with an OBPG that is extremely close to a source region may be difficult unless the transient thermal effect of OBPGs is properly corrected (Hirata, 2009). Typical

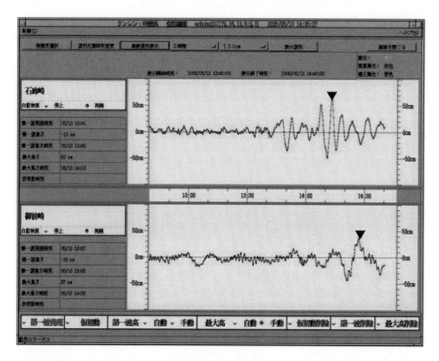

FIGURE 14.5 The JMA's man–machine interface screen image in the Sea Level Data Monitoring System. Shown are two sea-level recordings by different tide gauge stations. The ocean tide component has been removed (detided) for a more precise reading of the relevant tsunami wave parameters. Arrival time, initial wave amplitude, polarity, maximum amplitude, and corresponding time are picked and tabulated (on the left side of the corresponding detided sea level). The reversed solid triangles denote the maximum amplitude within the analyzed time window. *Source: Kamigaichi, 2009, Courtesy of Dr. Osamu Kamigaichi of JMA; Encyclopedia of Complexity and Systems Science, R. A. Meyers (Ed.), Tsunami forecasting and warning, 9592-9617; Reproduced with kind permission from Springer Science+Business Media.*

power spectral densities of observed seafloor pressure data from two pressure gauges are provided in Figure 14.6. The pressure spectra show the common shape in the infragravity wave band (Webb et al., 1991) below 0.02 Hz. Pressure fluctuations observed in this band are caused by long-wavelength surface gravity waves (Webb, 1998).

14.2.2. Japan Meteorological Agency Tsunami Monitoring

Currently, open ocean bottom pressure sensors for tsunami warnings are only watched on a computer for the following reasons:

1. The pressure sensors are useful for early detection of tsunamis, but to do this, the optimal filter must be tuned, especially for a nearby earthquake, to exclude apparent data changes due to shaking on the ocean floor.

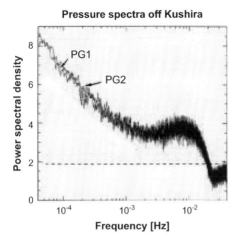

Pressure spectra off Kushira

FIGURE 14.6 Example of power spectral densities of observed seafloor pressure data from two pressure gauges, recorded by the permanent seafloor observatory of JAMSTEC off Hokkaido, Japan, in the Pacific Ocean. Both of the pressure data sampled at 10 s are used in the spectral calculations. *Source: Hirata et al., 2002; IEEE J. Oceanic Engg; Reproduced with kind permission of IEEE.*

2. The observed tsunami amplitude is usually much smaller than that predicted at the coast because it is before the amplification by the bathymetry change. Therefore, if the observed amplitude data are announced to the public, it can give a misleading impression that the tsunami is much smaller than JMA's estimate. JMA is considering the best way to communicate this information to the public.

JMA also uses pressure sensors to revise warnings. JMA disseminates the first warning as quickly as possible, using seismological data only before the confirmation of tsunami generation by the sea-level data monitoring. After the tsunami observation (and if it is necessary), JMA will revise (upgrade/downgrade) the first warning, depending on the sea-level change observation results. JMA adopts quantitative tsunami simulation techniques for the tsunami amplitude estimation. The observed off-the-coast tsunami amplitude is free from contamination by the complicated fine bathymetry change, and therefore it is more suitable for a comparison between the estimated and observed amplitude than the usual sea-level data (tide gauge data) measured at the coast.

However, with appropriate future modifications in the existing software for tsunami height computation for coastal regions, the difficulties that exist now for issuance of tsunami forecasts for coastal regions can hopefully be surmounted. To achieve improved reliability, a tsunami height computation scheme would take into account the complicated fine bathymetry and topography changes at the coast, as well as the complexities involved in the tsunami amplification and focusing (arising from the implications of Green's law). Despite the presently existing difficulties, Japan's ocean bottom seismograph (OBS) stations have been found to be very efficient and useful for early tsunami detection. For example, early detection of both the 1996 Irian Jaya (Indonesia) $M_W = 8.2$ (Figure 14.7) and the 2003 Tokachi-oki $M_W = 8.0$

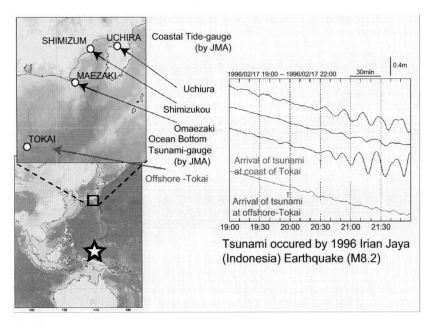

FIGURE 14.7 Early detection of the 1996 Irian Jaya (Indonesia) magnitude 8.2 earthquake-generated tsunami by JMA's single OBS station long before the tsunami's arrival at Japan's coastal locations as recorded by three coastal sea-level gauges. *Source: Courtesy of Dr. Osamu Kamigaichi of JMA.*

earthquake-generated tsunamis by JMA's single OBS station long before the tsunamis arrived at Japan's coastal locations are impressive examples of the practical utility of Japan's continuous real-time reporting submarine cable-based ocean-bottom recorders.

An important advantage of offshore tsunami observations is early tsunami detection. For deep-ocean cabled observatories, past events have demonstrated that for nearby earthquakes, the tsunami could be detected approximately 20 minutes earlier than at the nearest coastal tide gauge station.

14.2.3. Merits and Drawbacks of Submarine Cable-Mounted Systems

Submarine cable-mounted systems are preferred in situations where long-term continuous monitoring is a requirement. Whereas the availability of an ample supply of electrical power for long-term continuous operation is impossible with cableless installation, cable-mounted systems have clear advantages over autonomous systems that must rely on limited self-contained sources of power. Furthermore, a continuous supply of power to sensors and electronics eliminates many complications in sampling and communications. Close examination of geophysical events is made possible by cable-mounted systems because they permit a continuous stream of measurements sampled at high frequency to be transmitted to shore. Perhaps another important advantage of submarine cable-mounted systems is its freedom from vandalism, which is an issue of great concern in the use of DART buoy systems. They also operate for long periods of time without problems. For example, during the last 30 years of operation, there have been only two cases of necessary repairs (Dr. Masahiro Yamamoto's presentation at the July 2009 International Tsunami Symposium). Despite the considerably large initial cost during deployment, maintenance-free operation as claimed is indeed a point in favor of cable-mounted systems.

However, submarine cables have their limitations as well. For example, submarine cables, which are meant for operational tsunami measurements, are laid in trenches and then buried to protect them from fishing activities and other hazards. Further offshore, cables have been damaged by submarine landslides moving down the continental slope and rise. For example, 12 telegraph cables were cut off from eastern Canada during the 1929 Grand Banks earthquake ($M_W = 7.2$) submarine landslide and tsunami (Murty, 1977; Bryant, 2001; Clague, 2001; Fine et al., 2005). In this case, an orderly sequence of breaks occurred in these cables at distances up to 500 km from the epicenter. According to Heezen and Ewing (1952) and Heezen and colleagues (1954), the successive series of breaks in the cables was caused by the turbidity current generated by the slump on the continental slope. More recently, submarine landslides triggered by earthquakes damaged submarine cables in the Mediterranean Sea in 2003 and off Taiwan in 2006. However, submarine cables can be made less vulnerable to hazards through careful siting and installation.

14.3. THE RUSSIA TSUNAMI WARNING SERVICE

Russia has historically been the victim of several large- and medium-scale tsunamis because of the strong seismicity of the zone covering the Kuril Islands and the Kamchatka Peninsula in the western Pacific (Figure 14.8). Because the northeastern shelf of Sakhalin Island in the Sea of Okhotsk is the principal region of the oil and natural gas producing industries, terminals, and so on in the region off the Far Eastern coasts of Russia, a destructive tsunami would lead to a serious ecological disaster in this region.

14.3.1. Historical Background

The tsunami warning service in Russia was established in 1958, six years after an extremely destructive tsunami struck on November 4, 1952. With the goal of developing and exploiting technical means and methods of tsunami registration in the open ocean using the so-called hydrophysical method of tsunami forecasting, a special hydrophysical observatory was created at the beginning of the 1960s at Shikotan Island, situated on the southern end of the Kuril Islands, by S. L. Soloviev of the Sakhalin Complex Scientific Research Institute (Dykhan

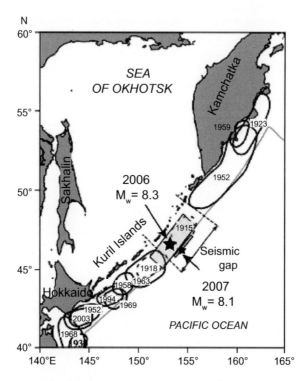

FIGURE 14.8 Source regions of the strongest earthquakes ($M_w \geq 8$) in the Kuril-Kamchatka zone for the period 1900–2005 plotted using different catalogue data: solid lines denote reliable contour lines; dashed lines denote unreliable contour lines; numbers are the year the earthquake occurred; straight dashed lines denote the limits of the seismic gap in the region of the Central Kuril Islands; asterisks denote epicenters of earthquakes on November 15, 2006, and January 13, 2007; gray rectangles denote the contours of source regions of these earthquakes; and the gray solid line is the axis of the deep Kuril-Kamchatka Trench. *Source: Lobkovsky et al., 2009; Original Russian Text published in Okeanologiya in 2009; Reproduced with kind permission from Springer.*

et al., 1983). Since 1965, twenty hydrophysical installations have been placed on the bottom near Shikotan Island. Each installation was connected to the observatory by underwater cable. Due to the rough natural conditions (frequent storms, drifting ice in the spring, and so on) and the experimental character of the installations, the duration of their life varied from several weeks to one year. Inevitable interruptions in bottom observations did not allow for the recording of very reliable tsunamis for a long time, although many small long waves of probably seismic origin were recorded (Jaque and Soloviev, 1971). The nineteenth installation was put into operation in the summer of 1979. The bottom water pressure sensor (vibrotron type) was placed at a depth of 113 m at a location 8 km to the southeast of Shikotan Island. The total lengths of the underwater and land cables connecting the sensor and observatory were approximately 20 km. Variations of pressure and sea level were transformed into frequency modulated electrical signals. At the output of the channel, they were transformed again into analog signals, and final registration was carried out in three frequency bands: water waves and swell, tsunamis, and tides (Dykhan et al., 1983). The February 23, 1980, earthquake ($M_w = 7.0$), which occurred to the southeast of Shikotan Island, was accompanied by a small tsunami that was recorded by tide gauges on the shores from Iturup Island in the north to Honshu Island in the south. The tsunami was also recorded by a bottom pressure sensor an hour earlier than the arrival of the tsunami at population centers. Personnel of the observatory could follow in real time all the details of tsunami propagation above the bottom sensor. Perhaps it was one of the first times in the world that a tsunami was detected and recorded offshore (Dykhan et al., 1983).

14.3.2. Service and Science Components of the Russian Tsunami Warning System

The Russian tsunami warning service is under the jurisdiction of two ministries: the Federal Service of Hydrometeorology and Environment Monitoring and the Russian Academy of Sciences (RAS). In Russia, the operational aspect related to tsunamis (as well as to all kinds of anomalous sea-level oscillations, including storm surges) is the mandate of the "SERVICE" wing. Accordingly, the organization of the Russian TWS comes under the jurisdiction of this wing. Likewise, the "SCIENCE" wing has the specific mandate to undertake and promote advancement of scientific studies related to all aspects of tsunamis (such as data processing, numerical modeling, long-term earthquake and tsunami forecast, dissemination of the acquired scientific knowledge in the form of scientific publications, etc.).

14.3.3. Tsunami Service Component

Tsunami monitoring, prediction, and warning services for the Pacific coasts of Russia are provided by the TWS centers of Russian Federal Service for Hydrometeorology and Environmental Monitoring (ROSHYDROMET) in Yuzhno-Sakhalinsk and Petropavlovsk-Kamchatsky working in close cooperation with the regional administrations of Ministry for Emergency Situations and the Ministry for Information Technologies and Communications of the Russian Federation, and SS of the Geophysical Service of the Russian Academy of Sciences (GS RAS). In addition, the Russian TWCs efficiently interact with the TWCs of other Pacific countries. Thus, the Russian TWS is an important part of the PTWS, operating under the umbrella of the IOC of UNESCO. The Russian subsystem as a whole and the Russian TWCs are guided by regulations and instructions provided at the federal and local levels. The TWS predicts the tsunami arrival times and the coastal impact (if possible) and uses this information to provide timely and effective information and warnings to minimize the hazards to human life and to provide adequate safety. The divisions involved in the TWS provide 24-hour operation, including continuous monitoring of seismicity and sea-level variations, situation analysis, declaring and canceling tsunami watches and warnings, and preparation and relaying of appropriate signals and messages in accordance with the established procedures.

14.3.3.1. Local Tsunami Procedures

In cases of tsunamigenic events in the near zone (i.e., local events), the parameters of the earthquakes are estimated by SS of the GS RAS located in Yuzhno-Sakhalinsk, Severo-Kurilsk, and Petropavlovsk-Kamchatsky. The initial tsunami warning is provided by the same SS. Criteria for the warning notification are based on the magnitude, M_s, and the location of the tsunamigenic earthquake. The magnitude criteria (i.e., magnitude threshold values for tsunami warning) presently accepted by the GS RAS are as follows (Russian Federation Report to IOC, February 2009):

- $M_s = 7.0$ for areas along the coasts of Kamchatka, the Kuril Islands, the Sea of Okhotsk and the Sea of Japan
- $M_s = 7.5$ for areas along the coasts of the Komandor Islands and Hokkaido Island
- $M_s = 8.0$ for areas along the coasts of the Andreanof Islands and Honshu Island

Likewise, a tsunami warning is canceled if:

- The tsunami has been recorded, but maximum wave heights are less than 0.5 m
- The tsunami warning has been declared, but tsunami signatures are absent in the data of the coastal tide gauges

The warning is canceled 0.5−1.0 hour after the latest estimated tsunami arrival time (TAT) to the settlements on the coast. The tsunami warning is canceled by the TWCs of the ROSHYDROMET in Yuzhno-Sakhalinsk and Petropavlovsk-Kamchatsky.

14.3.3.2. Distant Tsunami Procedures

Tsunami warnings for distant tsunamigenic events are provided by the TWCs in Yuzhno-Sakhalinsk and Petropavlovsk-Kamchatsky. The following actions are carried out after receiving information and corresponding parameters of a major distant earthquake from the SS of the GS RAS, foreign SS, and the PTWC:

- Estimation of tsunami threat for the Russian coasts based on the magnitude-geographical criterion.

- Calculation of tsunami arrival times to specific coastal sites.
- Sending "warning and watch" messages to the coastal hydrometeorological stations; activating tide gauge (sea-level) monitoring and eye-witness observations of sea level changes near the coast.
- Carrying out a situation analysis based on the entire set of information, including data from tide gauge observations from the PTWC and other foreign centers.
- Making the final decision about the actual tsunami threat for the Russian coast and declaring (if necessary) a tsunami warning.
- Transmitting tsunami warning emergency messages via communication channels according to the rules of notification to local and central authorities, all sectors of the population at risk, and foreign TWCs.

More precise definitions of tsunami parameters and threats to the Russian coasts are based on information about recorded tsunami wave heights at stations located near the source area or between the source area and the Russian coast, as well as on other information arriving from the foreign centers. On an experimental basis, a situation analysis was carried out each time PTWC provided a tsunami warning for the Pacific Ocean during 2005–2006. These analyses, in particular, included examination of the tide gauge data from Russian and foreign stations.

14.3.3.3. Russian National Sea-Level Network

Hydrometeorological stations located along the Russian coasts of the Pacific Ocean and marginal seas of the Russian Far East carry out sea-level observations (Table 14.1 and Figure 14.9). Starting in 2007, considerable improvements were made in the TWS observation network (Figure 14.10). In 2008, construction of an interregional center in Petropavlovsk was initiated for collecting, processing, and transmitting information about seismic events and tsunamis. Further, reconstruction works for installation of automated points at Kholmsk and Petropavlovsky Mayak had been carried out. The Hardware-Software Complexes (HSCs) of the Information Processing Centers at Yuzhno-Sakhalinsk and Petropavlovsk allow switching from the existing method of manual calculation of earthquake parameters to computer-based automated processing of the seismic signals simultaneously from a group of stations. Preliminary tests of the installed HSCs and new technologies have indicated the bright possibilities of reducing the time and increasing the accuracy of calculations of tsunamigenic earthquake parameters.

Andreev and colleagues (2009) reported the development of an automatic information and management system for the Russian TWC. This system has been developed and implemented as part of the Federal Targeted Program's "risk reduction and mitigation of consequences of natural and technology-related emergencies in the Russian Federation to 2010." The flowchart of the

TABLE 14.1 Marine stations of the Russian TWS.

Name of the Station	Type
Kamchatka Region	
Petropavlovsk-Kamchatsky	Telemetric tsunami station
Sakhalin Region	
Severo-Kurilsk	Telemetric tsunami station
Korsakov	Telemetric tsunami station
Kholmsk	Telemetric tsunami station
Kholmsk	Analog tide gauge
Malokurilskoe	Analog tide gauge
Uglegorsk	Analog tide gauge
Starodubskoe	Analog tide gauge
Primoriye Region	
Nakhodka	Analog tide gauge
Vladivostok	Analog tide gauge

Source: Courtesy of Dr. Tatiana Ivelskaya, Head, YSTWC, Russia.

computerized tsunami decision-making procedure is shown in Figure 14.11. The decision-making support system performs the following functions (Andreev et al., 2009):

- Presentation and monitoring of incoming messages and real-time data
- Automatic selection of notification protocol based on submarine earthquake data
- Implementation of logically related steps of event processing by duty ocean-ographer protocol in real time
- Calculations of tsunami wave heights and arrival times for threatened areas; presentation and monitoring of calculations
- Continuous analysis of sea-level data coming from automated posts and calculations to detect tsunami wave emergence and estimate its key parameters (i.e., first wave arrival time, amplitude, and period)
- Formation and automatic transmission of all output signals and messages by the notification protocol
- Forecasting tsunami impact on coastal areas based on submarine earthquake data
- Map-based presentation of spatially distributed data
- Automatic documentation of all steps of event data processing

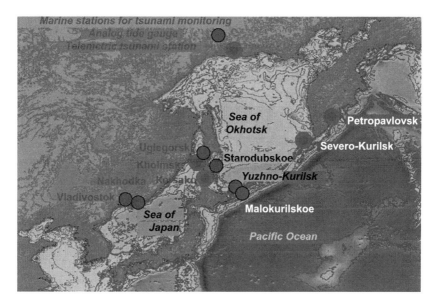

FIGURE 14.9 Sea-level stations of the Russian TWS network, 2008. *Source: Courtesy of Dr. Tatiana Ivelskaya, Head, YSTWC, Russia.*

FIGURE 14.10 Locations of 17 digital tide gauge installations for sea-level observations under the Russian TWS project, 2009–2010. *Source: Courtesy of Dr. Tatiana Ivelskaya, Head, YSTWC, Russia.*

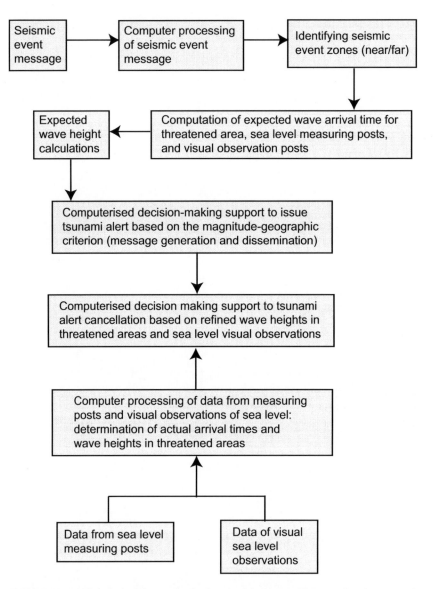

FIGURE 14.11   Flowchart of the computerized tsunami decision-making procedure incorporated in the automatic information and management system for the Russian tsunami warning center. *Source: Courtesy of Andreev et al., 2009.*

The computerized system for decision-making support in case of a tsunami threat consists of an integrated database, computation modules for estimating tsunami wave travel time and height, a duty oceanographer workstation, a control subsystem, a telecommunications subsystem, and an archived data viewer. The flowchart for operations of the computerized system of tsunami decision-making support is shown in Figure 14.12. A map-based data presentation of tsunami travel wave isolines is shown in Figure 14.13. Wave calculations for threatened areas are implemented as follows:

- The currently used methods of real-time prediction of tsunami characteristics are based on premodeling. Basically, premodeling consists of doing some of the tsunami characteristics calculations in advance (before an event occurs) and developing a database to be used for calculations in real-time operations.
- The database includes a set of elementary models of earthquake sources matching seismotectonic features of the water areas. For each model earthquake source, the expected wave heights are calculated. Results of calculations are entered into the database.
- When a message on a seismic event with date, time, and geographic coordinates of the earthquake epicenter and magnitude is received, the expected wave heights are calculated by interpolation of expected wave source parameters using the available database.

As indicated, the TWC's primary objective is to provide timely tsunami warnings and "all clears" information/advisories for the benefit of its citizens and hazard mitigation organizations. The SS is responsible for tsunami warning issuance in case of near-source quakes (i.e., when epicentral distance is up to 3000 km from Yuzhno-Sakhalinsk). However, the TWC issues warnings based solely on tsunami information after the occurrence of a major earthquake in the Pacific when the epicentral distance of the earthquake is more than 3000 km from Yuzhno-Sakhalinsk. The tsunami warnings are immediately transmitted to Civil Defence and Emergency Regional Headquarters and to the Central Telegraph Station of Yuzhno-Sakhalinsk. Subsequent to the confirmation of an earthquake occurrence, the TWC personnel computes tsunami arrival times for the coastal places of Sakhalin and Kuril Islands and then determines the approximate time that any danger is over, after which an "all clear" advisory is issued.

The present development and improvement of the Russian TWS are associated with installation of a telemetric station that includes a bottom pressure gauge (the so-called hydrophysical TWS subsystem) in the area adjacent to the Kuril Islands. The necessary condition for the effective subsystem activity is thorough knowledge about the following (Shevchenko et al., 2009):

- Statistical characteristics of background tsunami-range long-wave oscillations in the areas of telemetric station locations
- Frequency-sensitive features of these areas caused by their bathymetric topography
- Long-wave spectra variability depending on weather conditions

The preceding information is necessary to make efficient use of algorithms constructed for critical signal (tsunami) identification, automatic tsunami alert

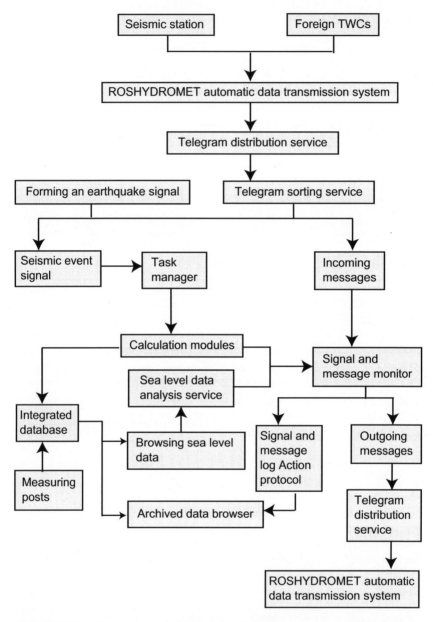

FIGURE 14.12 Flowchart for operations of the computerized system of tsunami decision-making support of the Russian tsunami warning system. *Source: Courtesy of Andreev et al., 2009.*

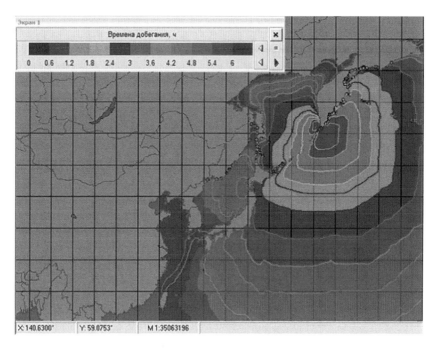

FIGURE 14.13 Map-based data presentation of tsunami wave isolines. Black point is the earthquake epicenter. *Source: Courtesy of Andreev et al., 2009.*

determination, and so on. To examine tsunami-range long-wave features, the Institute of Marine Geology & Geophysics of RAS, in cooperation with Nizhni Novgorod State University, installed bottom pressure gauges near Cape Kastrikum, Cape Van-der-Lind (the Urup Island), and Cape Lovtsov (the Kunashir Island) in July 2008.

14.3.3.4. Record of Tsunami Registration by the Sakhalin Tsunami Warning Center

For the period 1958–1998, the Russian TWC recorded 42 tsunamis in the regions of Sakhalin and Kuril Islands. Of these, 34 were generated in the near-source zone (Kuril-Kamchatka Trench, Japan and Okhotsk Seas), and 8 were generated in distant regions of the Pacific. Although not all major earthquakes near the Russian coasts have resulted in the generation of tsunamis, tsunami warnings are invariably issued whenever the earthquake magnitude is greater than or equal to a given threshold (the threshold is 7.0 for the Kuril-Kamchatka Trench, Japan and Okhotsk Seas). Because the warning center is located in a small town, the warning messages are communicated to the concerned agencies through telex and telegraph lines, as well as short-wave radio channels. For a nearby earthquake, the Sakhalin TWC issues tsunami warning messages within 6 to 10 minutes after receiving a warning from the seismic subsystems of Tsunami Services.

Russia has several SS, and its TWS is based primarily on the seismic data acquired by these stations. During the November 15, 2006, tsunami event, the Yuzhno-Sakhalinsk SS of Russia sent urgent information about a major earthquake to the Sakhalin TWC, which in turn announced the requisite tsunami warning for all the Kuril Islands. Further, the Russian TWC calculated the time of tsunami arrival at all the closest inhabited sites. During the earthquake event, the Sakhalin and the Pacific tsunami centers regularly exchanged operative information using a direct telephone connection, which facilitated rapid informed responses and decisions after the announcement of the tsunami warning.

14.3.4. Tsunami Science Component

Russia is considerably advanced in seismic research, and as a result, its scientists have been able to make very reliable earthquake predictions (primarily at the Institute of Oceanology, Russian Academy of Sciences (IO RAS)). For example, the Central Kuril seismic gap of about 400 km length, which last experienced a major earthquake in 1780 (Lobkovsky, 2005; Laverov et al., 2006a, b), was defined as the zone with the highest risk for a catastrophic event. Accordingly, they predicted the high probability of an imminent earthquake event in the region of the Central Kuril Islands one and a half years before the actual event occurred and before the likely earthquake source region was investigated. Based on the perceived intense earthquake strike and the attendant tsunami threat, they were able to model different scenarios for the formation and propagation of tsunami waves well in advance (Laverov et al., 2006a, b; Lobkovsky et al., 2006b, c; Lobkovsky and Kulikov, 2006). Thus, for the first time in the history of tsunami research, an earthquake/tsunami event was investigated long before it actually happened. For the worst-case scenario, the expected wave heights were more than 16 m for the coast of the Central Kuril Islands and more than 5 m for the coast of Sakhalin. The prediction came true, and a strong earthquake (on a somewhat reduced scale) eventually occurred on November 15, 2006 ($M_w = 8.3$), exactly within the Central Kuril gap at Simushir Island, as anticipated. The strongest oscillations, exceeding several meters, occurred near the source region of the Kuril Islands (Rabinovich et al., 2008). This seismically generated trans-Pacific tsunami appears to have been the strongest since the 1964 Alaska tsunami. The parameters of the observed tsunami were similar to those predicted, except that the northeastern coast of Sakhalin Island was mainly sheltered from destructive waves by the Simushir coast. Close on the heels of this powerful earthquake, the latest in the list of strong earthquakes ($M_w = 8.1$) in this region occurred on January 13, 2007 (Figure 14.14). These earthquakes also generated strong tsunamis recorded throughout the Pacific Ocean and impacted regions highly remote from the source (Rabinovich et al., 2008; Lobkovsky et al., 2009).

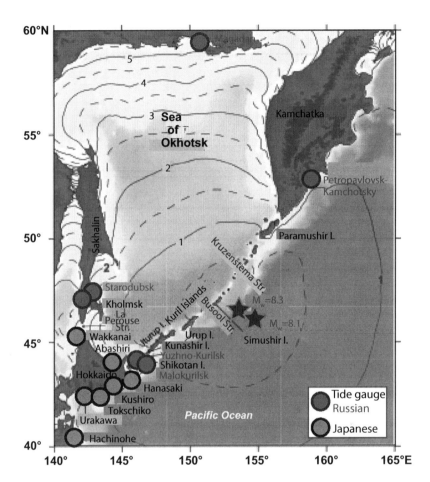

FIGURE 14.14 Map of the northwestern Pacific showing epicenters (stars) of the November 15, 2006 ($M_w = 8.3$), and January 13, 2007 ($M_w = 8.1$), Kuril Islands earthquakes and locations of tide gauge stations on the coasts of the Russian Far East and northern Japan. Solid and dashed curved lines denote the 30-minute isochromes of tsunami travel time from the November 15, 2006, earthquake source area. *Source: Rabinovich et al., 2008; Advances in Geosciences; Copernicus Publications on behalf of the European Geosciences Union.*

Therefore, the situation with respect to the Simushir earthquakes and tsunamis in 2006 and 2007 is quite unique: Strong earthquakes and seismically generated tsunamis were predicted by Russian scientists, two expeditions were carried out in the forecasted source zone during the period preceding these events, and the results of these expeditions were effectively used to construct prognostic models of earthquakes and tsunamis (Figure 14.15). Thus, the IO RAN forecast (Lobkovsky, 2005; Laverov et al., 2006a, b) and prognostic earthquake model used for the simulation of a hypothetical tsunami (Laverov et al., 2006; Lobkovsky and Kulikov, 2006) were quite precise. The real

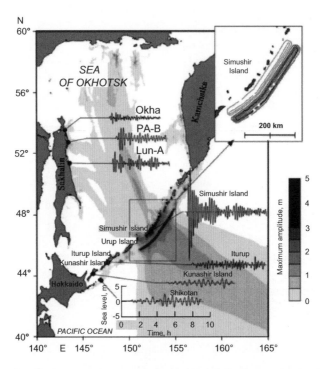

FIGURE 14.15 Maximum simulated amplitudes of a "hypothetical" tsunami with a source in the vicinity of the Central Kuril Islands. The initial tsunami source is shown in the inset (elevation is denoted with solid lines; depression is shown with dashed lines). Simulated tsunami records are shown for different coastal sites of Sakhalin and the Kuril Islands. *Source: Lobkovsky et al., 2009; Original Russian Text published in Okeanologiya in 2009; Reproduced with kind permission from Springer.*

earthquake had a slightly smaller displacement amplitudes magnitude and a shorter extension (apparently because, fortunately, the northern "keys" were not activated), and consequently the earthquake generated a less intense tsunami than was predicted in the "worst-case scenario." The results of the prognostic model calculations corresponded qualitatively to the actual sea-level observations at the coast and in the open ocean (a number of DART bottom-pressure stations operated by NOAA [United States] and by deep-ocean cable stations operated by JAMSTEC [Japan]), as well as to the preliminary field survey data collected in expeditions on the coast of the Kuril Islands during the summer of 2007 (Levin et al., 2008).

While the Russian TWC personnel had a clear idea of the probable tsunami amplitude and energy fluxes based on numerical simulations (Figures 14.16 and 14.17), the actual heights of tsunami waves at the coasts, which are crucial for tsunami risk estimates, were obtained from eyewitness accounts and instrumental observations (Figure 14.18) from coastal tide gauges. Whereas some of

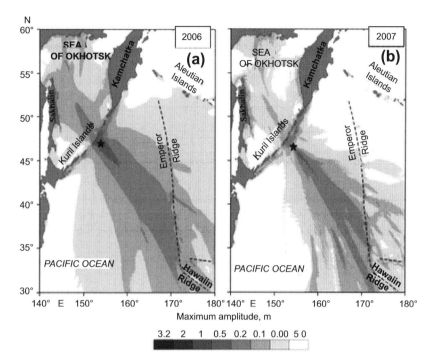

FIGURE 14.16 Numerical simulation of the maximum tsunami amplitudes for (a) November 15, 2006, and (b) January 13, 2007, earthquakes in the northwestern part of the Pacific Ocean and adjacent region of the Sea of Okhotsk. Asterisks denote the locations of the corresponding earthquakes. The thin dashed lines show the Emperor and Hawaiian submarine ridges. *Source: Lobkovsky et al., 2009; Original Russian Text published in Okeanologiya in 2009; Reproduced with kind permission from Springer.*

the tide gauges (read sea-level gauges) are conventional float-driven "pen-and-paper" analog ones, the rest are subsurface pressure-based digital gauges. However, there are locations that suffer from the absence of operating tide gauges (e.g., northeastern coast of Sakhalin). As usually happens during tsunami events, one of the gauges was damaged by the tsunami waves and did not function for a few hours. As predicted by the prognostic model, part of the tsunami energy propagated into the Sea of Okhotsk through the Bussol and Kruzenstern straits and caused noticeable oscillations in the Magadan region based on a tide gauge record.

Although the Russian tsunami warning schemes work effectively, there is much scope for greater expansion and improvement of its present sea-level gauge network. Because of the current, mostly archaic, older-generation sea-level gauge system, some of the sea-level records were found to have been plagued by relatively poor quality and low signal-to-noise ratios (SNR). Consequently, it was difficult to identify reliably the tsunami of January 13,

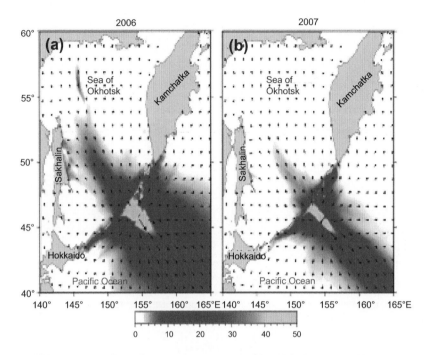

FIGURE 14.17   Energy flux for tsunami waves generated by (a) the November 2006 and (b) the January 2007 earthquakes in the Kuril Islands. *Source: Rabinovich et al., 2008; Advances in Geosciences; Copernicus Publications on behalf of the European Geosciences Union.*

2007, in the sea-level records from Petropavlovsk-Kamchatskiy, Magadan, and Starodubskoye. The sea-level gauge record at Malokurilsk, although significant and the most interesting one, was partly corrupted due to a malfunction. The poor recording was a result of problems with the ink that created gaps in the record. However, the initial six-hour interval of the record was processable, and it permitted the estimation of tsunami wave characteristics at this location (Lobkovsky et al., 2009). Unfortunately, the Russian sea-level gauge network is still under reconstruction.

In November 2006, inherently digital deep-sea bottom-pressure gauges with sampling intervals of 1 min were deployed at two deep-water locations (Malokurilskaya and Krabovaya bays on Shikotan Island) by the Institute of Marine Geology and Geophysics, Far Eastern Branch of the RAS. The January 13, 2007, tsunami was clearly recorded by these deep-sea pressure gauges. Despite these improvements, the Russian TWC's ability to accurately detect tsunami waves on the coasts of Russia and its capacity for providing operational corrections to tsunami warnings is still limited due to the lack of necessary instrument support, such as adequate number of digital sea-level gauges with real-time signal transmission capability. The closest sea-level gauges to the epicenters of the November 2006 and January 2007 earthquakes were located

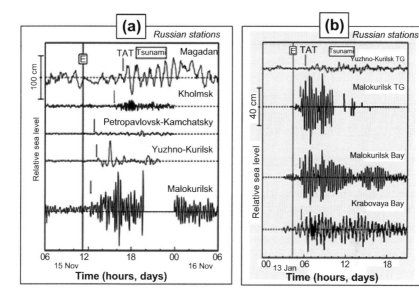

FIGURE 14.18 Tsunami records for (a) November 15, 2006, and (b) January 13, 2007, events obtained at the Far Eastern coast of Russia. The subsurface pressure tsunami records were obtained in Malokurilskaya and Krabovaya bays on Shikotan Island. Arrows indicate the time of arrival of the first tsunami wave. The solid line with the "E" denotes the time of occurrence of the corresponding earthquakes. *Source: Rabinovich et al., 2008; Advances in Geosciences; Copernicus Publications on behalf of the European Geosciences Union.*

500−800 km away. The lack of earthquake and tsunami monitoring systems on a real-time basis, the lack of information concerning tsunamis at coasts closest to the epicenter of earthquakes, and the impossibility of performing visual observations in the dark created extreme difficulties for the operational tsunami warning services in Russia during the 2006 and 2007 events (Lobkovsky et al., 2009). In view of this, deployment of adequate numbers of modern digital sea-level gauges and seismometers at sufficiently close spatial intervals at the coasts and in the open ocean with the capability for real-time reporting of data remains to be a top priority with the Russian TWC. The technical review by Rabinovich (2006) provides considerable input required for the revival and reconstruction of the Russian TWS system into a modern one.

Recent developments in Russia suggest that the "science" of tsunamis is going to be used more efficiently for tsunami warnings. For example, Babajlov and colleagues (2009) implemented a project to design a new generation of the TWS for the Far East coast of the Russian Federation. The aim of the project is to develop a technology that will allow constructing a database of possible occurrences of disastrous waves along the coast in a series of numerical experiments. At the first stage of this study, a collection of basic model sources of tsunamigenic earthquakes is defined. The model sources are used to calculate the initial elevation fields on the ocean surface. The next stage employs the numerical modeling to simulate the propagation and trans-formation of tsunami waves on the way from the source toward the coast. This information is presented as a decision support system used by those responsible for initiating the disaster mitigation procedures such as evacuating residents and instructing vessels to move away from the dangerous areas of the coast.

The project involves the numerical solution of a large number of instances of wave hydrodynamics problems. At the same time, the interpretation of the results requires nontrivial postprocessing and building specialized information systems. The numerical solution of the wave hydrodynamics problems for multiple combinations of parameters of an earthquake source with a high spatial resolution constitutes the major part of the computational requirements of the project. The amount of computations needed implies the use of high-performance computers and requires adaptation of numerical algorithms to specific computing platforms.

Another example of the progress made in the "science" of tsunamis is a scheme for tsunami hazard zoning for the southern Kuril Islands, which was developed by Kaystrenko and colleagues (2009). The main problem with statistical estimation of tsunami risks for the Russian Far East is a poor representation of historic tsunami data. For instance, there are few observa-tional tsunami data regarding a sparsely populated area along the Tatar Strait coast. At the same time, the Japan Islands are known to be a region with long-term reliable historical records regarding tsunamis. Kulikov and Fine (2009) presented results of the tsunami risk estimation for the region of the Nevelskoi Channel (the Tatar Strait) based on numerical modeling that has been applied to extend tsunami statistics from the Japan Sea.

Tsunami hazard is a complex parameter that involves many factors and is often determined to have the probability of one or more unfavorable conse-quences. The basis for the tsunami hazard theory is a probability model of tsunami-mode. Kaystrenko and colleagues showed that a sequence of tsunamis has the Poisson characteristic, and the probability that n tsunamis will occur at a given location with the wave height exceeding a given threshold value h is given by the formula:

$$P_n(> h) = \frac{(\varphi(h) \cdot t)^n}{n!} e^{-\varphi(h) \cdot t} \qquad (14.1)$$

The average frequency $\varphi(h)$ of tsunami as the function of "threshold" h has the name of a recurrence function. This function is monotone nonincreasing, and at $h > 0.5$ m, it is well approximated by the exponent:

$$\varphi(h) = f \cdot e^{-\frac{h}{h^*}} \qquad (14.2)$$

The parameter h has the physical sense of the frequency of a great tsunami, and it is regional, slowly changing from one point to another point on the coast, and H^* is a characteristic height of a tsunami for any coastal point. Using the least-squares method, the model parameters h and H^* together with their dispersions can be estimated for all the points with a good representative series of heights of the historical tsunami. The needed parameters H^* for all the coastal points can be calculated using numerical modeling incorporating a fine-grid size. The correlation between natural tsunami data and calculated tsunami heights for selected 18 points was from 0.85 (August 11, 1969, tsunami) up to 0.98 (May 22, 1960, tsunami). Such a good correlation allowed the use of a numerical tsunami height for calculation of the parameters H^* along the coast. Finally, the tsunami hazard zoning scheme for the Southern Kuril Island coast with $f = 0.17$/year and distribution of H^* was developed. It was found that all the tsunami heights H_T with the recurrence period T can be calculated as $H_T = H^* \ln (f\, T)$.

14.4. THE CANADA TSUNAMI WARNING SYSTEM

Significant parts of the almost 27,300-km-long British Columbia coast of Canada (straight distance 775 km only), which consists of a complex network of inlets, straits, passes, sounds, narrows, and islands (Figure 14.19), are susceptible to the effects of tsunamis generated within the Pacific basin. Major Cascadia earthquakes accompanied by destructive tsunamis have an average recurrence of 500 years in this region (Clague and Bobrowsky, 1999; Clague et al., 2003). Trans-Pacific tsunamis caused by major earthquakes in the Pacific "Rim of Fire" can also significantly affect the British Columbia coast (Murty, 1992; Clague, 2001). The March 1964 Alaska earthquake with magnitude $M_w = 9.2$ produced a catastrophic tsunami that swept along the British Columbia coast, causing considerable damage (Murty, 1977; Clague, 2001; Clague et al., 2003). This area is also at high risk for destructive tsunamis caused by local major earthquakes (Rapatz and Murty, 1987; Hebenstreit and Murty, 1989).

The existing Canadian TWS with its Permanent Water Level Network (PWLN) underwent a major upgrade in the year 1998, with higher-precision sea-level gauges having 1-minute sampling capability. Three of the sea-level gauge stations (Tofino, Winter Harbour, and Langara) have been selected for use as tsunami warning stations. Each of these stations is equipped with RST 2000 Mobile Satellite (MSAT) packet data Mobile Earth Terminals (MET). Should a tsunami event occur, a detection algorithm in the Sutron data logger would cause the MET to transmit an emergency message (Figure 14.20). The gauge computes the mean per-minute change from the last three minutes of observations and compares this value with a threshold value. If the threshold value is exceeded for three consecutive minutes, a tsunami alarm is generated. The algorithm has been designed to provide the lowest possible detection threshold without triggering false alarms due to ordinary atmospherically generated seiches. A host computer at the Institute of Oceanographic Sciences (IOS) receives the alarm message and automatically issues a pager call to alert

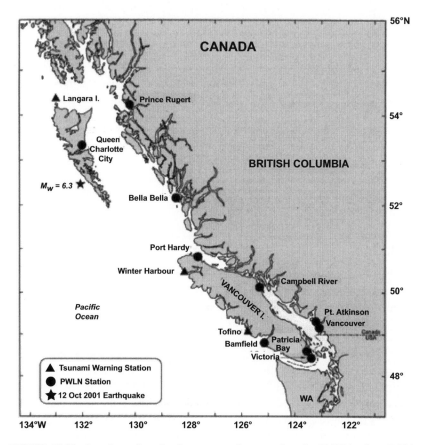

FIGURE 14.19 Locations of sea-level gauges on the approximately 27,300-km-long British Columbia coast of Canada (straight distance 775 km only), consisting of a complex network of inlets, straits, passes, sounds, narrows, and islands. Tsunami warning stations are marked by triangles, and the other Permanent Water Level Network (PWLN) stations are indicated by circles. The epicenter of the October 12, 2001, Queen Charlotte earthquake ($M_w = 6.3$) is indicated by a star. *Source: Rabinovich and Stephenson, 2004; Natural Hazards; Reproduced with kind permission of Springer.*

the response personnel to investigate the tsunami event. Data can then be requested by the host computer and forwarded to the pager for ongoing analysis and response.

Under normal conditions, the sea levels at every tsunami- and PWLN-station are sampled and stored every minute. At all stations except Langara, these data are transmitted by telephone modems every three hours to the host computer for storage. At Langara, the gauge operation is checked daily, and data backups are made monthly (to floppy disk) and mailed to the IOS. In normal operation, the MSAT communication mode (see Figure 14.20) is not used, but an "event" will trigger its operation. The communication link can also

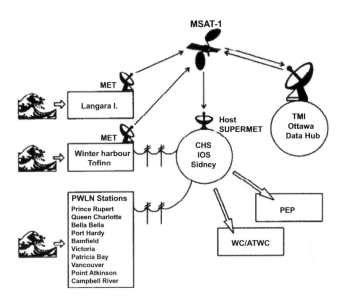

FIGURE 14.20 Present scheme of tsunami monitoring on the coast of British Columbia, Canada, based on tsunami warning and PWLN stations. MSAT: Mobile Satellite; MET: Mobile Earth Terminal; CHS: Canadian Hydrographic Service; PEP: Provincial Emergency Program; PWLN: Permanent Water Level Network; WC/ATWC: West Coast/Alaska Tsunami Warning Center. *Source: Rabinovich and Stephenson, 2004; Natural Hazards; Reproduced with kind permission of Springer.*

be transferred to MSAT at any time by response personnel using the telephone modem or during a brief time window every 15 minutes when the remote MSAT unit powers up to look for incoming messages (see Figure 14.20).

The new hardware has been in operation since early 1999. During this period, the data return has been in excess of 99% from the network, with most stations providing 100% data return. The percent network has been found to be reliable, and the data it provides are accurate and timely (Rabinovich and Stephenson, 2004). When a potentially tsunamigenic earthquake occurs, the Alaska TWC provides information on the location and magnitude of the event, as well as expected tsunami arrival times at various locations around the Pacific coast. Following these predicted arrival times, the emergency response personnel expect to begin receiving information on the observed/recorded tsunami. The response staff must be able to quickly and accurately estimate the wave heights of arriving waves, determine if the measured signals are from a tsunami or merely atmospherically generated seiches, and assess the danger of the arriving waves. That information must then be quickly passed to the ATWC and to local emergency response personnel. The Provincial Emergency Program (PEP) is responsible for receiving watch and warning messages from the West Coast/Alaska TWC and relay those messages, with refinements for the Canadian coast, to all sectors of the population at risk in the province (Rapatz and Murty, 1987). The initial tsunami warning is based on seismic information. While there have been instances of false tsunami warnings for British Columbia associated with the 1994 Shikotan earthquake ($M_w = 8.2$), which resulted in lost productivity valued at a few million dollars, sometimes even relatively weak local earthquakes have produced significant tsunami waves for the Pacific coast of North America capable of devastating communities in the near field (Gonzalez et al., 1995).

14.5. THE AUSTRALIA TSUNAMI WARNING SYSTEM

Australia is vulnerable primarily to distant tsunamis generated in any part of the Pacific Ocean and to a limited extent by local tsunamis generated by underwater slippage of sediments off the continental shelf in some areas. Usually there would be insufficient time for a warning system to operate due to the very short tsunami arrival times from nearby sources.

After the Indian Ocean tsunami of December 26, 2004, the Australian government committed AU$ 68.9 million to the development of the Australian tsunami warning system (ATWS). The development and implementation of the ATWS is a four-year joint-agency project involving the Bureau of Meteorology, Geoscience Australia, and Emergency Management. The Joint Australian Tsunami Warning Center (JATWC), operated by the Australian Bureau of Meteorology (Bureau) and Geoscience Australia (GA), was formally established on July 1, 2007. The JATWC now provides Australia with an independent capability to detect and warn about impending tsunamis. The Australian public is now regularly advised of nil or potential threats from tsunamis being generated by undersea earthquakes in neighboring regions.

The JATWC is composed of two hubs: seismic detection and monitoring, based at GA in Canberra, and tsunami warning and sea-level verification, based at the Bureau's National Meteorological and Oceanographic Center (NMOC) in Melbourne. The Australian seismic network has already been addressed, so the next section addresses only the sea-level network.

14.5.1. Sea-Level Network

A network of coastal sea-level stations and deep-ocean tsunami detection buoys (Figure 14.21) that are operated by the Bureau of Meteorology are used to verify the existence of a tsunami. The composite network provides coverage for tsunamis originating from both the Pacific and the Indian Ocean basins.

The coastal sea-level station network uses an existing network of air-acoustic sensor-based SEAFRAME gauges that were designed initially for climate monitoring. The expanded network incorporates microwave radar sensors for sea-level monitoring. All of the new stations are configured for 1-minute data reporting intervals, and the real-time data are reported in a table-driven code

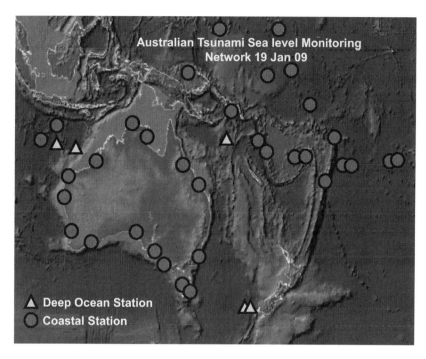

FIGURE 14.21 Network of Australia's coastal sea-level stations and deep-ocean tsunami ·detection buoys used for sea-level monitoring and verification of the existence of a tsunami. The data are provided in real time on the GTS. *Source: National Report of Australia to IOC-UNESCO, February 2009; Courtesy of IOC-UNESCO; Reproduced with kind permission of IOC-UNESCO.*

form (CREX) to the ATWS. From there, the 1-minute data are distributed at 3-minute intervals on the GTS. Countries that are unable to access the GTS can download the data via an ftp file.

While early deep-ocean station deployments met with considerable success, a series of failures or technical problems were encountered in the second half of 2008, and in January 2009 four of the five deployed stations (*Australian National Report to IOC,* February 2009) were affected.

14.5.2. The Tsunami Warning Scheme

The seismic detection and monitoring hub based at GA in Canberra automatically issues seismic solutions for potentially tsunamigenic earthquakes to the tsunami warning and sea-level verification hub in Melbourne. After further assessment, the GA issues a manually assessed seismic solution. Electronic notification is relayed between the two organizations via a dedicated fiberoptic cable. Receipt of the manual seismic assessment automatically triggers tsunami travel time software at the Bureau. Predicted tsunami amplitudes from deep-ocean model tsunami scenarios are then used to determine whether a nil threat bulletin or a potential threat bulletin will be issued through the Bureau's dedicated dissemination systems.

Once the bulletin has been issued, a monitoring process begins using sea-level gauges across the Pacific and the Indian Oceans. If no tsunami is detected by the gauges, then the JATWC issues a cancellation bulletin. If either a tsunami is confirmed or no signal has been detected by a gauge when there is only 90 minutes of travel time remaining, the JATWC will issue a tsunami warning for either a marine or a land inundation threat for the relevant states and territories, as determined by the tsunami model scenario. This could lead to an evacuation and other actions in consultation with the state and territory emergency response agencies. Information from the PTWC and JMA is used both as a backup if GA's facilities were to become inoperable and also as a reference in issuing the JATWC bulletins.

The Bureau of Meteorology has national responsibility for issuing JATWC tsunami advisories and warnings specific to Australia and its territories. This information is provided to the operational centers of state and territory emergency management organizations via the Bureau's meteorological and severe weather (multihazard) communications network in the Bureau's state and territory offices. The Bureau is also responsible for promulgating any public suggestions and warnings relevant to an event. Tsunami bulletins and warnings are distributed to the media and public by a variety of electronic media such as the Internet, fax, and ftp. Because of Australia's geographic position bordering the Indian Ocean and the Pacific Ocean, the Australian TWS is expected to deliver significant benefits to both of these oceanic regions.

14.5.3. Tsunami Risk Assessment Program

Tsunami risk research in Australia is being conducted across a broad range of disciplines and organizations, primarily three areas (*Australian National Report to IOC,* February 2009):

1. Geologic studies of tsunami sources and evidence of prehistoric tsunami
2. Tsunami wave propagation and inundation
3. Vulnerability of communities and the environment to tsunamis

Several organizations are conducting geologic and tectonic research to form the basis for probabilistic tsunami hazards/risk models. These include James Cook University (field studies of prehistoric tsunami in Australia and Southeast Asia), Wollongong University (field studies of prehistoric tsunami in Australia), and Macquarie University and GA (earthquake sources at the plate boundaries). Tsunami wave propagation modeling studies are also being undertaken by several centers such as the Bureau (real-time monitoring and tsunami warning), the University of Queensland (numerical models of earthquake, tsunami generation, and propagation), GA together with the Australian National University (tsunami inundation model for shallow water and the resulting run-up onto the coast), and the University of Melbourne (dynamic modeling of the effect of tsunamis on critical infrastructure).

A first-generation scenario database (T1) that was completed in 2006 proved to be a very useful tool for forecast guidance and general event analysis. However, it was somewhat limited in terms of its domain, model run-time, and rupture construction. An enhanced database (T1.1) was developed to replace it and transferred into operational use at the JATWC in early 2009. The domain of T1.1 covers the entire Pacific and Indian Oceans, and includes source locations at 100-km intervals along all subduction zones in these basins. The South Sandwich subduction zone is also included, with scenarios generated on a separate domain. Each source location has four scenarios associated with it, with earthquake magnitudes of $M_w = 7.5$, 8.0, 8.5, and 9.0. This results in a total of 1865 scenarios. The rupture modeling for T1.1 is an improvement over T1, with large ruptures being represented as the sum of a number of smaller rupture elements and each scenario being run for 24 hours of model run-time. The scenarios are linearly scaled in order to produce guidance for intermediate magnitude events.

The tsunami warnings are obtained by considering predefined zones around the Australian coastline and determining which of the three specific levels of warning is appropriate, depending on the values of maximum modeled wave amplitude that occur in each zone. The three levels are no threat, marine threat, and land threat (Greenslade et al., 2009).

Although much progress was made, several problems and challenges persist in the real-time systems. One is heavy fouling of the moorings with fishing line or nets and/or the surface buoys dragged off-station by vessel tie-ups. Some other problems include intentional vandalism and theft of surface instrumentation, antennas, solar cells, and so on. Such problems are quite serious. For example, in the case of the Indian Ocean tsunameter network consisting of 26 tsunameters deployed until July 2009, 7 buoy systems were damaged due to vandalism and 8 systems became inoperative due to technical problems (Greenslade and Jarrott, 2009).

When a tsunameter network is a multination network (e.g., Indian Ocean tsunameter network), coordinated deployments and sharing of real-time data become important. In situations where some of the participating countries are lesser developed than the others, assistance from donor countries and agencies becomes necessary. In such cases, the departure of donors implies an issue of sustainability. Multination participants mean different manufacturers and different station data formats. Warning centers currently access only local (national) stations and the DART™ product types. DART data are accessible via the NOAA-NDBC Website and through GTS transmission. Only when data sharing is fully established will countries be able to move from a philosophy of national independence (with the consequences of oversampling and higher costs) to a paradigm of interdependence, with prospects of communal network optimization (Greenslade and Jarrott, 2009).

14.6. THE PUERTO RICO TSUNAMI WARNING SYSTEM

Puerto Rico is a U.S. Commonwealth Island nation (Figure 14.22) that is located in the Caribbean Sea, which has large tsunami inundation zones. Historical tsunami records from sources such as the National Oceanic and Atmospheric Administration's (NOAA) National Geophysical Data Center

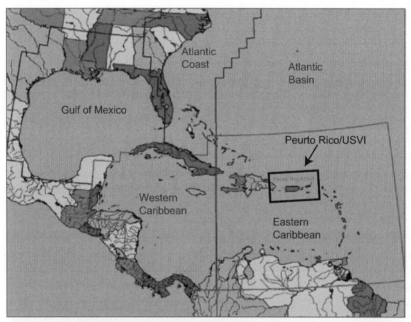

FIGURE 14.22 Map indicating the different regions considered in the Caribbean Tsunami Warning System. The Puerto Rico USVI region falls currently under the responsibility of the Puerto Rico Seismic Network and the NOAA NWS West Coast Alaska Tsunami Warning Center. The tsunami warning guidance for the rest of the Caribbean is currently provided by the NOAA NWS Pacific Tsunami Warning Center. *Source: Courtesy of von Hillebrandt-Andrade et al., 2008.*

(NGDC) show that over 75 tsunamis with validity greater than 1 have been observed in the Caribbean over the past 500 years (Prof. Christa G. von Hillebrandt-Andrade; Manager of NOAA NWS Caribbean Tsunami Warning Program; Personal communication, 2010). These represent approximately 7-10 percent of the world's oceanic tsunamis. Earthquakes, landslides, and volcanic tsunami sources have all impacted the region. Since 1842 at least 3510 people have lost their lives to tsunamis in the Caribbean, which is six times more than in the northeastern Pacific. For example, in 1867 and 1918 local $M_w = 7.3$ earthquakes generated tsunamis that impacted the Virgin Islands and eastern Puerto Rico and Western Puerto Rico, respectively, which generated significant loss of lives and property (von Hillebrandt, 2009). Over the past years there had been an explosive population growth and influx of tourists along the Caribbean coasts that increase the tsunami vulnerability. In the case of the December 2004 Indian Ocean tsunami, a day after the earthquake in Sumatra, tsunami waves were detected in the sea-level records from Punta Guayanilla, Magueyeres Island, and San Juan in Puerto Rico (Figure 14.23). Observed maximum tsunami wave heights at these stations are: Punta Guayanilla, 3.0 cm; Magueyeres Island, 7.5 cm; and San Juan, 4.5 cm (Rabinovich et al., 2006).

Until recently, the agency responsible for monitoring tsunamis in Puerto Rico was the Puerto Rico Seismic Network (PRSN) of the Department of Geology of the University of Puerto Rico, located at Mayaguez. The State Emergency Management Agency was responsible for issuing evacuation orders, while the National Weather Service forecasted and disseminated the emergency messages regarding earthquake and tsunami information within its Area of Responsibility (AOR) which included Puerto Rico (PR), the US and British Virgin Islands (VI) and Eastern Dominican Republic (DR). The PRSN operated a network of 25 seismic and 7 sea-level gauges and weather stations in PR, VI and the DR, together with Caribbean and Atlantic. Typical installations of coastal sea-level gauges and a weather station are illustrated in Figure 14.24.

A GOES (Geostationary Operational Environmental Satellite) receiver and central recording system installed at the Puerto Rico Seismic Network recorded and analyzed the data (seismic, sea-level, and meteorological) from the Puerto Rico network as well as other sea-level gauges that were operated by NOAA in Puerto Rico, Virgin Islands, Caribbean and Atlantic. The data gathered were analyzed along with earthquake information to decide whether or not tsunami information, warnings, watches or advisories should be issued and when to upgrade or cancel them if they had been issued. A siren functioning as an All Hazards Broadcast System was also established at Mayaguez. Tsunami warning, tsunami watch, tsunami advisory or tsunami information statement were issued based purely on seismic data analysis or sea-level record (von Hillebrandt-Andrade et al., 2008).

In 2010 the Caribbean Tsunami Warning Program (CTWP) was established by the National Weather Service (NWS) of NOAA as the first step of a phased approach for a Caribbean Tsunami Warning Center (CTWC). Pictures of the

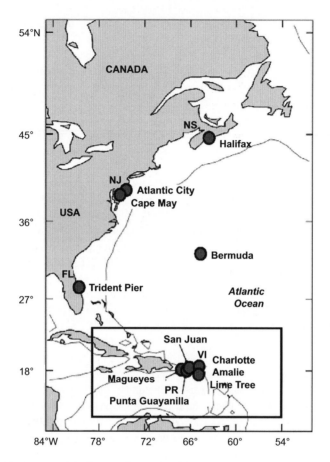

FIGURE 14.23 Map of the Puerto Rico (PR) region in the western part of the North Atlantic Ocean showing sea-level stations at Punta Guayanilla, Magueyeres Island, and San Juan in Puerto Rico, that recorded the 2004 Sumatra tsunami. The other two sea-level stations located close to Puerto Rico are in the U.S. Virgin Islands (VI). *Source: Rabinovich et al., 2006; Surveys in Geophysics; Reproduced with kind permission from Springer Science and Business Media.*

exterior of the Puerto Rico Seismic Network (University of Puerto Rico at Mayagüez) where the NOAA NWS Caribbean Tsunami Warning Program is jointly located are shown in Figure 14.25. Operations area of the Puerto Rico Seismic Network is illustrated in Figure 14.26. The office currently gives support and guidance for tsunami observations, including seismic and sea-level systems, tsunami forecasting, communications and education, and preparedness. Its staff is fully bilingual in English and Spanish and provides briefings and materials for use by the general public, government, scientists and broadcast media. CTWP works closely with the NWS Pacific and West Coast and Alaska Tsunami Warning Centers and the Puerto Rico Seismic Network

FIGURE 14.24 (a) Vieques, Puerto Rico coastal sea-level station with meteorological sensors operated by the Puerto Rico Seismic Network. (b) Fajardo coastal sea-level station operated by the Puerto Rico Seismic Network. Two gauges are used for redundancy. *Source: von Hillebrandt-Andrade et al., 2008; picture taken by Javier Santiago.*

that are currently providing tsunami warnings and guidance to different areas of the Caribbean and adjacent regions. It also contributes to meeting the objectives of the UNESCO Intergovernmental Oceanographic Commission's (IOC) Intergovernmental Coordination Group for Tsunamis and Other Coastal Hazards Warning System for the Caribbean Sea and Adjacent Regions (CARIBE EWS) and works closely with other international, regional and national observational, monitoring and preparedness organizations.

Although the CTWP was just established in 2010, the history of the efforts to establish a Caribbean Tsunami Warning System began in the 1990's when scientists from the region alerted on the lack of a tsunami warning system in the region. Initial regional efforts were coordinated through the IOC (UNESCO) Sub-Commission for the Caribbean and Adjacent Regions (IOCARIBE) and then after the 2004 Indian Ocean Tsunami, through the CARIBE EWS. Since its establishment in 2006, the CARIBE EWS has held five sessions until 2010 in which recommendations for the strengthening of the early warning system, including the establishment of a Caribbean Tsunami Warning Center, have been made.

According to Professor Christa G. von Hillebrandt-Andrade, who took over charge in 2010 as the Manager of the NOAA NWS Caribbean Tsunami Warning Program, there is a committed plan to strengthen the organization in the areas of seismology, oceanography, electronics, information technology, education and outreach, and administration. The tsunami forecast and warning

FIGURE 14.25 Pictures of the exterior of the Puerto Rico Seismic Network where the NOAA NWS Caribbean Tsunami Warning Program is jointly located. The different dishes are for GOES, VSAT and Satellite TV. In addition, on the rooftop are various antennas for VHF and Spread Spectrum radio communications. *Source: Courtesy of Prof. Christa von Hillebrandt-Andrade, Manager, NOAA NWS Caribbean Tsunami Warning Program; picture taken by Javier Santiago.*

area of responsibility would include the Caribbean and Adjacent regions. The CTWP is jointly located with the Puerto Rico Seismic Network (PRSN) at the University of Puerto Rico at Mayagüez.

The National Weather Service is one of the offices of the NOAA of the Department of Commerce. The mission of the NWS is to provide weather, hydrologic, and climate forecasts and warnings for the United States, its territories, adjacent waters and ocean areas, for the protection of life and

FIGURE 14.26 Operations area of the Puerto Rico Seismic Network, University of Puerto Rico at Mayagüez. *Source: Courtesy of Prof. Christa von Hillebrandt-Andrade, Manager, NOAA NWS Caribbean Tsunami Warning Program.*

property and the enhancement of the national economy. NWS data and products form a national information database and infrastructure that can be used by other governmental agencies, the private sector, the public, and the global community.

14.6.1. Tsunami Warnings

Tsunami warnings are issued when a potential tsunami with significant widespread inundation is imminent or expected. Warnings alert the public that widespread, dangerous coastal flooding accompanied by powerful currents is possible and may continue for several hours after the arrival of the initial wave. Warnings also alert emergency management officials to take action for the entire tsunami hazard zone. Appropriate actions to be taken by local officials may include the evacuation of low-lying coastal areas and the repositioning of ships to deep waters when there is time to safely do so. Warnings may be updated, adjusted geographically, downgraded, or canceled. To provide the earliest possible alert, initial warnings are normally based only on seismic information.

14.6.2. Tsunami Watches

Tsunami watches are issued to alert emergency management officials and the public of an event that may later impact the watch area. The watch may be upgraded to a warning or advisory. It may also be canceled based on updated information and analysis. Therefore, emergency management officials and the public should prepare to take action. Watches are normally issued based on seismic information without confirmation that a destructive tsunami is underway.

14.6.3. Tsunami Advisories

Tsunami advisories are issued in the event of the threat of a potential tsunami that may produce strong currents or waves that could be dangerous to those in or near the water. Coastal regions historically prone to damage due to strong currents induced by tsunamis are at the greatest risk. The threat may continue for several hours after the arrival of the initial wave, but significant widespread inundation is not expected for areas under an advisory. Appropriate actions to be taken by local officials may include closing beaches, evacuating harbors and marinas, and relocating ships to deep waters when there is time to safely do so. Advisories are normally updated, upgraded to a warning, or canceled.

14.6.4. Tsunami and Earthquake Information Statements

Tsunami and earthquake information statements are issued to inform emergency management officials and the public that an earthquake has occurred or that a tsunami warning, watch, or advisory has been issued for another section of the ocean. In most cases, information statements are issued to indicate that there is no threat of a destructive tsunami and to prevent unnecessary evacuations as the earthquake may have been felt in coastal areas. An information statement may, in appropriate situations, caution about the possibility of destructive local tsunamis. Information statements may be reissued with additional information, although normally these messages are not updated. However, a watch, advisory, or warning may be issued for the area, if necessary, after an analysis is done or updated information becomes available.

14.7. THE KOREA TSUNAMI WARNING SYSTEM

The Korean peninsula is vulnerable to tsunamigenic earthquakes that might cause damage along the coasts of Korea and Japan. Two earthquake-generated regional tsunamis in 1983 and 1993 caused severe damage along the eastern coast of Korea. Further, the December 2004 Indian Ocean tsunami and the earthquake of Fukuoka in 2005 accelerated the endeavor for the establishment of an earthquake monitoring and tsunami early warning system for Korea. As a result, the Korea Meteorological Administration (KMA) established an integrated seismic network and a TWS in 2005.

The National Earthquake Center of the KMA is responsible for issuing the tsunami advisory or warning based on the information of the preliminary epicenter and magnitude of earthquakes. This center would disseminate the tsunami information, which includes the earthquake information as well as the arrival time and the height of the tsunami along the coast, to the key national emergency management agencies, the media, and local authorities via a short message service (SMS), a multimedia message service (MMS), fax, e-mail, or a computer message, taking advantage of Korea's advanced IT infrastructure. Local meteorological agencies have the responsibility to monitor the sea-level variations and report them to the headquarters of KMA. The KMA in turn would analyze the sea-level data to confirm the tsunami. The criteria adopted by Korea to issue national tsunami advisories or warnings are as follows (*National Report of the Republic of Korea to IOC-UNESCO,* 2009):

> *Advisory:* Earthquake magnitude ~7.0–7.5
> Expected tsunami height ~0.5–1.0 m
> *Warning:* Earthquake magnitude ≥7.5
> Expected tsunami height ≥1.0 m

14.7.1. Seismic Networks

KMA operates a national seismic network known as the Korea National Seismographic Network (KNSN), which consists of 12 broadband seismometers, 32 short-period seismometers, 109 accelerometers, 3 borehole seismometers, and an earthquake analysis system. KMA established a geomagnetic observatory in 2008. It exchanges seismic data with the JMA and National Research Institute for Earth Science and Disaster Prevention of Japan (NIED) in real time in order to mitigate tsunami disasters in Korea and Japan.

The KNSN was designed to provide an automated solution of seismic events for immediate response to tsunamigenic earthquakes. It is composed of velocity recording seismometers and accelerometers that run continuously and automatically provide digital data. In the event of an earthquake, the seismic signals are transmitted through KMA's own network system to a central processing station. The recorded seismic waveforms are analyzed and estimated automatically. In addition, KMA established the National Earthquake Information System to analyze the earthquake data, construct a seismic database, and issue timely warnings.

When an earthquake occurs, the seismic data gathered from each station are automatically analyzed within five minutes. This way, a tsunami warning can be issued quickly if the earthquake is found to be tsunamigenic.

14.7.2. Sea-Level Networks

The National Oceanographic Research Institute (NORI) operates 37 sea-level stations around Korea (Figure 14.27) whose data are available in real time on the NORI website, which is updated at 1-minute intervals. The KMA operates a sea-level monitoring system at Ulleung Island, which is located between Korea and Japan.

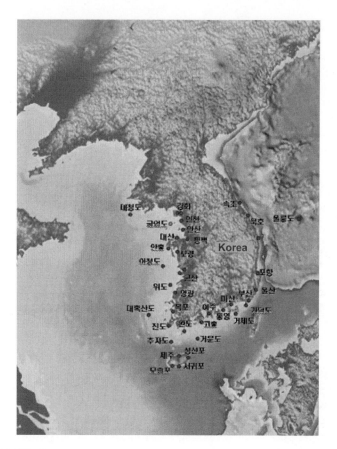

FIGURE 14.27 Network of Korea's coastal sea-level stations used for sea-level monitoring and verification of the existence of a tsunami. The data is provided in real time through the website of the National Oceanographic Research Institute of Korea. *Source: National Report of Republic of Korea to IOC-UNESCO, 2009; Courtesy of IOC-UNESCO.*

14.8. THE CHILE TSUNAMI WARNING SYSTEM

Historically, several tsunamigenic earthquakes have occurred along the Chilean continental margin due to an active subduction zone. These events have caused destructive effects to several coastal communities in the Pacific Basin. Since the inception of the Chile TWS, considerable efforts have gone into the improvement of technology and operational methodology to issue tsunami warnings in the shortest possible time.

Chile is a member of the International Coordinating Group of the TWS in the Pacific. The Hydrographic and Oceanographic Service of the Chilean Navy, known as SHOA (in Spanish, Servicio Hidrografico y Oceanografico de la Armada), located at Valparaiso, is in charge of identifying and characterizing the tsunamigenic events in Chile. The criteria for declaring a potential tsunami emergency is an earthquake with $M_W \geq 7.5$. The SHOA operates around the clock and sends out tsunami threat information to other sister agencies such as the National Emergency Office of the Interior's Department (ONEMI; in Spanish, Oficina Nacional de Emergencia del Ministerio del Interior) and to naval authorities as high-priority messages. Tsunami informs, watches, or warnings are issued when an earthquake is strong (i.e., >6.5 on the Richter scale). The emergency situation is terminated when SHOA detects no sea-level anomalies at the near-field stations. The usual communication media used between SHOA and ONEMI are fax, e-mail, radio, and telephone. The Chilean TWS consists of a network of SS, coastal sea-level gauges (Figure 14.28), and a DART system located north of Chile. The communication system used is satellite Iridium telephone.

14.9. THE NEW ZEALAND TSUNAMI WARNING SYSTEM

New Zealand is vulnerable to tsunamis because it has many volcanoes along its coastline, and it is close to faulting and continental plate subduction zones. Because of its proximity to tsunami generation sources, it would be almost impossible for New Zealand to issue timely warnings for local tsunamis. However, timely warnings can be realized in cases of distant tsunamis (>3 h travel time) and regional tsunamis (1−3 h travel time). Because a tsunami that is generated in conjunction with a nearby large earthquake or undersea land-slide may not provide sufficient time to implement official warning procedures, persons in coastal areas who experience strong earthquakes (e.g., hard to stand up) or observe strange sea behavior (such as the sea level suddenly rising and falling, or hear the sea making loud and unusual noises or roaring like a jet engine) have been advised not to wait for an official warning. Instead, they have been advised to accept nature's signs as a warning and immediately evacuate and proceed to high ground or inland locations.

14.9.1. Sea-Level Networks

Currently, a network of real-time tsunami gauges is under installation around the New Zealand coasts and on nearby offshore islands. Upon completion, the network will consist of 20 tsunami monitoring stations (Figure 14.29a). This will include 18 stations designed, owned, and operated by New Zealand. An additional two stations at Norfolk and Macquarie Islands, which will be designed, owned, and operated by Australia, will complete the network. Six New Zealand stations and one Australian station have been installed in North Cape, Tauranga, Gisborne, Napier, Wellington, Chatham Island, and Macquarie Island.

At each New Zealand station, sea levels will be measured by two subsurface (i.e., submerged in the sea) pressure sensors (two are used for the sake of

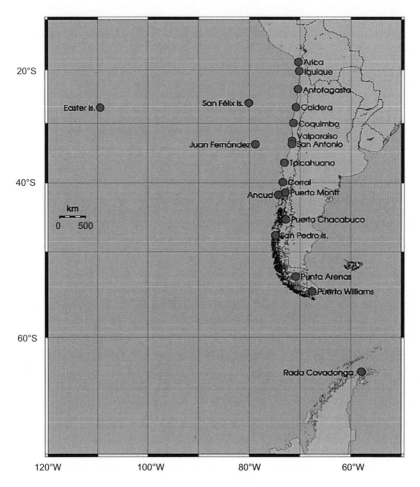

FIGURE 14.28 Network of Chile's sea-level stations used for sea-level monitoring and verification of the existence of a tsunami. The communication system used is satellite Iridium telephone. *Source: National Report of Chile to IOC-UNESCO, 2009; Courtesy of IOC-UNESCO.*

redundancy). Sea-level measurements, sampled at 10 Hz, are transmitted to the GeoNet Data Management Center in Lower Hutt. Sea-level data are available to TWCs in real-time via the GTS as well as over the Internet via Seedlink (a seismic data exchange protocol). Real-time raw and detided time-series are displayed on the GeoNet website at http://www.geonet.org.nz/tsunami/gauges and are freely available for download via the GeoNet ftp site: ftp://ftp.geonet. org.nz/tsunami. Since the installation of the New Zealand tsunami monitoring network, two tsunamis have been detected on the gauges around New Zealand. Another open-coast network of 21 sea-level gauges, operated by another organization known as National Institute of Water and Atmospheric Research

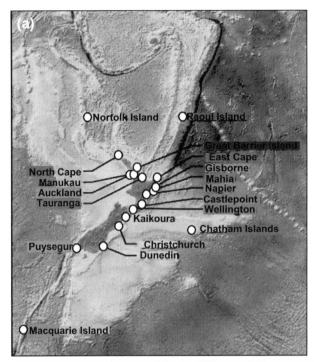

FIGURE 14.29   (a) New Zealand's tsunami monitoring network; (b) NIWA open coast sea-level network. *Source: National Report of New Zealand to IOC-UNESCO, January 2009; Courtesy of IOC-UNESCO. Reproduced with kind permission from IOC-UNESCO.*

(NIWA), exists around New Zealand (Figure 14.29b), mostly recording at 1-minute intervals. The system is not real-time, but loggers can be interrogated on demand. A daily update on sea levels, tides, storm surges, and tsunamis from selected sites can be found at http://www.niwa.co.nz/services/free/sealevels.

14.9.2. Tsunami Warnings

Initial tsunami advisories or warnings are issued by the Ministry of Civil Defence & Emergency Management (MCDEM) as a default action when the information received meets or exceeds specific thresholds (i.e., criteria). However, when an event does not hold a threat for New Zealand but the information otherwise available is considered to potentially lead to public concern, a National Advisory would be issued. Following the issue of a national tsunami advisory or warning, local authorities are responsible for activating local public alerting mechanisms, following their own procedures, while national agencies activate response plans relevant to their areas of business. MCDEM maintains a Memorandum of Understanding (MoU) with key media (radio and TV) for the public broadcasting of warnings (New Zealand National Report to IOC-UNESCO, 2009). National tsunami advisories and warnings are disseminated to all local authorities, key national agencies, and the media via the National MCDEM Warning System via SMS, e-mail, and fax. Tsunami advisories and warnings are followed up by continuous subsequent advisories/warnings until a formal cancellation is issued via the National MCDEM Warning System.

14.10. INDIA'S EARLY-WARNING SYSTEM FOR TSUNAMIS AND STORM SURGES

The consequences of the December 2004 global tsunami, which originated in the Indian Ocean, quickly became a tragic lesson for the majority of the countries in the region. As a result, many coastal economies around the Indian Ocean are now more aware of the necessity to make well-coordinated efforts to deal with tsunamis. The scale of this hazard exceeded all historical cases of catastrophic tsunamis. Therefore, scientists and several national governments are more intensely focused on the problem of how to prevent the consequences of such marine catastrophes, and several individual nations with sea boundaries have begun to install TWS systems.

India, one of the major victims of the December 2004 tsunami, took considerable initiative to put in place a National Early Warning System for Tsunamis & Storm Surges for future benefits. The Indian Ocean rim countries are likely to be affected again in the future by tsunamis that could be generated in the two known tsunamigenic zones: the Andaman-Nicobar-Sumatra Island arc in the Bay of Bengal and the Makran subduction zone north of Arabian Sea. These zones have been identified by considering the historical tsunamis, earthquakes, their magnitudes, the locations of the area relative to a fault, and also by tsunami modeling. Despite the infrequent occurrence of tsunamis (about six events reported in the twentieth century) in the Indian Ocean, they could occur at any time and could be very devastating. The east and west coasts of the Indian Ocean and the island regions are likely to be affected by tsunamis generated mainly by subduction zone—related earthquakes from the two potential source regions.

In India, a government-funded organization known as the Indian National Centre for Ocean Information Services (INCOIS), located at Hyderabad (Andhra Pradesh) and in partnership with institutions from Ministry of Earth Sciences (MoES), Department of Space (DOS), Department of Science & Technology (DST), Center for Scientific & Industrial Research (CSIR), Ministry of Home Affairs (MHA), and Ministry of External Affairs (MEA), is responsible for providing all information to the public related to the surrounding seas, including tsunamis and storm surges. After the December 2004 tsunami, India successfully set up the first TWC in the Indian Ocean in record time (Nayak and Kumar, 2008).

14.10.1. Sensor Networks

The Indian Tsunami Early Warning System consists of a real-time network of SS (Figure 14.30), bottom pressure recorders (BPRs) (Figure 14.31), and coastal and island sea-level gauges (Figure 14.32) to detect tsunamigenic earthquakes and monitor tsunamis. The traditional methods of tsunami warnings practiced in most parts of the world are based primarily on seismic information obtained immediately after the earthquake and on calculations of the time of wave propagation and its amplitude at each specific point. However, the efficiency of such methods decreases, and occurrences of false alarm increase when there is a lack of data about the parameters of tsunamis in the source. For example, during the Shikotan tsunami on October 4, 1994, which was a catastrophic event for the Southern Kuril Islands and Hokkaido, the PTWS announced a tsunami alarm on the Hawaiian Islands and warned of the possible approach of a tsunami with a height of a few meters. Significant finances were spent on the evacuation of thousands of people, and two persons died during this evacuation. Sadly, the alarm was false; the actual height of the tsunami wave was less than half a meter (Kulikov et al., 2005). The seismic network of the Indian TWC is configured to minimize such false alarms. The Indian seismic subsystem receives real-time seismic data from the national seismic network of the Indian Meteorological Department (IMD) and other international seismic networks and is capable of detecting all earthquake events occurring in the Indian Ocean in less than 15 minutes of occurrence (Nayak and Kumar, 2008).

As noted earlier, direct measurements of tsunamis obtained from coastal sea-level recorders contain fluctuations that are strongly distorted with respect to the form of the wave in the open ocean. The arrival of the tsunami wave to the shallow-water zone and the reflection from coasts lead to an increase in

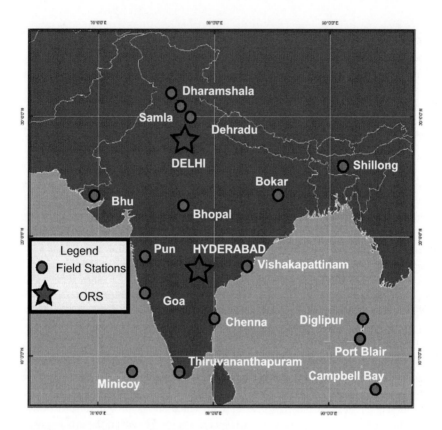

FIGURE 14.30   Real-time broad-band seismic stations of India's ocean observation network. *Source: Courtesy of Dr. T. Srinivasa Kumar, INCOIS, Hyderabad, India.*

amplitude, but the spectrum of the signal (waveform) is transformed due to the resonance properties of the shelves, bays, and straits. It is a well-established fact that quality data for tsunamis in the open ocean can be obtained using the data from BPRs. Thus, the BPRs installed in the deep ocean are the key sensors to confirm the occurrence of a tsunami. However, data from coastal and island sea-level stations are vital to monitor the arrival and progress of tsunami propagation in areas of human settlement. Thus, the Indian TWC incorporates data from both BPRs and coastal sea-level stations. Whereas the BPRs have been installed by National Institute of Ocean Technology (NIOT), the coastal sea level stations are managed by the Survey of India (SOI) to monitor the progress of tsunami wave arrivals.

14.10.2. Tsunami Warning Centers

A state-of-the-art early warning center has been established at INCOIS with all the necessary computational and communication infrastructure that enables

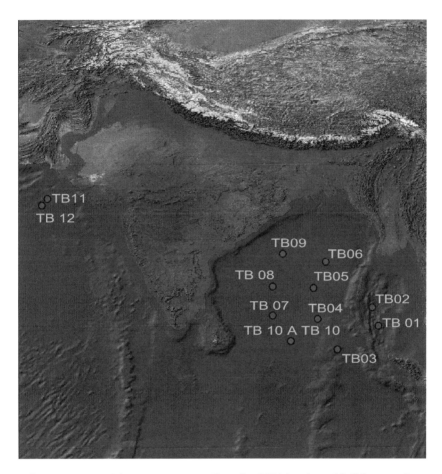

FIGURE 14.31   Real-time Bottom Pressure Recorder (BPR) locations of India's ocean observation network. *Source: Courtesy of Dr. T. Srinivasa Kumar, INCOIS, Hyderabad, India.*

reception of real-time data from all the sensors, analysis of the data, and generation and dissemination of tsunami advisories following a standard operating procedure. A host of communication methods have been employed for timely dissemination of advisories. Seismic and sea-level data are continuously monitored in the Early Warning Center using a custom-built software application that generates alarms and alerts in the warning center whenever a preset threshold is crossed. The data is organized in a central database on a storage server and affords analysis and retrieval. A display facility (Figure 14.33) allows visualization of various data streaming in, prerun model outputs, and vulnerability maps. Tsunami warnings/alerts/watches are generated based on preset decision support rules and disseminated to the concerned authorities for necessary action following a Standard Operating Procedure. The database is also

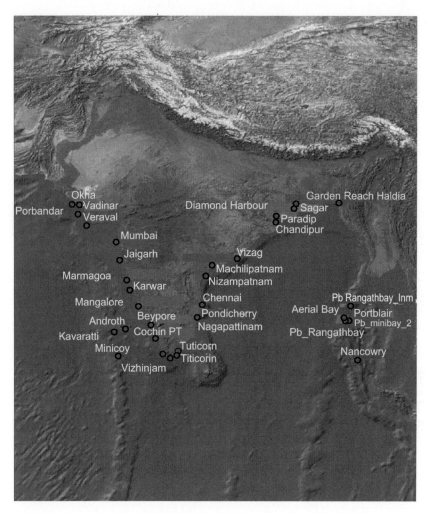

FIGURE 14.32   Tide station locations of India's ocean observation network. *Source: Courtesy of Dr. T. Srinivasa Kumar, INCOIS, Hyderabad, India.*

linked to the dedicated tsunami website, through which data/information/advisories are made available to the users. The efficiency of the end-to-end system was proved during the undersea earthquake of 8.4 M that occurred on September 12, 2007, in the Indian Ocean (Nayak and Kumar, 2008).

14.10.3. Standard Operating Procedures at Early Warning Centers

The criteria for generation of different types of advisories (warnings, alerts, and watches) for a particular region of the coast are based on tsunami travel time.

FIGURE 14.33 Snapshot of India's Tsunami Early Warning Center at INCOIS, Hyderabad. *Source: Nayak and Kumar, 2008; The first tsunami early warning centre in the Indian Ocean; Reproduced with kind permission from Tudor Rose Publishers.*

The warning criteria are based on the premise that coastal areas falling within 60 minutes of travel time from a tsunamigenic earthquake source need to be warned based solely on earthquake information. This is because enough time may not be available for confirmation of the expected sea-level anomaly based on measurements from BPRs and coastal/island sea-level gauges. Those coastal areas falling outside the 60 minutes of travel time from a tsunamigenic earthquake source would be put under a watch status and would be upgraded to a warning only upon confirmation of an anomalous sea-level change (rise or fall). This implies that the possibility of false alarms is higher for those areas that are located close to the earthquake source. However, for other areas, the question of a false alarm does not arise because the warnings are issued only after confirmation of an anomalous sea-level change. In order to minimize false alarms even in the near-source regions, alerts are generated by analyzing the prerun model scenarios, so warnings are issued only to those coastal locations that are at risk. The coastal areas, where tsunami threat perception exists, will be categorized under different risk zones: major tsunami, medium tsunami, and minor tsunami (Nayak and Kumar, 2008).

14.10.4. Vulnerability Maps

The *tsunami N2 model* has been used for the purpose of preparing vulnerability maps indicating the areas likely to be affected by flooding and the extent of inundation of seawater into the land. Information from remote sensing and field investigations are being integrated in GIS for modeling and mapping of the inundation of seawater to determine setback lines, to plan coastal defense, and so forth.

14.10.5. Performance of the System: Case Study of September 12, 2007

INCOIS generated a database of model scenarios considering various earthquake parameters. The prerun scenario for the September 12, 2007, earthquake event in Sumatra was used to calculate the estimated travel time and run-up heights at various coastal locations and sea-level sensor locations (tide gauges and BPRs). The directivity map generated from the picked scenarios indicated that the southeast and southwest Indian coasts were likely to be affected by a minor tsunami (~20 cm) and Andaman and Nicobar Islands (~10 cm). Whereas observed tsunami heights at Padang, Port Blair (Andaman Archipelago), and Chennai (southeast Indian coast) were lower than the estimated tsunami heights by 20, 2, and 2 cm, respectively, those at Coco's Island, Sabang, and Male (Maldives Archipelago) were higher than the estimated tsunami heights by 10, 10, and 1 cm, respectively. However, those at the offshore BPR sites were very realistic (within ±1 cm). Although no comparisons were reported by INCOIS for tsunami heights from the west coast of India, real-time reporting on the Internet by a sea-level gauge at Verem in Goa (west coast of India) that is maintained by the NIO revealed the tsunami height at this location to be up to 29 cm (Prabhudesai et al., 2008), which is larger by 9 cm compared to that predicted for the southwest Indian coast. The differences observed at the coastal regions could perhaps be because of the inability of the model to incorporate high-resolution bathymetry near the coast. In this case, the model cannot resolve small local effects and resonant amplifications. Nested grids (Cherniwsky et al., 2007) with fine-resolution coast and bottom topography are needed to resolve such problems so the fast changes in tsunami heights at the coastal region, resulting from the law of conservation of energy (Greens's theorem), can be well accounted for.

Except for these differences in tsunami heights at the coastal locations, the end-to-end system of the Indian TWS performed well, thereby enabling reception, display, and analysis of the real-time and model data sets, as well as generation and dissemination of timely and accurate advisories following the standard operating procedure. The information was effectively used to provide necessary advisories to the concerned authorities, thus avoiding unnecessary public evacuation for this event (Nayak and Kumar, 2008).

14.10.6. Contributions to the Indian Ocean Region

The Indian TWC is equipped with world-class computational, communication, and technical support facilities, and it possesses all the essential ingredients of a modern TWC. The instrumentation of the Indian TWC is established in such a way that it is capable of detecting tsunamis originating from both the known tsunamigenic sources in the Indian Ocean (i.e., the Andaman-Nicobar-Sumatra Island arc and the Makran subduction zone north of Arabian Sea). The Indian TWC has robust application software based on geospatial technologies to generate and disseminate timely tsunami advisories to the Indian Ocean countries. INCOIS has also set up the warning center infrastructure so as to have the desired capabilities of a Regional Tsunami Watch Provider. India has begun providing regional tsunami watch services from its national system for the Indian Ocean region. At present, it provides earthquake source information to potential tsunami threats and travel times. Shortly, the relevant information, such as earthquake parameters, travel times, run-up heights, and potential threat zones, will be provided. The warning center can also support the generation of inundation maps and risk and hazard assessments.

14.10.7. National Institute of Oceanography Stations

Although the responsibility of providing ocean information services to the people of India officially rests with INCOIS, Joseph and Prabhudesai (2005) from the NIO India, proposed a cost-effective cellular-based Internet-accessible real-time reporting system that would enable people all over the world to view in real/near real time a graphical display of sea-level, sea-state, and surface meteorological changes from coastal and island locations in India. Subsequently, based on this idea, a group of scientists and technologists at NIO (an oceanographic research laboratory under the Indian Council of Scientific & Industrial Research (CSIR)) promptly designed such a system for sea-level monitoring and established it at Verem Naval Jetty in Goa, with logistical support from the Indian Navy (Prabhudesai et al., 2006). The station (Figure 14.34) began operations on September 24, 2005, and provided real-time sea-level changes in graphical format (measured sea-level, astronomical tide, and residual) that could be monitored over the Internet (http://inet.nio.org). Subsequently, similar subsurface pressure sensors as well as microwave radar-based systems were established at Kavaratti and Andrott Islands in the Lakshadweep Archipelago, located at the eastern Arabian Sea.

The installations at the coastal and island locations enabled the detection and close monitoring of the arrivals and persistence of tsunamis at these two distant locations in real time on the Internet after the occurrence of an $M_w = 8.4$ earthquake in Sumatra on September 12, 2007, was announced (Prabhudesai et al., 2008). Sea-level records at 5-minute sampling intervals (measured sea-level, astronomical tide, and residual) provided a clear indication of the tsunami signal (Figure 14.35). Residual and wavelet power spectra provided additional information on this tsunami (Figure 14.36). The tsunami arrived at Kavaratti Island and Goa after traveling nearly 5 hours 15 minutes and 8 hours, respectively, from the source region in Sunda Trench. The arrival of

FIGURE 14.34 NIO's cellular based Internet-accessible real-time reporting sea-level station established at Verem Naval Jetty, Goa. Sea level is obtained based on subsurface pressure measure measurement. *Source: Prabhudesai et al., 2008; Reproduced with kind permission from Indian Academy of Sciences.*

a detectable tsunami signal first at Kavaratti Island and then 2 hours 45 minutes later at the shallower Goa coastal region of the mainland indicates the importance of having high-sensitivity real-time monitoring and Internet-accessible sea-level stations on India's island locations for effective tsunami warning purposes for the mainland.

In quick succession, an Internet-accessible real-time reporting surface meteorological station was established at Puducherry (formerly, Pondicherry) on the southeast coast of India as an integral part of the ocean information services scheme of INCOIS for providing advisories to fishermen and coastal communities at Puducherry, besides the use of data from the network for a variety of scientific studies. A nongovernmental organization, known as Pondicherry Multipurpose Social Service Society (PMSSS), provided its space

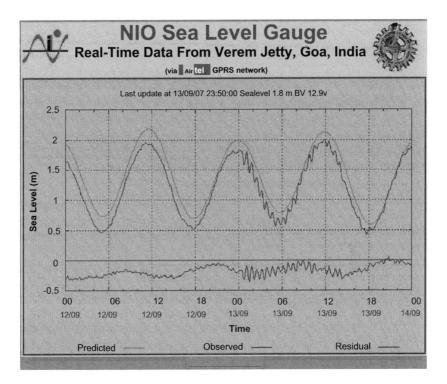

FIGURE 14.35 Real time graphical display of 12 September 2007 weak tsunami at Goa, India from NIO's sea-level gauge. *Source: Prabhudesai et al., 2008; Reproduced with kind permission from Indian Academy of Sciences.*

and logistical support for this installation. Since its establishment, INCOIS has routinely used the data for the benefit of the fishing community of Puducherry. In subsequent years, a network of Internet-accessible real-time reporting sea-level, sea-state, and surface meteorological stations (Figure 14.37) were established at several locations on the Indian coasts and islands under financial support primarily from NIO and in part from INCOIS. As of August 2010, the network consists of eleven sea-level stations, three surface wave stations (significant wave height and direction), and eleven surface meteorological stations.

The in-house designed and developed sea-level stations of the network have incorporated subsurface pressure sensor and two variants of downward-looking aerial microwave radar sensors, incorporating cellular-based and Internet-enabled real-time reporting capability. The microwave radar gauges have been installed from coastal jetties and open waterfronts, with relatively fewer logistical problems. The NIO network allows the real-time display of other related information such as surface gravity waves as well.

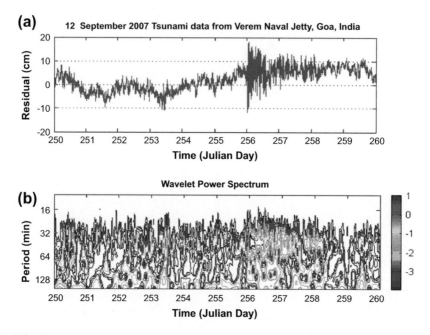

(a)

12 September 2007 Tsunami data from Verem Naval Jetty, Goa, India

(b)

Wavelet Power Spectrum

FIGURE 14.36 Detailed information on the real-time-reported September 2007 tsunami at Goa from sea-level residuals and wavelet power spectrum.

Unlike float-driven gauges and guided air-acoustic gauges that require stilling-wells (with their inherent slow responses, nonlinearity, and waveform distortions in the presence of large-amplitude waves), sea-level gauges incorporated in the NIO network have fast and linear responses. The gauges are installed under "open" environments (i.e., without the use of "tide gauge huts") and are powered from 12-volt batteries that are charged through solar panels. Sea-level data are acquired at fast sampling intervals (every 5 minutes). A set of Internet-timed synchronized time-tagged datasets are uploaded to an Internet server with the use of a cellular modem. The preference for cellular-based data reporting stems from the ubiquity of cellular base stations throughout the country, the relatively small size of the cellular modems, and state-of-the art information accessibility at significantly low cost. Graphical displays of the measured sea levels, astronomical tides, and residuals (i.e., anomalous sea-level oscillations) can be viewed in real/near real time from the Internet. Data loss due to power or instrument failure is largely circumvented by the induction of a redundant gauge at the same station. Maintenance of accurate time stamps of the dataset through Internet time synchronization of the gauges' real-time clock (RTC) twice a day using Internet network time protocol (NTP) allows error-free estimation of the arrival times and speed of propagating oceanic events such as coastal waves, storm surges, tsunamis, migrating eddies, and so on. The AWS records Internet-time synchronized vector averaged winds, air temperatures,

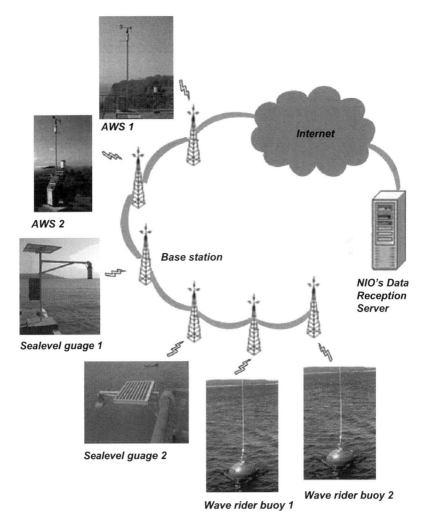

FIGURE 14.37 Schematic illustration of NIO's in-house designed network of cellular based Internet-accessible real-time reporting sea-level and surface meteorological stations. *Source: Prabhudesai et al., 2010. OCEANS' 10 – IEEE Seattle Technical Conference; September 2010; IEEE.*

relative humidity, solar radiation, barometric pressure, and rainfall at 10-minute intervals.

As indicated earlier, real-time and near-real-time sea-level and surface meteorological information from the NIO network can be accessed through the link http://inet.nio.org, where the viewer is introduced to the network through a screen display of stations (Figure 14.38). By clicking on the "Location Map" icon on the main screen, a screen display of location maps of stations that are

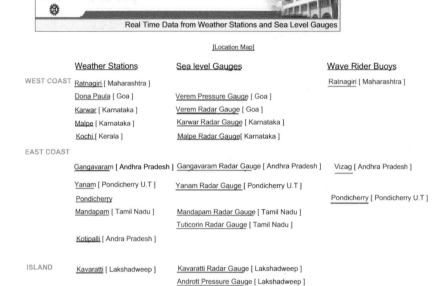

FIGURE 14.38 Main screen display of stations available on the NIO-network of sea-level, sea-state, and surface meteorological stations until July 2010. *Source: Prabhudesai et al., 2010. OCEANS' 10 – IEEE Seattle Technical Conference; September 2010; IEEE.*

available on the NIO network of sea-level, sea-state, and surface meteorological stations appears on the computer console (Figure 14.39). By clicking on any desired location on the display of stations, the viewer gets access to the respective real-time information (Figures 14.40, 14.41, and 14.42). At present, the only real/near-real time visual presentation of sea-level, sea-state, and surface meteorological time-series information from India that can be accessed through the Internet from anywhere is from the NIO network. This network—the first of its kind in India—is likely to be of practical value as a real-/near-real-time visual information dissemination system that anyone can access during an oceanogenic eventuality.

India's Early Warning System for Tsunamis & Storm Surges can be considerably improved by inducting the real-time data stream (sea-level data at 5-minutes interval, significant wave height and direction data at 30-minutes interval, and meteorological data at 10-minutes interval) from the NIO network into the forecasting system of INCOIS. NIO is considering further expansion of its network in the near future. Although the NIO network is primarily meant for high-resolution data acquisition for its research programs, real-/near-real-time graphical display of sea-level, sea-state, and surface meteorological information on the Internet from a network of coastal and island stations will be of

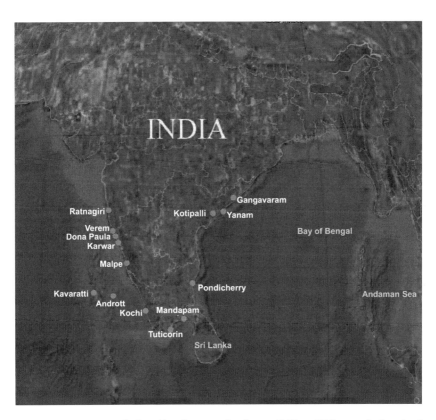

FIGURE 14.39 Screen display of location map of stations available on NIO-network of sea-level and surface meteorological stations until July 2010. *Source: Prabhudesai et al., 2010. OCEANS' 10 – IEEE Seattle Technical Conference; September 2010; IEEE.*

considerable value for the general public as well as disaster management agencies. Real-time reporting capability to the network yields several benefits, including the following:

- The possible use of data for running automated real-time numerical models for operational forecast
- Validation of models and remotely sensed satellite data
- The availability of real-time data from a network of stations, thereby giving the disaster response team access to the latest in situ information from several locations
- Periodic arrival of the data stream from all stations at a single central server, thus yielding backup for the data from all the stations
- Remote monitoring of proper working condition of individual stations
- Implementation of repair/maintenance in the shortest possible time, thereby minimizing breaks in the time-series data stream

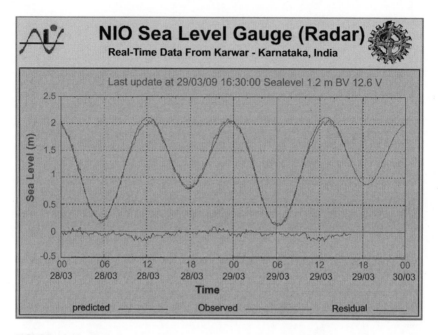

FIGURE 14.40 Typical screen display of real-time sea-level data from a selected location available on NIO-network. *Source: Prabhudesai et al., 2010. OCEANS' 10 – IEEE Seattle Technical Conference; September 2010; IEEE.*

The data and information gathered by the network will thus benefit the coastal communities that are at risk from tsunamis, storm surges, and sea-level changes in general.

Effective visual reporting of dangerously anomalous sea-level variability and surface meteorological conditions is of paramount importance to the safety of the people. When such anomalous sea-level oscillations are anticipated either because of earthquakes or intense meteorological disturbances or both, the Internet-accessible graphical display of sea-level anomaly and meteorological conditions at multiple locations available from the NIO network can also be brought to the attention of all sectors through various modes of communication, such as personalized devices (mobile telephone, e-mail, SMS, etc), mass media (radio, television), and community media (loudspeakers, sirens, etc.), as well to ensure that the desired objective is effectively met.

The network is also useful for providing the requisite real-time data to aid navigation, high-quality digital data required for climate change monitoring and related studies, constructing maps of tsunami risk along the coast (local tsunami zoning), and issuing detailed wave height and wave arrival information for incoming waves. The high-resolution surface meteorological data gathered from such a network will also help the corresponding agencies and researchers

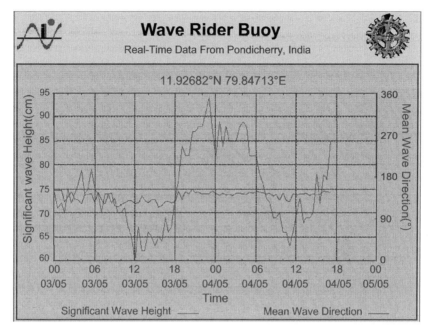

FIGURE 14.41 Typical screen display of real-time sea surface wave data from a selected location available on NIO-network. *Source: Prabhudesai et al., 2010. OCEANS' 10 – IEEE Seattle Technical Conference; September 2010; IEEE.*

to improve weather and extreme weather forecasts and to better understand the phenomena.

In contrast to the limited bandwidth provided by INSAT transmitters, coastal observations at high bandwidths at significantly low cost have become realizable using cellular GPRS networks. The NIO network allows, for the first time in India, Internet-based real-/near-real-time tracking and monitoring of sea-level, sea-state, and surface-meteorological conditions along the Indian coasts and islands and from almost anywhere in the world—an issue of considerable practical significance during natural disasters such as storms, storm-surges, and tsunamis.

14.11. MALAYSIA'S MULTIHAZARD EARLY WARNING SYSTEM

Malaysia is relatively free from severe natural hazards such as earthquakes, tsunamis, tropical storms, and typhoons. However, this country has often been a victim of several types of geological and weather-related hazards such as flash floods, monsoonal floods, landslides, and wind gusts in association with severe thunderstorms. The abnormal weather and the resulting flash flood in Johor in December 2006/January 2007 is a recent example of such events. In Malaysia,

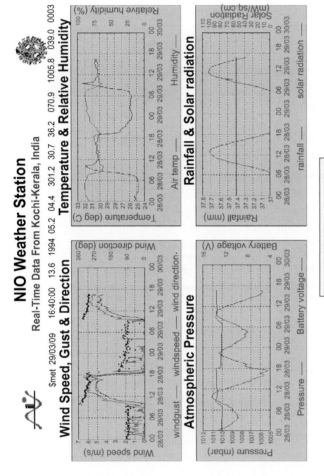

FIGURE 14.42 Typical screen display of real-time surface meteorological data from a selected location available on NIO-network. *Source: Prabhudesai et al., 2010. OCEANS' 10 – IEEE Seattle Technical Conference; September 2010; IEEE.*

the Malaysian Meteorological Department (MMD) is the official agency responsible to address issues related to meteorological and seismological events and the official source of severe weather warnings. It is also its goal to comprehensively serve the nation's needs for seismological information and tsunami warning services. The high-resolution data gathered from a network of environmental real-time monitoring stations are expected to be useful to generate more accurate, timely, and reliable forecasts and warnings of weather, water, and related environmental elements and to improve the delivery of environmental information and services to the public, governments, and other users. The environmental information obtained is also expected to be useful to provide scientific and technical expertise and advice in support of policy and decision making.

14.11.1. Seismic Networks

To effectively realize its goals, the MMD has set up a multihazard early warning center and enhanced its observation program. It has also initiated several environmental and ocean prediction models for weather and is contemplating building a comprehensive national earthquake and TWS that will consist of a national seismic network comprising several SS. The SS consist of sensors, power supplies, data loggers, digitizers, and so on. The seismic data acquired through the seismic networks are transmitted to the MMD's operation center via satellite. The seismic waveform data received in real time from the seismic network are displayed at the monitoring center for visual inspection by scientists to set the stage for analysis and interpretation. On the basis of international and regional cooperation, Malaysia is receiving seismic waveform data from USGS, IRIS, CTBTO, Indonesia, Singapore, and Australia through the Internet and VSAT in near real time. In turn, Malaysia is contributing real-time seismic waveform data from its broadband stations for regional and international exchange.

14.11.2. Sea-Level Networks

As part of the tsunami network, sea-level data are gathered from a network of deep-ocean bottom pressure recorders. The bottom pressure data that are tel-emetered from the buoy via INMARSAT C are received at an earth station and immediately archived at the National Operations Centre for processing and dissemination. The Malaysian coastal sea-level station network currently consists of 6 gauges. However, an additional 15 gauges are planned to be installed at various locations. The sea-level network was very valuable in the real-time reporting of the July 17, 2006, tsunami and the September 12, 2007, tsunami that occurred in the Indian Ocean.

Malaysia's multihazard early warning system includes a network of four coastal camera networks as well. There is a proposal to install another

14 new coastal camera systems. Like many other countries, Malaysia has also incorporated a network of siren warning systems to alert the coastal communities in the event of an impending disaster from a natural calamity.

14.11.3. Automatic Weather Station Networks

The aim of the automatic weather station (AWS) network is to monitor weather phenomena for forecasting, climatology, and research. Since 1995, a total of 180 AWS systems have been established over the country. Real-time monitoring is realized using the latest Web technologies. The network incorporates both broadband and GPRS data transmission to ensure real-time data streaming. The telecommunication infrastructure is proposed to be upgraded in terms of speed (upgrading Internet access speed for MMD headquarters from 3 Mbps to 8 Mbps, and for regional forecast offices from 256 kbps to 1.5 Mbps).

14.12. THE SINGAPORE TSUNAMI WARNING SYSTEM

In Singapore, the Meteorological Services Division (MSD) of the National Environmental Agency is the designated national organization that is responsible to assess tsunami risks and issue warnings and advisories to the relevant government agencies. MSD maintains a seismic monitoring system that as of February 2009 consists of five sensors, including one global seismic network (GSN) station at Bukit Timah Dairy Farm (BTDF). With the completion of the first phase of the seismic upgrade in March 2006, data from the GSN station are exchanged in real time with Malaysia, Indonesia, and Australia. The second phase of the seismic upgrade involves the upgrading of obsolete data loggers and the installation of one broadband and two strong-motion stations. The broadband station will be installed with STS-2 and strong-motion sensors. This station is expected to serve as a backup to the BTDF station. The upgrade of seismic equipment was completed in 2007, and three new stations are scheduled to be completed in the near future.

The Maritime and Port Authority (MPA) maintains a network of 12 sea-level stations around the coastal waters of Singapore. Real-time sea-level data from the Tanjong Pagar station are available to the warning center at MSD for the purpose of sea-level and tsunami monitoring. Sea-level data from this station have been available to the international community via GTS since November 2008.

Along with many nations around the Indian Ocean and major research centers in the world, Singapore is actively contributing to the development of an early warning system for the South and Southeast Asian regions. A two-year study, commissioned by MSD to assess and evaluate the possible impacts of tsunamis on Singapore, was completed in August 2008. During the study, the MSD worked in tandem with the researchers from the local institutes—the

National University of Singapore and the Nanyang Technological University—to assess the possible impacts of tsunamis on Singapore and to develop tsunami modeling capabilities (Lesley Choo Yap Chui Wah, *Singapore National Report to IOC,* February 2009). Under the project, an Operational Tsunami Prediction and Assessment System (OTPAS) was developed to integrate source identification, modeling modules on propagation and inundation, real-time data from monitoring stations, system infrastructures, and telecommunications. The system can assimilate near-real-time data into numerical models to produce forecasts of tsunami arrival times, heights, and inundation/run-up to assist the duty forecaster at MSD to make rapid assessments and issue tsunami warnings to the relevant response agencies and the public. A graphical user interface (GUI) of the OTPAS allows monitoring of tsunamigenic events in the domain of interest. The GUI is capable of GIS functions, and it provides tools for data visualization, analysis, and animation. The system has been in operation since August 2008. The system shall continue to undergo refinement in the update of modeling modules, adaptation of GUI to the users' requirements, incorporation of worldwide seismic sources, and enhancement of modeling codes for more efficient simulation runs.

MSD is coordinating a task force involving key government agencies having the responsibility of tsunami monitoring and disaster mitigation. Based on the results of the modeling studies, the national tsunami response plan focuses on warning and evacuation efforts to the public at unprotected beach areas and low-lying southern islands, and workers at waterfront and coastal facilities. The plan also has a provision for managing the consequences under the worst-case scenario of a destructive tsunami resulting in severe structural damages and loss of lives.

Although Singapore's current intention is to rely more on real-time information available on the Web, it is deploying DART-type national monitoring systems that are designed to detect and report tsunamis on their own without any remote control from ground stations. Efforts are underway for the development of a real-time data-driven model that would rapidly predict the tsunami wave amplitudes at the coast by using the data monitored continuously by the deep-ocean moored buoys (Tkalich et al., 2007). Such buoys are already available off Phuket (Figure 14.43). This country is currently paying more attention to tsunami research activities such as tsunami propagation modeling (e.g., Dao and Tkalich, 2007) than on operational programs such as real-time reporting from a large network of seismometers and sea-level gauges.

14.13. EUROPEAN UNION INITIATIVE

In view of providing adequate tools for rational and sustainable development of the coastal zones, it is important to start programs for systematic evaluation of the tsunami hazard in the region and to adopt economically sustainable

FIGURE 14.43   Deep ocean buoy networks off Phuket deployed by Thailand. *Source: Tkalich et al., 2007, Journal of Earthquake and Tsunami, Reproduced with kind permission from World Scientific Publishing Company.*

programs to assess vulnerability to tsunami waves and to make estimates of risk as regards properties and infrastructures as well as human lives. The descriptions presented in the previous sections have shown that the United States, Japan, Russia, Australia, Canada, Puerto Rico, Korea, Chile, New Zealand, India, and Malaysia have established reasonably elaborate TWS systems.

The member states that participated in IOC/UNESCO in 2005 decided to develop the North-East Atlantic and Mediterranean TWS (NEAMTWS). As a contribution to this effort, the **NA**tional **T**sunami **NE**twork of **G**reece (NATNEG) is currently under development. The network's primary goal is to develop the national TWS along the Hellenic Arc, which is the most tsunamigenic area in the NEAM region, but also to create databases that may support the TWS, as well as promote basic and applied research on tsunamis. At the initial phase of its development, a schedule for the instrumental part of NATNEG may include (Papadopoulos, 2009) a broadband seismograph network from the Institute of Geodynamics, National Observatory of Athens (NOA) and a network of pressure-type sea-level gauges, installed along the Hellenic Arc, that would be telemetrically connected with NOA for real-time transmission of sea-level signals at a sampling rate of about 30 sec. The NATNEG may also include a Decision Matrix, which is an empirical tool that supports the evaluation of the possibility for tsunami generation as soon as the focal parameters of an earthquake have been determined, regardless of whether sea-level or other oceanographic signals are available. The NATNEG will be complemented by earthquake and tsunami catalogues, which already exist, as well as by a database of predetermined tsunami numerical simulations for the scenario tsunami generation, propagation, and inundation that could support the decision making in parallel to the Decision Matrix.

Another European country that took initiative in the establishment of a TWS system under the EU program is Portugal, which, due to its geographical location, is the first country to be hit by a tsunami that is generated in the Gulf of Cadiz. This fact makes it the natural candidate to host a regional TWS for this area. The two components set up under IOC/UNESCO umbrella are the Intergovernmental Coordination Group for the North-East Atlantic and the Mediterranean and Connected Seas Tsunami Warning System (ICG/NEAMTWS). One of the two European projects under this umbrella is GITEC-TWO (Genesis and Impact of Tsunamis on the European Coasts-Tsunami Warning and Observations). The present status of the ongoing implementation of the NTWS in Portugal is as follows (Matias et al., 2009):

- Real-time collection and processing of seismic data
- Real-time acquisition and transmission of coastal sea-level data to the data collection center
- Completion of operationalization of software tools for assimilation of seismic and sea-level data and preparation of messages

The purpose of the system is to issue to the Portuguese Civil Protection Authorities the first message based only on seismic information, after 5 minutes of the earthquake origin, as indicated in the preceding first entry. Using an extensive set of precomputed tsunami scenarios allows the NTWS to inform also on the predicted arrival times and amplitudes of tsunami waves at pre-defined critical points along the coast. Next, per the second and third entries, the NTWS will be able to access in real time the sea-level measurements and confirm, change, or cancel the previous tsunami message. At present, only coastal sea-level gauges are available. When fully operational with expansion of the present network, the Portuguese NTWS will be able to send messages to other countries in the Atlantic, providing the services required by a Regional TWC. The development of the Portuguese TWS is a joint effort between volunteer researchers from a few Portuguese institutions.

The contributions so far provided by international projects financed by the European Community (e.g., EU tsunami research project TRANSFER (Tsunami Risk ANd Strategies For the Europen Region)) resulted in substantial progress in improving basic assessment of hazards (by increasing knowledge of tsunami sources and by testing and refining numerical tools for tsunami simulations and inundation map computations) in a number of test areas in the Mediterranean, including the Messina Strait in Italy (Tinti et al., 2009a, b). In view of its vulnerability to tsunamis, the city of Rhodes, which occupies the northeast edge of the island, has been selected as the master test site of the TRANSFER. One of the major components in that research project, as regards the master test site, is the production of risk maps based on several methods. Greece made substantial contribution to the project TRANSFER (Fokaefs et al., 2009). Greece researchers selected an extreme tsunami wave on the basis of two independent approaches. The first was based only on tsunami descriptions contained in historical documentary sources. The second was based on inundation results coming from numerical simulation of some potential tsunamigenic sources. Finally, they adopted an extreme wave of 5−6 m in height with three different directions: east, north, and west. Maps illustrating a number of attributes, such as tsunami inundation, local topography, land use/land cover, importance of facilities, and so on, have been used to characterize and map the expected tsunami impact on a scale of 1:5000, which was practically applicable. The tsunami impact for each of the three scenarios included three different items: damage potential, vulnerability of population, and risk of life.

Under the framework of an Italian national project and the European project TRANSFER, Bressan and colleagues (2009) undertook the development of a real-time detection algorithm, which is important for the implementation of a TWS. The detection algorithm's goal is to distinguish the first tsunami wave from the background signal. The algorithm detects a tsunami if the instantaneous signal parameter exceeds a set threshold and is significantly different from the previous background signal parameter.

Real-time tsunami detection presents very different challenges; depending on whether the algorithm must be applied for either an offshore sea-level station or for a coastal sea-level station. In particular, the algorithm parameters are site-specific for coastal sea-level signals because time-series in that case are characterized by oscillations driven by the coastal topography. It was found that the presence of resonant waves with periods in the tsunami range complicated the algorithm application. To safely detect a tsunami, the algorithm had to be adjusted to target only tsunamis with amplitudes larger than the natural frequency oscillations. On the other hand, the offshore sea-level signal is mainly composed by white noise and the tide. The main issue in this case is that the tsunami signal may not have a very large amplitude with respect to noise. However, if the offshore station is located near the tsunami source, the seismic signal is relatively large and could mask the tsunami signal.

In Europe, France is hardly affected by local tsunamis (generated by cliff failure or submarine landslide), and therefore no procedures for local tsunamis have been implemented. Currently, there are four sea-level gauges installed at the following locations (French National Report to IOC, 2009):

- Papeete (Tahiti harbor)
- Rikitea harbor (Gambiers Islands)
- Taiohae (Nuku-Hiva) in Marquesas Archipelago
- Tahauku bay (Hiva-Oa) in Marquesas Archipelago

Two new sea-level stations will be installed in 2009, one in Tubuai (Austral Island) and one at Rangiroa atoll (Tuamotu Archipelago). Although French Polynesia has only a limited number of sea-level gauges, it routinely receives sea-level data from about 30 other sea-level stations from the Pacific Ocean region (14 from Australia, 14 from Chile, and the others from Central Pacific) via WMO/GTS (Figure 14.44a). The transmission delays involved in data reception through GTS (TELEX), via the WMO (World Meteorological Organization) network, are in the range of 5–60 minutes.

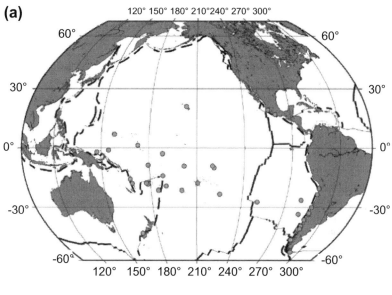

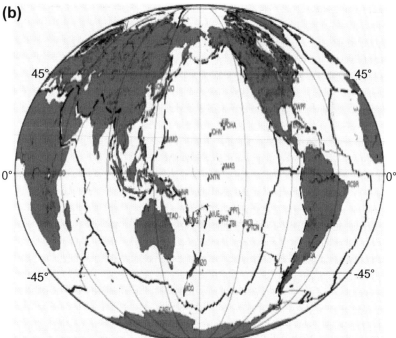

FIGURE 14.44 Map showing (a) sea-level stations and (b) seismic stations in the Pacific Ocean from where French Polynesia receives sea-level data in near-real time via GTS (TELEX) and seismic data in real time via SEEDLINK protocol. The plate boundaries (marked by thick dark broken lines) are indicative of the sensitiveness of Pacific and Indian Oceans to seismic tsunamis. *Source: French National Report to IOC-UNESCO, 2009; Courtesy of IOC-UNESCO.*

Seismic data are received in real time from about 40 stations (Figure 14.44b). The seismic data received are used to calculate an average earthquake magnitude (M_m), the scalar seismic moment (M_o), the slowness of earthquake (E/M_o) (Newman and Okal, 1998), and the moment tensor of big earthquakes.

14.14. TSUNAMI WARNING SYSTEMS AROUND THE AFRICAN CONTINENT

We have seen that sea-level monitoring network coverage is quite good around Europe, the Pacific rim, and North America. However, there are gaping holes around Africa. Even where sea-level gauges exist, the equipment is often old. Under EU funding as well as under the umbrella of Global Sea Level Observing

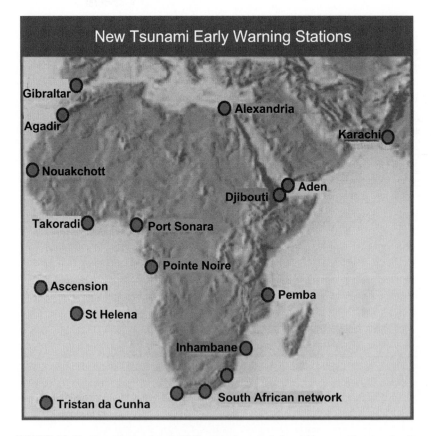

FIGURE 14.45 Recently-developed TWS network around the African continent under the UNESCO/EU initiative. *Source: Woodworth et al., 2007b; Planet Earth, Reproduced with kind permission from Natural Environment Research Council.*

System (GLOSS) and Ocean Data and Information Network for Africa (OdinAfrica), Proudman Oceanographic Laboratory (POL) in the United Kingdom designed and established a network of "tsunami enabled" sea-level stations in Africa and in the western Indian Ocean, and is currently working on establishing similar systems along European coastlines (Woodworth et al., 2007).

In Africa, two types of coastal sea-level gauges are installed: a radar gauge and a subsurface pressure gauge. The former is clamped to a sturdy beam bolted to a harbor wall. The gauge hangs out over the sea and measures the time it takes for a pulse to reflect back from the sea surface. The subsurface pressure gauge is both a backup to the radar gauge and the main tsunami sensor. It is installed safely below the low-tide level. Unlike the radar gauge, the pressure gauge can sample changes in sea level as rapidly as required to measure individual short-period waves.

The crucial part of the tsunami system is getting the information from the gauges and back to warning centers as quickly as possible. Because the Indian and Atlantic Oceans are much smaller relative to the vast Pacific Ocean, tsunami travel times in the former are much shorter than in the Pacific. Thus, tsunami alert systems for the Indian and the Atlantic Oceans would need to be substantially faster than the PTWS. Because of this need for speed, POL developed a solution that uses the same satellite technology that broadcast journalists use to send live reports from war zones. The system provides an "always on" broadband connection almost anywhere on the Earth's surface, using a service known as Broadband Global Area Network (BGAN). POL and IOC have linked up with INMARSAT, which provides BGAN service. A modem hooked up to the sea-level gauge allows transmission of sea-level data to a satellite and the Web. A recently developed TWS network around the African continent (Figure 14.45) is expected to be expanded to a much wider network in the near future.

Chapter 15

Technological Challenges In Detecting Tsunamis

The 2004 Sumatra event proved that the coasts of several countries were at high risk for tsunamis and that a precise sea-level network situated offshore and along the coasts to provide continuous monitoring of tsunamis and other marine hazards was mandatory. To reduce the impact of tsunamis for coastal areas, help prevent false alarms, and mitigate the tsunami hazard, it is important to establish a reliable network of deep-ocean and island/coastal sea-level gauges/ tsunami recorders. Such gauges are also essential for providing the requisite data to aid navigation, storm-surge monitoring, climate change monitoring and related studies; constructing maps of tsunami risk along the coast (local tsunami-zoning); estimating resonant characteristics/properties of local topography (Rabinovich, 1997), and issuing detailed wave height and wave arrival information for incoming waves (Mofjeld et al., 1999a, b, 2000).

Providing real-time reporting capability to the network of sea-level stations would give the tsunami response team access to the latest sea-level information. The data and information gathered by the network would be helpful to the coastal communities that are at risk from tsunamis and sea-level changes in general. Meteorological data from a vast network of real-time reporting meteorological stations are equally important to identify whether the currently prevailing tsunami is of geophysical origin or meteorological origin (i.e., to distinguish between a "geophysical" tsunami and a "meteorological" tsunami). The high-resolution meteorological data gathered from such a network would also help the corresponding agencies and researchers to improve weather and extreme weather forecasts and to better understand the phenomena.

The primary purpose of the majority of sea-level gauges already in place all over the world is measuring relatively low-frequency processes, such as tides and seasonal to climatic sea level variations. When relatively short-period and high-amplitude oceanic waves such as tsunamis and storm surges occur, the measuring equipment needs to be capable of recording such sea-level vari- ability without waveform distortion or amplitude attenuation. Quite often, sea- level changes or disturbances can be difficult to measure. Most of the in situ sea-level measuring equipment depends critically on environmental conditions (for example, the presence or absence of water currents, vertical temperature

Tsunamis: Detection, Monitoring, and Early-Warning Technologies. DOI: 10.1016/B978-0-12-385053-9.10015-8

gradients, suspended sediment, floating objects, etc.), as well as the nature of the measurement site (e.g., open-sea conditions, inside protected harbors, constricted hydraulic inlets, etc.) and the available local infrastructure. Sea-level gauges are often installed at locations that are not optimal for recording tsunamis. Unlike old-fashioned gauges, which were designed exclusively for tidal measurements for navigational applications, the new sea-level gauges need to be "tsunami-enabled" to ensure preservation of the waveform and amplitude of the tsunami signal.

An important component of the assessment and, thus, mitigation of the effects of the tsunami impact is a computer-aided simulation. One of the most important issues of the tsunami modeling is gaining some insight into the basic characteristics of a tsunami source. Preservation of tsunami waveforms is vital in tsunami research because the source processes (e.g., slip distribution on fault planes) of the earthquakes are estimated using tsunami waveforms based on tsunami inversion techniques. For quantitative estimation of the source process, the observed tsunami waveforms are compared with computed ones to find the best fault model (Tanioka and Ioki, 2009). Recently, Namegaya and colleagues (2009) estimated the vertical seafloor deformation caused by the 2007 Niiga-taken Chuetsu-oki earthquake (M_{JMA} 6.8), which occurred in the coast of the Japan Sea, by using inversion of the tsunami waveforms. Voronina (2009) proposed a new method for reconstructing a tsunami source. In this approach, the inverse problem of inferring the initial sea displacement is considered as a usual ill-posed problem of the hydrodynamic inversion of tsunami sea-level records. The forward problem—the calculation of synthetic sea-level records from the initial water elevation field—is based on a linear shallow-water system of differential equations in the rectangular coordinates.

In terms of spatial coverage, the network should have gauge stations located at the closest distance from the region of highest tsunami risk. According to the modern requirement for operational tsunami service, the spatial distribution of sea-level gauges at seismically active coasts should be around 100–150 km. The location of the sea-level station should be such that it is not too far inland, and it should not be strongly shielded from the arriving tsunami waves. The station should be sufficiently sturdy and be of minimal size so it will not be easily damaged by the force of the surge (both geophysical and atmospheric). Frequently, sensors and gauge shelters are swept away or damaged by strong waves during storms (Rapatz and Murty, 1987).

In several cases, poor instrument quality (insufficient accuracy) has created tsunami spectra (both "normal" and "storm") at high frequencies (>0.02 Hz) that resembled "white noise" (Rabinovich and Stephenson, 2004). New digital sensors that can measure sea-level variations with high precision and fast sampling intervals are needed to detect and record even weak tsunamis. Uninterrupted time-series data from such sea-level stations would allow the identification and separation of seismically generated tsunami waves and atmospherically generated seiche oscillations, which is a key problem for disaster warning personnel and an important scientific problem (Rabinovich and Monserrat, 1996; Gonzalez et al., 2001).

An ideal sea-level gauge incorporated into the tsunami monitoring network should be efficient enough to detect and record even small tsunamis—both far-field (generated at far distances) and near-field (generated locally). In general, our ability to detect tsunami waves in a sea-level gauge record depends strongly on the signal-to-noise ratio (i.e., the ratio between the tsunami and background oscillations). For the 2004 Sumatra tsunami, this ratio in the Indian Ocean ranged from 40:1 to 20:1, so tsunami detection from the Indian Ocean records was straightforward (Merrifield et al., 2005; Rabinovich and Thomson, 2007). In contrast, the tsunami signal-to-noise ratio in the North Pacific and North Atlantic records ranged from 4:1 to 1:1, thereby making tsunami detection from these records more difficult (Rabinovich et al., 2006). This would mean that the gauge should have excellent accuracy and resolution and should also be free from electrical noise. If these criteria are achieved, even weak tsunami signals in the sea-level record will be clear and distinguishable from the topographi-cally induced background noise. The ultimate accuracy limit of onshore measurements in the tsunami frequency band will depend on the location selected, and therefore preliminary measurements are required to establish this limit (Zielinski and Saxena, 1983).

Another aspect that must be considered is the response of the sea-level detecting device to "normal tsunamis" and "meteo-tsunamis." Infragravity waves (IG-waves) generated by nonlinear interaction of wind-waves and tsunami-like, atmospherically induced seiches can be obstacles to identifying weak tsunamis (Rabinovich and Stephenson, 2004). In fact, one of the main purposes of any preliminary analysis of sea-level data (detiding, detrending, and low-pass and high-pass filtering) is to reduce the background noise level and thereby improve the tsunami-to-noise ratio (Candella et al., 2008). Gauges that could not measure the waves in the tsunami frequency band would record strongly distorted signals. Digital recordings of sea levels have significantly increased the accuracy of long-wave collections and made data processing much easier and more efficient.

Close sampling intervals are an important prerequisite to distinguishing and separating out the tsunami signal from the sea-level record and to minimizing underestimations of tsunami heights. Sahal and colleagues (2009) reported that the long sampling rate of the French sea-level gauge records (10 minutes) did not allow for a proper evaluation of the impact of the tsunami induced by the May 21, 2003, Boumerdes-Zemmouri [Algeria] earthquake [$M_w = 6.9$] on the French Mediterranean coasts, since the measured amplitudes were underestimated in contrast to the observed phenomena based on surveys. Their studies underlined the need for networked sea-level gauges with high-resolution records and short sampling rates. The periods of the tsunami signal at a given location are related, to some extent, on the topographic admittance of that location. The net effect of reducing data sampling intervals is a loss of important information about the waves in the tsunami frequency band and possibly serious aliasing problems (Emery and Thomson, 2001). For example, Rabinovich and Stephenson (2004) noted that the major topographic admittance peaks at most stations on the coast of British Columbia (Canada) are located at periods less than 12 minutes. If the sampling interval is set to 6 minutes, the Nyquist period will become 12 minutes. Thus, to adequately distinguish tsunami signals of 12 minutes and less, the sampling period must be less than 6 minutes at these locations. As noted earlier, periods of landslide-generated tsunamis can be as small as 2 minutes. It is obvious that long sampling intervals can lead to either totally masking the desired tsunami signal or a marked distortion of the wave properties. Waves are not well represented if the sampling interval is not sufficiently shorter than half the actual wave periods that are to be measured. The resulting aliasing can significantly affect the statistical results, especially maximum wave heights (Emery and Thomson, 2001). More specifically, the longer the sampling interval, the greater the attenuation of the recorded wave heights.

The influence of sampling intervals on recorded tsunami wave parameters depends on two primary factors: the frequency and energy content of the incoming waves and the frequency response of the observational site (Candella et al., 2008). The first is known fairly well for each event, while the second depends on the local topographic admittance function, which cannot be estimated without high-quality observations with short sampling intervals (Rabinovich and Stephenson, 2004) or detailed numerical modeling (Raichlen et al., 1983). In any case, closer (i.e., shorter) sampling intervals are preferred to record sea-level data for investigating tsunamis and other sea-level signals such as IG-waves. While in the past higher (i.e., longer) sampling intervals reduced some expenses and simplified data storage, today cost-effective memory modules with very large capacities (e.g., MMC cards and pen drives) are commonly available and effectively incorporated into sea-level gauges (Prabhudesai et al., 2008; 2010), so limiting the data sampling interval because of a logger's memory capacity is not justified.

An important technical requirement for realistic measurements of tsunamis is the linear response of the measurement device over the full range of wave periods of interest. It is well known that the tsunami wave periods could be in the range of approximately 2 minutes to an hour or more, depending on the source characteristics and the local topographic resonant characteristics. Thus, the tsunami measuring device should have a linear response in the wave-period range, even for the largest tsunami wave height expected. In many cases, tsunami heights were greatly underestimated because of these problems (Kulikov et al., 1996).

Maintaining accurate time stamps for the sampled dataset is another important criterion in sea-level measurements for tsunami source studies. With large errors on the time-stamp of the sea-level records, it would be difficult to decide the best fault model (Tanioka and Ioki, 2009). Incorporating a GPS-derived or an Internet-accessed time stamp in the sea-level data logger is a practical solution to this problem (Prabhudesai et al., 2010).

While an ideal network should be completely maintenance-free (which is an impractical proposition), a near-ideal network should consist of field stations that have a minimum amount of maintenance and maximum trouble-free operation. A display of the battery voltage at the field station (vital to the smooth operation of the gauge) will help minimize the downtime of the data logging (field) instruments. Because any single measuring/communication instrument can fail, redundancy should be a critical design feature to minimize total data loss when they are needed the most. Experience has shown that in the case of real-time reporting stations, missing data at the reception (i.e., monitoring) stations due to interruption of the communication network can be remedied by sending a sufficiently long "last few records" during each transmission sequence. With the effective application of such "temporal redundancy," the data collection process can continue without any interruptions. Although this is an effective method to minimize loss of data due to interruption of the communication network, the method is not technically the best solution because it creates an unnecessary waste of bandwidth. Implementation of an optimal method for achieving temporal redundancy without sacrificing bandwidth is desirable.

Due to infrequent occurrence of large tsunamis, the midocean measurement program should be designed for the detection of much more probable, small tsunamis. This requirement calls for high resolution in midocean measurements. Onshore measurements are usually associated with high background noise. However, this can be reduced by locating sensors in more suitable locations and/or using an array of sensors and attendant signal processors to take advantage of the spatial characteristics of the noise.

Real-time sea-level data reporting from offshore regions has an important benefit in terms of efficient short-term tsunami forecasts for coastal regions. For instance, a method offered by Korolev (2009a) requires only seismological information about earthquake epicenter coordinates and open sea-level data for tsunami forecasts in any specific point. This method is based on a fundamental property of symmetry of a Green function of a wave equation (known in acoustics and seismology as the *reciprocity principle*). It involves creating a transfer function that permits forecasting tsunami waveforms in any specific point using data from sea-level stations. Waveforms from auxiliary circular sources with centers that coincide with earthquake epicenters are computed for the desired forecast points on the coast as well as the offshore sea-level measurement stations. The transfer function is created as the ratio of spectra of waveforms in appropriate points. The advantage of this technique is that the time required for creating the transfer function is appreciably less than the time of tsunami propagation to a point, the data from which are used for the forecast. The convolution of transfer function with sea-level spectrum gives the forecasted tsunami waveform. The duration of the forecasted tsunami is equal to the duration of the sea-level data time-series. In that way, Korolev's method for short-term tsunami forecasting can work in real-time modes. The advantage of this method is that one can do away with the requirement of precreated databases of synthetic mareograms if a transfer function is created during an event. Thus, Korolev's method can be used in areas for which databases of synthetic mareograms are not suitable. Therefore, the method can be used by both regional and local tsunami warning services if they can receive sea-level information in real-time mode.

The challenges of supporting operational tsunami detection include installation and maintenance of reliable and accurate coastal and offshore tsunami observing systems, complexity of real-time tsunami forecasting, and the impact of communication technology and data management. These and the preceding challenges indicated are largely met in the real-time Deep-Ocean Assessment and Reporting of Tsunami (DART) system implemented by NOAA. In this system, 1-minute averaged sea-level data are acquired, which improves the reliability and timeliness of warnings to the coastal communities. In addition to the real-time data, the DART bottom-pressure recorders (BPR) provide high-frequency data stored internally on a flash card for later retrieval. These high-resolution continuously recorded data are integrated over 15-second intervals (Stroker et al., 2009). Operational and postevent high-resolution long-term records are invaluable and are directly usable for many applications, including investigations into small tsunamis, the study of tsunami physics and propagation across ocean basins, global tide and inundation modeling, tidal corrections to hydrographic data, comparisons of bottom pressure tides with altimeter tides, and in tsunami and climate research, which lead to the development of additional operational and climate applications (Stroker et al., 2009; Allen et al., 2009).

Chapter 16

Sea-Level Measurements From Coasts and Islands

In the distant past, tsunami measurements were made solely on the basis of the location of watermarks on buildings and trees and eyewitness accounts. However, this situation changed slowly and steadily with the development of sea-level measuring instruments, initially known as tide gauges and known today as sea-level gauges. Consequently, most of the worldwide sea-level measurements available at present have been obtained from a variety of conventional and modern sea-level gauges. While the former were primarily designed for navigational and hydrographic survey applications, the latter came into existence primarily as an effort toward improved measurements to observe the global climate change signal more accurately. This chapter addresses the rapid progress made in the technology of sea-level measurements, as well as its reporting in real time or near real time.

Over the years, many devices have emerged for measurements of sea-level oscillations. The devices of the old generation were known as tide gauges because they were primarily meant for measuring tides. Subsequently, it was realized that tides are not the only factor that causes oscillations and variability in sea levels. Other factors such as swells, seiches, Kelvin waves, storm surges, meteorological and geophysical tsunamis, and remote forcing from far-off locations all contributed significantly to sea-level variability; thus, these devices are now aptly termed sea-level gauges.

Tsunamis: Detection, Monitoring, and Early-Warning Technologies. DOI: 10.1016/B978-0-12-385053-9.10016-X

While global observations are needed to assimilate data sets into numerical models to gain insight on the principal factors that control the transport of tsunami energy and its distribution and propagation characteristics throughout the world oceans, the requisite high-quality observations were not available until recently. Thus, the old-generation gauges were not of much use to the study of tsunamis. For example, the first known global tsunami, which is associated with the Krakatau volcano explosion of August 27, 1883, was recorded by 35 gauges, including those in La Havre (France), Kodiak Island (Alaska), and San Francisco (Symons, 1888; Pelinovsky et al., 2005). However, the quality of the analog records available for the 1883 tsunami did not allow for a thorough quantitative analysis of the tsunami wave properties and global propagation patterns (Rabinovich and Thomson, 2006). This was the situation in the case of the 1960 Chilean tsunami as well, where the tsunami was measured by about 250 tide gauges in the Pacific Ocean (Berkman and Symons, 1960) and very strong tsunami waves (over 3–4 m) were observed at many far-field sites (10,000 km from the source area). However, all these were analog records in which the crests and troughs of the largest waves (i.e., those most important for tsunami analysis) were frequently clipped off because the instruments of the day were not designed to measure such strong oscillations (Rabinovich and Thomson, 2006). Although tsunamis were recorded by several gauges along the coasts of Thailand and Indonesia, most gauges were archaic analog devices that were not designed to measure tsunamis. For this reason, the derived wave characteristics (especially arrival times) are not fully reliable (Rabinovich and Thomson, 2006).

Fortunately, frequent modernization in tune with advancement in technology resulted in progressively improved instrumentation for such measurements. In this effort, the Global Sea Level Observing System (GLOSS) of the UNESCO's Intergovernmental Oceanographic Commission (IOC) has played a key role, with the primary intention of obtaining high-quality sea-level measurements for studies of global climate change. The new-generation gauges have proved to be capable of recording even very weak tsunami signals, as seen from the recent experience (Woodworth et al., 2005; Prabhudesai et al., 2008). To put the historical progress achieved in sea-level instrumentation development scenario in perspective, it is appropriate to address all the devices, which have been used at one time or the other in all parts of the world. There has also been an evolution with the electronic circuits and data loggers that monitor and record the physical changes detected by the sensing technology. These include basic temperature compensation or more sophisticated microprocessor-controlled sensors that manage temperature-compensation, linearization, and so forth. In the following section, the international conventions used in sea-level measurements and various methodologies used for measurement of sea-level oscillations, together with the advantages and disadvantages of the different methods, are addressed so the technological advances that have taken place in this area over the years can be fully appreciated.

16.1. CHART DATUM

In coastal and island waters, *sea level* is defined as the distance of the sea surface above a recognized reference datum known as chart datum (CD). Thus, coastal and island sea-level elevations are always expressed with reference to CD. By international agreement, the level used as CD should be just low enough so low waters do not go far below it. The CD is, therefore, a safe low-water level in order to maintain the minimum depth useful for guiding a vessel safely to port.

The device that carries out sea-level measurements with reference to CD has been conventionally called a *tide gauge* rather than *sea-level gauge* primarily because at most coastal locations the astronomically induced and topographically modified tide is the largest part of the sea-level oscillations for most of the time. The observed sea level can be viewed as oscillating about a mean sea level (MSL). These low-frequency oscillations in sea level are measured relative to the land, to which the benchmarks are permanently attached. In the past, measurements of low-frequency sea-level oscillations were routinely made only from coastal and estuarine waters. In recent years, such measurements from beyond the coast began gaining greater importance.

Measurements of low-frequency oscillations in coastal and island sea levels have a long history. Sea levels have been measured almost as long as human development. For all nations the "datum of national leveling network" has been estimated from long-term sea-level measurements. The MSL is an important survey parameter because it is used as a reference level for heights.

The technologies used over the years for automated measurements of sea-level oscillations include stilling-well-based floats and pulleys attached to chart recorders or digital encoders, subsurface mounted pressure sensors and bubbler gauges, acoustic reflection devices, downward-looking aerial microwave reflection devices, microwave interferometers, satellite-borne altimeters, and methods that used differential global positioning technology. In most of these measurements, tide staff readings have been used for setting the initial value for the recording devices and for periodic visual checking of the instrument readings. Tide staffs of various kinds and recommended ways of their use are explained in the IOC manuals and guides.

16.2. FLOAT-DRIVEN GAUGES

The oldest standard recording instrument for automatic tracing of low-frequency sea-level oscillations on graph paper is a mechanical device known as a *float-driven gauge,* or simply *float gauge.* In this device, sea level is detected by a "float" resting on the water surface that is confined within a vertical pipe known as "stilling-well." The "well" is usually attached to a purpose-built in-water structure established in navigational ports. The stilling-well, which is set in the water and connected to the sea by an orifice

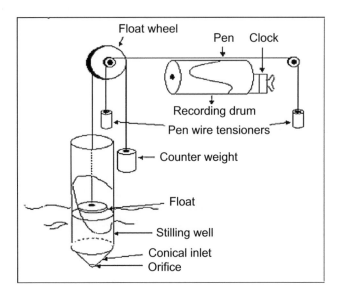

FIGURE 16.1   A conventional float-driven gauge.

(Figure 16.1), was conventionally used in most of the places worldwide for sea-level measurements. However, in some countries (e.g., Japan) tide-wells of a different structure are also used (Satake et al., 1988), wherein the stilling-well is generally dug into a wharf and hydraulically connected to the sea through an intake pipe (Figure 16.2). Various configurations of such wells and their intake pipe or pipes have been in common use in Japan (Figure 16.3). The pipe is narrowed down at the outer part to reduce unnecessary surf beats during severe winter weather along the Japan Sea coast. At times, a damper is also inserted in the pipe, or a net or other obstacle is attached to the exit point to further reduce the surf beats problem.

An important component in this gauge is a cylindrical or conical float, which rests on the surface of the water within a large cylindrical pipe (7 to 15 m long) of 60 to 80 cm diameter with a tapered bottom end, known as a stilling-well. The well is often attached to a vertical structure or a concrete wall that extends well below the lowest level to be measured. A hole located at the bottom end, whose diameter is 0.1 times that of the pipe, is known as an *orifice*. The orifice to well diameter ratio of 0.1 provides an area ratio of 0.01. By convention, the orifice is usually located at a point at least 1 m below CD and at least 1 m above the seafloor, after allowing for wave-induced fluctuations in sea surface elevation. This orifice forms a low-pass hydraulic filter that allows only the low-frequency oscillations within the well and suppresses the wave-induced high-frequency oscillations. By the action of this hydraulic filter, the float rests on an approximately quiet water surface. For most wave regimes this ratio was assumed to give satisfactory results. However, in regions where the waves are

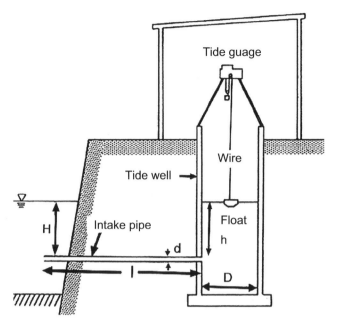

FIGURE 16.2 Tide-well dug into a wharf and hydraulically connected to the sea through an intake pipe. *Source: Satake et al., 1988; Journal of Marine Research; Reproduced with kind permission from Journal of Marine Research, Yale University.*

very active, the area ratio was sometimes reduced to 0.003. Copper material, as a result of its ability to inhibit marine growth (Huguenin and Ansuini, 1975), is often used at the orifice region to prevent its fouling by marine growth and the resulting blocking of the orifice.

A float is used to detect the dampened motion inside the well, thus the name *float-driven* gauge. Usually the float is cylindrical or conical in shape, and its diameter is half that of the well. The float, which is large enough to give sufficient force to overcome friction in the recording system, is weighted at the bottom so it floats upright and is balanced by a suitable mechanism that keeps its top clear of the water. This arrangement allows the float to rise and fall with the sea-level oscillations, with a much-reduced disturbance from the high-frequency wind-wave action. The float, which is attached to a counterweight via a taut flexible wire that passes over a pulley, energizes a pen via a set of gears. Alternatively, the float can also be linked to a sprocket wheel via a precisely perforated metallic band. A steady spring provides torque to the band storage wheel equal and opposite to that provided by the weight of the unwound band and the residual weight of the partially immersed float. As the float rises and falls with the tide, the pen moves back and forth on a sheet of paper rolled over a cylindrical drum. A mechanical-clockwork drives this roll of recorder paper at a constant rate. While the clockwork gives the recorder

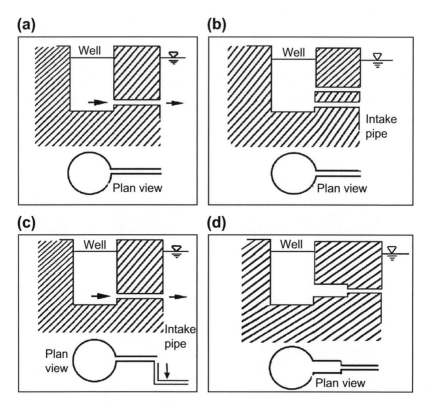

FIGURE 16.3 Various configurations of tide-wells that are dug into wharfs and hydraulically connected to the sea through intake pipes. *Source: In part from Satake et al., 1988; Journal of Marine Research; Reproduced with kind permission from Journal of Marine Research, Yale University.*

paper a continuous motion in the X-direction (time-axis), the float drives the pen over the paper in the Y-direction (sea-level-elevation axis), thereby continuously tracing the sea-level oscillations on the graph paper repetitive. The first such self-recording gauge that we know of, which was designed by Palmer, began operating in the Thames estuary in the midnineteenth century (Palmer, 1831). Photographs of two popular recording devices of float-driven gauges are shown in Figure 16.4. The sea-level record plotted on the paper came to be known as a *marigram*. Because the occurrences of high and low water associated with the tidal rhythm progressively shift every day by approximately 1 hour, the same paper roll is used to plot the tides for several consecutive days. Continuous recordings of the sea level ensures the availability of time-series data on changes in sea level over long periods of time. A typical marigram from the Bombay Harbor sea-level station in India, one of the oldest stations in the world, is shown in Figure 16.5. Here, the original paper

FIGURE 16.4 Two popular recording devices of float-driven gauges. (1) Horizontally positioned spool with mechanical clock and (2) tidal graph being traced. (3) Vertically positioned spool that gives one-month reading. *Source: Clock and tidal graph photos courtesy of Proudman Oceanographic Laboratory, United Kingdom.*

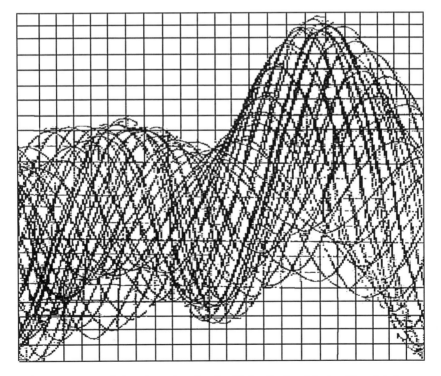

FIGURE 16.5 A typical marigram from Bombay Harbor Sea Level Station. The original paper chart is 160 cm × 90 cm in size. *Source: Courtesy of Bombay Port Trust, Mumbai, India.*

chart is 160 cm × 90 cm, and the automated measurements using the float-driven gauge are related to permanent benchmarks on land.

Analysis of data from marigrams is cumbersome because of the time required for manual smoothing (i.e., manually tracing a smooth thin line through approximately the mean of the high-frequency fluctuations on the tidal graph), outlier removal, and digitization. Despite these difficulties, such records proved to be valuable in the extraction of the December 26, 2004, Sumatra tsunami signal from several locations all over the world.

16.2.1. Shaft Encoders and Microprocessor-Based Loggers

The trend of modernization has influenced the age-old float-driven gauge as well. The modernization program sought to complement or replace the conventional mechanical recording portion and the marigram with an electronic counterpart known as a *shaft encoder*. Most shaft encoders are designed to mount directly on the shaft of a pulley. This scheme eliminates the need for external gears and chain interfaces during installation. In a shaft encoder, the angular motions of the float pulley, resulting from the vertical oscillation of the water-level within the stilling-well, are converted into electrical signals. At preset time intervals, the signals are output to a data logger for storage of the time-tagged data in a convenient medium. Float-operated shaft encoders with integral data loggers (different makes) are shown in Figure 16.6. The float-operated Thalimedes Shaft Encoder with an integral data logger, in combination with conventional mechanical chart recorders of different makes (http://www.ott.com/web/ott_de.nsf/id/pa_thalimedes_e.html), is a cost effective upgrade from a mechanical system to digital technology, and it is designed for continuous, unattended monitoring of water-level. The inbuilt buffered data logger offers many features such as event controlled recording, 1 minute … 24 hour storage interval etc. Data downloading or configuration can easily be done directly via IBM-compatible notebooks, palmtops or with OTT's rugged VOTA Multifunctional Field Unit. RS232 output / SDI12 interface are provided for remote control and data transmission via serial modem (land line), GSM (cellular), and so on. This allows communication worldwide.

Two classes of shaft encoders are available: those that provide analog signal output and those that provide digital signal output. While a circular potentiometric encoder provides an analog signal corresponding to the circular motion of the pulley, an optical/magnetic encoder provides a train of pulses (i.e., digital output) as the shaft rotates. The pulse train is generated with the use of a code-wheel or a code-strip. A combination of a rotating magnet and an integrated Hall-Effect integrated circuit (IC) provides a digital signal directly. While the code-wheel provides a rotary position, the code-strip provides a linear position.

A typical code-wheel developed by Hewlett Packard for use with a shaft encoder functions in conjunction with an LED (light emitting diode) based source/detector module that incorporates a lens in front of the LED. This

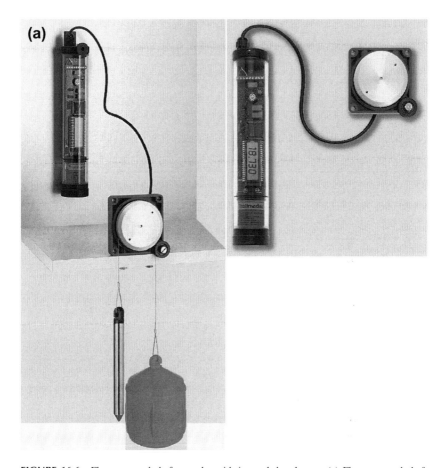

FIGURE 16.6 Float-operated shaft encoder with integral data logger. (a) Float-operated shaft encoder with integral data logger. (b) Float-operated shaft encoder with integral data logger in combination with a mechanical chart recorder (in an open housing). (c) Float-operated shaft encoder with integral data logger in combination with a mechanical chart recorder (in a closed housing). *Source: Courtesy of Helmut Hohenstein, OTT Hydromet Gmbh; Reproduced with kind permission of OTT Hydromet GmbH, P.O. Box: 2140 / 87411, Kempten / Germany.*

module contains a single LED as the light source. The generated light is collimated into a parallel beam by means of a single lens that is located directly over the LED. Opposite the emitter is an integrated photodetector array circuit. This IC consists of multiple sets of photodetectors and the signal processing circuitry necessary to produce the digital waveforms. The lens produces a highly collimated light beam. While the float moves in synchrony with the water-level oscillations in the well, the code-wheel/code-strip moves through a narrow passage located between the emitter and the detector. This motion causes the light beam to be interrupted by the pattern of spaces and bars on the code-wheel/code-strip. The photodiodes, which detect these interruptions, are

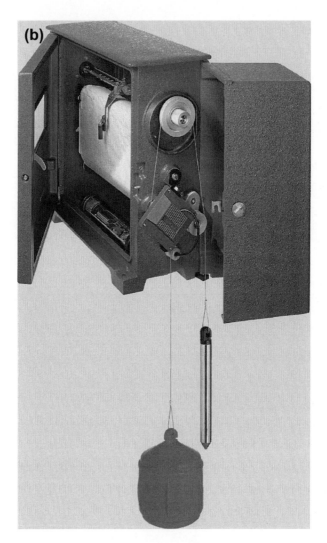

FIGURE 16.6 Continued

arranged in a pattern that corresponds to the radius and count-density of the code-wheel/code-strip. These detectors are also spaced such that a light-period on one pair of detectors corresponds to a dark-period on the adjacent pair of detectors. The photodiode outputs are fed through the signal processing circuitry. Two comparators receive these signals and produce the final outputs for channels A and B. As a result of this integrated phasing technique, the digital output of channel A is in quadrature (i.e., 90 degrees out of phase) with channel B. Thus, coupled with the code-wheel, the light emitter/detector module translates rotary motion into a two-channel digital output. The

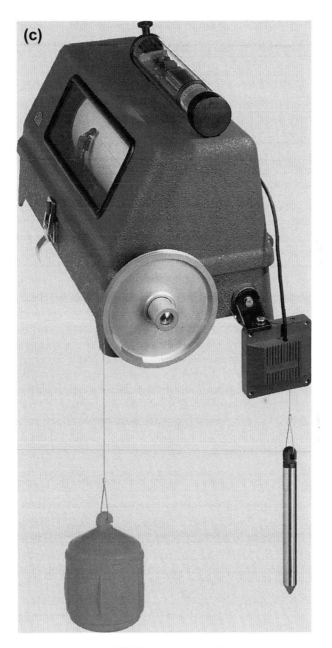

FIGURE 16.6 Continued

clockwise/counterclockwise direction of rotation of the code-wheel, which corresponds to the rising and falling phases of the water levels in the tide-well, is determined based on the phase of channel A relative to that of channel B. In the case of a code-strip, the detector circuitry translates linear motion into a digital output.

A microprocessor-based instrumentation attached to the encoder converts the angular motion of the pulley to the corresponding linear motion of the float that is attached to the pulley. The advantage with the shaft-encoder-based electronic recording system is considered to be elimination of errors that are brought about when data are read or transferred manually. The float-operated shaft encoder is considered an ideal device for cost-effective modernization of existing conventional float-driven sea-level stations. This is because the device can be easily combined with the existing conventional mechanical chart recorders, independent of the installation conditions.

16.2.2. Calibrating Float-Driven Gauges

Float-driven gauge installations require a calibration device such as a well-dipper or a water-sensing "plumb bob" that is mounted to a graduated tape. The well-dipper is just a calibrated rule with an electrical contact, and it is routinely used in many sea-level stations to set the initial water level during replacement of the paper chart. Regular visits to the sea-level stations are required to ensure that the sea-level record is accurate as verified with the well-dipper/plumb bob device. Recent innovations in the technology of sea-level monitoring have produced an automated version of the plumb bob calibration device that is combined with a shaft encoder. In this device (Figure 16.7) the "human" component required for the operation of the plumb bob is replaced by a DC motor, which drives a reel-mounted punched and graduated tape that engages a multipin pulley mounted on the shaft encoder. The shaft encoder maximizes its sampling rate as the DC motor lowers the plumb bob toward the water surface. The motor control algorithm ensures that the device immobilizes for several seconds immediately upon the touch of the plumb bob with the water surface in the tide-well. This delay allows the user to take a visual reading of the tape in cases where the measurement was initiated manually. Thus, in this device the DC motor performs the traditional human function of lowering the plumb bob and taking the tape readings periodically with a resolution claimed to be better than a millimeter.

16.2.3. Drawbacks of Conventional Tide-Wells

Although conventional tide-well (stilling-well) systems are robust and relatively simple to operate, they are known to have a number of disadvantages, such as large size and difficulty in installing. Until recently, inaccuracies in sea-level measurements were not a serious consideration, since the data records

FIGURE 16.7   A motor-controlled plumb bob calibration device for use with a tide-well. *Source: Technical Data Sheet of AMASS Data Technologies Inc.; Reproduced with kind permission from AMASS Data Technologies Inc.*

were primarily used for navigational and hydrographical survey applications, where an accuracy of 10 cm was considered sufficient. A closer examination revealed that the traditionally used stilling-well systems have a number of disadvantages. For example, they suffer from inherent problems, including slow and nonlinear responses to large-amplitude short-period waves (Cross, 1968; Noye, 1974; Braddock, 1980; Shih and Porter, 1981; Loomis, 1983), such as those associated with tsunamis and storm surges, as a result of which they respond differently to tides, tsunamis, storm surges, and so on. These problems differ from station to station, depending on the geometry of the stilling-well's orifice and the intake pipe. For example, Satake and colleagues (1988) examined the responses of stilling-well-based sea-level gauges (of the types used in Japan, as shown in Figures 16.2 and 16.3 by in situ measurements at 40 stations in northern Japan. In their experiments, the recovery times for a 1-m water-level difference were found to vary from station to station and range from 65 to 1300 s. The recovery times were also computed hydraulically on the basis of the geometry/structure of the stilling-well system. The ratio of the observed time to the computed one was found to be in the range between 1 and 10, which is attributable to environmentally induced changes within the tide-well system. While the original geometry itself is problematic, attachment of shells or deposition of sediments can clog the intake pipe and alter the effective diameter or surface roughness of the intake pipe, thereby further deteriorating the response characteristics. Temporal changes in responses resulting from such changes in the geometry of the intake pipe were evident in repeated measurements by Satake and colleagues (1988), in which the recovery time roughly doubled in two and a half years. Because of these effects, tsunami waveforms obtained from these tide gauge records in the case of the 1983 Japan Sea earthquake were found to have been significantly distorted (Figure 16.8).

Because in situ sea-level records obtained from tide gauges are usually the only instrumental data of tsunamis (except satellite altimetry data), and the tsunami waveforms recorded by tide gauges are used to quantify tsunamis or to study earthquake source processes, contamination of tsunami signals by the nonlinear responses of tide gauges is of serious concern in tsunami research. It has been suggested that a more suitable tide gauge system for tsunami measurement would be characterized by a short recovery time and a digital

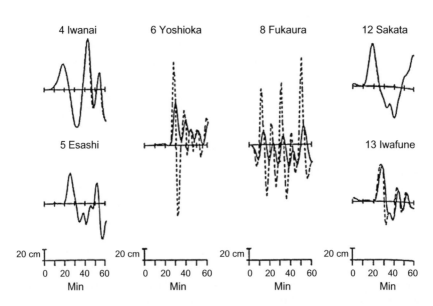

FIGURE 16.8 Significantly distorted tsunami waveforms obtained from stilling-well gauge records in the case of the 1983 Japan Sea earthquake. Waveform distortion arose from nonlinear responses of stilling-wells. Solid traces are the recorded tsunamis, and the dashed traces are the corrected tsunami waveforms after accounting for the gauge response. *Source: Satake et al., 1988; Journal of Marine Research; Reproduced with kind permission from Journal of Marine Research, Yale University.*

recording system, rather than a paper chart. A tsunami gauge, instead of a tide gauge (Curtis, 1986), is desirable to effectively handle problems specific to tsunamis, such as quantification of tsunamis or realistic operation of tsunami warning systems.

Accurately measured tsunami waveforms are important for investigating earthquake source problems, since a tsunami has a potential advantage for studying seismic sources because the propagation effects can be more accurately evaluated than with seismic waves (Satake, 1987). It is worth noting that none of the presently available seismic models (e.g., Global CMT (Harvard), Japan Meteorological Agency (JMA), USGS, NEIC (National Earthquake Information Center), and GSRAS (Geophysical Service, Russian Academy of Sciences)) define the spatial extent of the source area, which is normally identified on the basis of aftershock distributions (Rabinovich et al., 2008). The contradictions in estimated seismic parameters, including seismic moments and source sizes, prompted scientists to use inversion of the "observed tsunami waveforms" to reconstruct the initial tsunami sources (Fujii and Satake, 2007; Tanioka et al., 2008). Thus, using sea-level gauges that do not contaminate the tsunami waveforms is vital in the field of tsunami research.

Also, examination of past tsunami records obtained from float-driven gauges indicated that stilling-wells caused large underestimation of the tsunami signal (by a factor of 2 to 5) relative to those inferred from inland inundations (Titov and Synolakis, 1997; Merrifield et al., 2005). Clear evidence of large underestimations of tsunami signals by float-driven gauges was obtained from the December 26, 2004, tsunami recorded from Thailand. Intercomparison of simultaneous measurements of sea levels from the float gauge and a depth gauge (echo-sounder) of the Belgian yacht *Mercator* that was anchored about 1.6 km off the Phuket coast indicated that the tsunami height measurements by the former were less than one-third the latter. Seeling (1977) described a unique design of a stilling-well where water-level oscillations inside the well are linearly related to those outside the well. However, this design was too sensitive to marine growth and proved to be too demanding, so the design was not accepted (Lennon and Mitchell, 1992). Another drawback associated with a conventional stilling-well is a significantly large drawdown error induced by flow and wave motions at the measurement site (Shih and Baer, 1991). These errors arise because the stilling-well is basically a pressure-balancing system that balances the pressure of the water column above the inlet (inside the well and external to it). Any dynamic process, which perturbs the pressure-field at the inlet of the well, will cause a corresponding difference in the mean water level inside the well, thereby introducing an error in the sea-level measurement. The relative height of the water level inside the well versus the desired sea level outside the well was shown, through empirical studies, to be a function of the design of the well. The dynamical contamination of stilling-wells could be reduced by the use of appropriate front ends at the inlet of the well (see Shih and Baer, 1991).

Apart from the preceding dynamic effects, local water density changes have a profound influence on the accuracy of sea-level measurements using stilling-wells (Lennon, 1970). For example, the well acts as a freshwater trap in regions where the salinity varies in correlation to the tide (e.g., at the mouth of an estuary). In such regions, it is almost impossible to have equal density profiles inside and outside the well and generally gives rise to overestimation in the measurement of the sea level. A three-year study by Joseph and colleagues (2002) of water density anomalies in a conventional tide-well at Marmugao, Goa, India, indicates that the average water density in the well is consistently lower than that of the external ambient waters. The tide-well is situated at the mouth of the Zuari estuary, and anomalies were observed at all periods except during peak summer months and the onset of the summer monsoon season. These anomalies led to considerable overestimation of sea levels by tide-well-based gauges. The limitation of the conventional tide-well could be minimized by incorporating arrays of perforations on its entire submerged portion. Density measurements during different seasons indicate that these perforations give rise to good mixing between the waters inside and outside the tide-well, thereby improving the accuracy in sea-level measurement.

Finally, in the context of tsunami measurement, several instances were reported in which float-driven gauges clipped off the crests and troughs of the largest waves (i.e., those most important for tsunami analysis) because of the inability of these gauges to measure such strong vertical oscillations (Berkman and Symons, 1960; Takahashi and Hatori, 1961; Rabinovich et al., 2006).

16.3. ELECTRIC STEP GAUGES

An electric step gauge is a modern version of a combination of the age-old tide staff (Figure 16.9a) used for visual reading of sea-level with reference to CD, and the resistance-wire gauge (Ayers and Cretzler, 1963) that provides an electric output. In the electric step gauge, a row of horizontal electrodes mounted on an electric insulator staff that is vertically driven into the seabed is used to detect the position of the water surface (Figure 16.9 (b)). The electric step gauge works on the principle that if an electrode at the end of an insulated wire is submerged in water, the electrical resistance to the water is significantly lower than that when the electrode is above the water surface. This allows detection of the water surface with reference to the level of the highest submerged electrode while scanning the electrodes from the top of the staff to its bottom. In this device, the electrodes, preferably made of manganese bronze, are scanned at a fast rate (at least twice every second) and averaged in real time to yield tidal height records free from the disturbances of wind-waves. Though the electrodes are usually 5 cm apart, the averaging process permits measurements of tidal heights with a resolution better than 0.5 cm in most applications (Botma and Leenhouts, 1993).

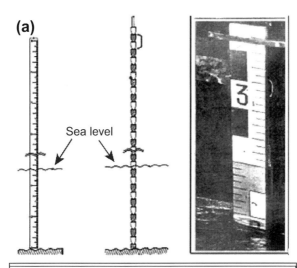

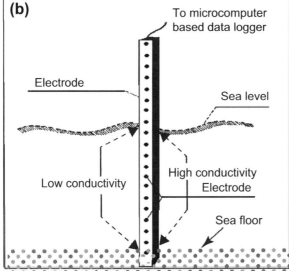

FIGURE 16.9 (a) Different types of tide staffs that provide visual indication of coastal sea-levels relative to a reference datum; (b) An electric step gauge that provides automated measurements of sea-level relative to a reference datum.

An advantage with the electric step gauge is that it is installable on simple mechanical structures and does not require hydraulic damping devices for waves. Because the measurements are digital, it is more or less free from drift effects and temperature influences.

A source of error associated with the electric step gauge, however, is water piling at the gauge staff due to flow and waves. Also, a layer of thick oil on the water surface, a possibility in oil berths, can be a serious impediment to the reliable functioning of this gauge, since it inhibits proper electrical contact of the electrodes with seawater.

16.4. AIR-ACOUSTIC GAUGES

Air-acoustics for sea-level measurements, which emerged in the early 1990s, are based on the use of sound wave transmission through air and subsequent reception after reflection from the air-water interface. An air-acoustic gauge estimates sea-level using time-domain reflectometry technique. In this method, the vertical distance, D, between the sea surface and the face of a piezoelectric transducer that is positioned in the air is estimated by measuring the average time elapsed between transmission and reception of a series of acoustic pulses. The value of D is estimated from the measured elapsed time and the known or assumed value of the velocity of sound in air. The acoustic transducer is rigidly mounted at a known datum above the highest expected water level. Sea-level elevation is estimated by subtracting the measured distance from this datum level, which is usually the CD.

In operation, a piezoelectric transducer transmits a series of acoustic pulses vertically downward. The pulses are reflected by the sea surface and are returned to the transducer, which converts them into analogous electrical pulses. The electronic circuitry associated with the transducer determines the average time interval, t, elapsed between emission and reception of the pulses. The value of D is given by the expression $D = (ct/2)$, where c is the velocity of sound in the air. Because the transducer is located above the sea surface, the sea-level elevation, H_t, from the CD is given by the relationship $H_t = (H - D)$, where H is the altitude of the transducer face above the CD.

An important consideration in air-acoustic sea-level gauges is the sensitivity of sound velocity to atmospheric pressure, ambient temperature, and moisture. Sound velocity c in air (m/s) is given by Dupuy (1992):

$$c = 331.2\left[1 + 0.97\left(\frac{u}{p}\right) + 0.0019\,t\right] \tag{16.1}$$

where u is relative moisture, p is the atmospheric pressure (mb), and t is the air temperature (°C). Sensitivity of c to ambient temperature is more significant than to atmospheric pressure and moisture. Application of an air-acoustic gauge for sea-level measurement was restricted until corrective measures were devised to compensate for errors due to temperature sensitivity.

Errors in sea-level measurement arising from sound velocity variations in the air can be corrected by measuring the temperature profile along the acoustic path and the use of a suitable calibration curve. The system can also be made self-calibrating by including a small acoustic reflector such as a perforation or an insert in the acoustic tube at a known distance in the air column. By relating the time of flight from the sea surface to that from the fixed reflector, a self-calibration and direct compensation for variations in sound velocity between the acoustic transducer and the fixed reflector can be achieved. The reflector, which functions as the calibration reference point, acts as a discontinuity that causes a decrease in acoustic impedance as the acoustic pulse passes it, resulting in another reflection, which propagates back toward the transducer. In this setup, the sensor is regarded as the acoustic head plus its individual calibration tube. The common calibration accuracy of such a self-calibration system has been claimed to be 0.025%. The self-calibration technique just described still does not account for variations in sound velocity between the fixed reflector and the sea surface. Thermal expansion of the calibration tube also introduces uncertainty because the stability of calibration relies on the stability of the calibration length in an operational mode.

16.4.1. Unguided Air-Acoustic Gauges

Measurements of sea-level elevation can be made using unguided signals from an acoustic transducer mounted vertically above the sea surface (Figure 16.10a). The unguided acoustic sea-level gauge consists of an acoustic transducer for generating and receiving acoustic signals, a suitable means for calibration, fixtures to erect the system at the site, and an electronic circuit for generation, processing, and displaying the sea-level data. A digital-to-analog converter or a purely digital circuitry is used for generation of electrical signals of suitable pulse width and frequency. A switching circuit isolates the transmission and reception of signals, and a power amplifier amplifies the received signals.

An unguided air-acoustic gauge can present several problems. One is that the amplitude of acoustic energy reflected back to the acoustic transducer varies considerably with changes in the angle of the liquid surface relative to the axis of the acoustic beam, such as that caused by waves and swells. At angles exceeding about 20 degrees from the vertical, the signal return from the water surface can be below the lowest signal-to-noise ratio that is acceptable for reliable measurement of water level. This problem arises primarily from

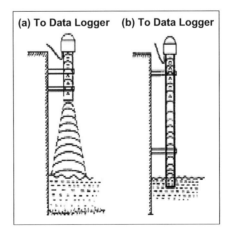

FIGURE 16.10 Air-acoustic gauge using (a) unguided signal transmission and reception and (b) guided signal transmission and reception.

scattering and absorption of the acoustic energy during perturbation of the water surface, thereby causing the return signal to drop below a usable level. In practice, the maximum working angle for most of the presently available acoustic probes is about 30 degrees. Even at certain angles that are considerably less than this, there can be blind spots where the return signal drops to a very low level (Sinclair, 1994). Further, an unguided acoustic signal path may suffer from missing pulse returns from a wave-laden sea. Floating debris can also interrupt the acoustic path.

16.4.2. Guided Air-Acoustic Gauges

Unguided and guided air-acoustic gauges have many features in common. The component that differentiates the latter from the former is a long and narrow sounding tube that is protected by a stilling-well. Several problems with the unguided air-acoustic gauge can be circumvented by guiding the acoustic pulse along this sounding tube (Figure 16.10b). Flow- and wave-induced errors arising from the use of a sounding tube are reduced by attaching a suitable hydromechanical front end at the lower end of its protective well.

The guided air-acoustic gauge consists of an acoustic transducer or a pair of acoustic transducers for generation and reception of acoustic signals, a mechanical system that consists of an acoustic signal guide tube, some kind of a means for calibration, stilling-well and fixtures to erect the system at the site, and an electronic circuit for generation, processing, and displaying the sea-level data. A stilling-well is used merely to protect the delicate acoustic signal guide tube against the impact of ocean waves and currents and to minimize the effects of water currents and waves on the water level inside the guiding tube. The entire assembly has provisions for fixing the gauge to a suitable structure for measuring sea levels. The acoustic transducer/transducer-pair is positioned at the upper end of the acoustic signal guide tube with its lower open end immersed below the lowest tide level.

The sounding/guiding tube serves several purposes. It helps to isolate the transducer from other sources of interference and confines the ultrasonic beam so it is directed vertically toward the region of the water surface directly below the acoustic transducer. Furthermore, because the sounding tube is positioned inside a stilling-well (which is a low-pass hydraulic filter that allows only the low-frequency oscillations within the well and suppresses the wave-induced high-frequency oscillations), the water surface inside the sounding tube is substantially damped of waves. In addition, the guide tube prevents bubbles and minimizes agitation effects. The sounding tube helps to increase the range of operational tilt angles over which the transducer can be used. Surprisingly, this tube has been found to cause a considerable increase in the amount of energy received by the transducer from reflections off the water surface.

Another advantage arising from the use of the tube is that it is easy to provide a reference height by mounting some form of reflector at a known height within the tube. In this way, the transducer will receive one reflection from the liquid surface and another one from the reference reflector, against which the vertical height of the acoustic transducer from the water surface can be calibrated. This, in principle, enables the ultrasonic gauging system to compensate for temperature variations that can affect the speed of acoustic propagation along the sounding tube.

Sound velocity variations arising from vertical differential temperature variations in that portion of the sounding tube that lies between the calibration reference reflector and the water surface continue to be a source of error, which is not accounted for in the preceding intrinsically automated self-calibration procedure using a single reflector/calibrator near the top end of the tube. Such an acoustic sea-level gauge is known to exhibit errors in the measurement of sea levels (Vassie et al., 1992). These errors arise from temperature-gradient-induced anomalies of sound velocity along the entire length of the air column in the sounding tube of the acoustic gauge. Intercomparison measurements by Joseph and colleagues (1997) of sea levels against a temperature-compensated pressure gauge, and air temperature profile measurements in the protective well of the air-acoustic gauge using a chain of temperature probes also indicated a strong temperature-gradient-induced noise in the sea-level data measured by the guided air-acoustic gauge (Figure 16.11).

Bower and colleagues (1992), Sinclair (1994), and Pathak and Ramadass (2002) described methods for temperature compensation of guided air-acoustic gauges through the use of a multiplicity of predetermined acoustic reflectors at several points along the sounding tube to provide for full-length calibration. Pathak and Ramadass (2002) described a scheme (Figure 16.12) wherein the

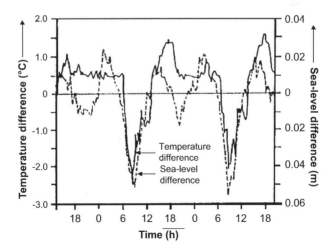

FIGURE 16.11 Typical temperature dependence of sea-level measurements from a self-calibrating guided air-acoustic gauge (NGWLMS) deployed in the Zuari estuary, Goa, from June 6 to June 8, 1997. *Source: Joseph et al., 1997.*

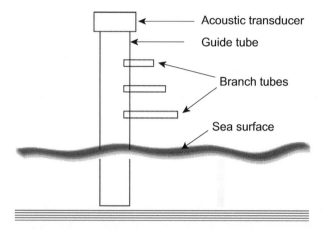

FIGURE 16.12 The acoustic reflectors used for temperature-compensation of a guided air-acoustic gauge, consisting of resonant side-branches located orthogonal to the sounding tube. *Source: Pathak and Ramadass, 2002 [http://www.uspto.gov]; Reproduced with kind permission from the United States Patent and Trademark Office.*

acoustic reflectors consist of resonant side-branches located orthogonal to the sounding tube. The side-branches are designed to respond to a specific frequency such that the sound pulse with appropriate center frequency is predominantly reflected by the branch, thereby providing a means for in situ calibration of the acoustic gauge by measuring the effective velocity of sound at different portions of the sounding tube. The length of the side branch, which plays an important role in its effectiveness as a calibrator, is an odd multiple of the quarter wavelength for a given transmission frequency of sound and is given by the expression (Pathak and Ramadass, 2002):

$$L = (2n - 1)\left(\frac{\lambda}{4}\right); \ n = 1, 2, 3, \ldots\ldots \quad (16.2)$$

where;

L: length of the side-branch tube, including the Raleigh end-correction [m]
λ: wavelength of sound [m]; ($\lambda = C/F$)
C: velocity of sound in air [m/s]
F: frequency of transmitted sound [Hz]

By using properly tuned resonating side-branches and signals of different frequencies for calibrating and for measuring the sea level, the limitations existing in in situ calibration can, in principle, be overcome. These side-branch tubes achieve maximum reflection of the acoustic signal. The improved signal-to-noise ratio leads to improvement in the accuracy of measurement. The diameter, d, of the guide tube is determined by another formula that takes into account the wavelength of the sound signals for realizing plane wave propagation within the guide tube for achieving higher accuracy and is given by the expression (Pathak and Ramadass, 2002):

$$d < 0.586\,\lambda \qquad\qquad (16.3)$$

where λ is the wavelength used. It is possible to have more branches connected to the sounding tube of the acoustic gauge at different distances from the acoustic head. Using Eqs. (16.2 and 16.3), each branch is designed to respond to a specific sound frequency such that the sound pulse with appropriate center-frequency is predominantly reflected by the branch without loss of energy of the acoustic pulse to the surrounding, with an improved signal-to-noise ratio. Thus, the effective velocity of sound in air can be estimated for different portions of the sounding tube. By using properly tuned resonating side-branches and signals of different frequencies for calibration and for measurement of sea level, it is possible to overcome some of the limitations of the existing in situ calibration methods.

This modified sounding tube can also allow good mixing of the water body in the sounding tube with that external to it if the closed-ended, resonating side-tubes described by Pathak and Ramadass (2002) be converted to open-ended side-tubes. Because resonance is equally possible with both closed-ended and open-ended tubes, this simple modification is likely to provide an easy solution to the lower-density water-trapping problem suffered by the air-acoustic sea-level gauges if arrays of perforations are also provided on their protective-well.

Although the preceding method sounds simple, there are some practical difficulties in its implementation. For example, the length of the resonating side-branch tube used for in situ calibration of the sea-level gauge is related to the wavelength of the acoustic transmission pulse. Because of this, it becomes necessary to attach a series of resonating side-branch tubes of differing lengths at several points on the sounding tube at different distances from the acoustic head for efficient full-length in situ calibration of this acoustic gauge during different phases of the tide. Another disadvantage with this methodology is that it is not possible to achieve the mandatory plane wave propagation condition for the sounding tube for all the different acoustic transmission frequencies because of the relationship between the wavelength of the acoustic trans-mission signal and the diameter of the guiding tube. Also, the presence of the protruding resonating side-branches necessitates a much larger diameter stil-ling-well because of the presence of the protruding side-branches.

Recent experience in South African waters was not in favor of the air acoustic gauge. Although the gauges were self-calibrating, their performance was far beyond specification; they were erratic and difficult to tune (Farre, 2004). Recent studies conducted by Gomez and Maldonado (2004) in Spain provide a somewhat more optimistic picture. Although their experience indicated that these gauges do not work properly if installed in the open air and when the distance from the acoustic head to the seawater is more than 9 m or less than 2.5 m, the performance of the acoustic gauge improves when its acoustic tube is painted white.

Another source of systematic errors in sea-level measurements using a guided air acoustic gauge appears to be trapping lower-density water in its long and narrow sounding tube. The air-acoustic gauge's long and narrow sounding tube, in the present form, is devoid of any lateral perforations on its submerged portion and therefore can trap a lower-density water body even if there is good mixing within the protective well.

Errors arising from temperature effects and trapped fluid in the sounding tube would be concerns when considering the overall accuracy attainable from guided air-acoustic gauges. This may require the use of postacquisition adjustments on the data for examining the sea-level signal with greater precision, particularly low-frequency signals. Averaging of a number of measurements will have a stilling effect (Porter and Shih, 1996). Removal of outliers from the acquired data sample sequence and recomputation of the mean water level would give improved accuracy in sea-level measurement.

A serious limitation of the air-acoustic gauge in the context of tsunami is its inability to record large-amplitude signals. For example, it has been reported that in the tsunami record from an acoustic gauge at Port Elizabeth (Republic of South Africa), the crest of a tsunami wave was chopped off (Rabinovich and Thomson, 2006).

16.5. DOWNWARD-LOOKING AERIAL MICROWAVE RADAR GAUGES

The microwave radar gauge represents an alternate noncontact device that is capable of remote measurement of sea-level elevation from the air. The basic premise involves transmitting microwave pulses toward the sea surface and receiving the reflected/backscattered microwave energy. The reflected microwaves are analyzed to estimate the two-way distance traveled by a given pulse. The principle of sea-level measurement based on microwave radar-sensor technology is shown in Figure 16.13. It can be readily seen that the operation of this device is based on the time-of-flight principle used with the air-acoustic gauge. The time of flight can be determined by measuring the phase of the return wave and knowing the frequency of the microwave signal that was transmitted. Further, the time of travel can be measured using the well-known digital sampling techniques. Depending on the device model, the transmission frequency ranges from almost 10 GHz to almost 24 GHz, and the beam width is almost ±5 degrees (for example, OTT microwave radar designed for sea-level measurement application operates at about 24 GHz; wavelength = 1.25 cm). Microwave signals in the form of linear sweep (chirp) are normally used for achieving better precision in distance measurement. The microwave-level gauge typically includes a microwave feed-horn directed on the sea surface, electronics housing spaced apart from the feed-horn, and a microwave waveguide coupled to the housing and the feed-horn for carrying microwave radiation between them. Coaxial or other appropriate cabling can also be used as

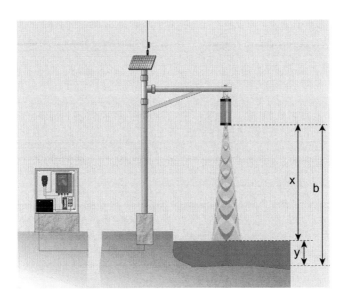

FIGURE 16.13 The principle of sea-level measurement based on microwave radar-sensor technology. *Source: Courtesy of Helmut Hohenstein, OTT Hydromet Gmbh; Reproduced with kind permission of OTT Hydromet GmbH, P.O. Box: 2140 / 87411, Kempten / Germany.*

a substitute for the wave-guide. A microwave transducer in the housing couples to the wave-guide and sends and receives microwave signals. A microprocessor in the housing measures the distance x in air based on a microwave echo from the air-water interface and a microwave echo from the feed-horn. A known microwave distance measurement technique is frequency modulated continuous wave (FMCW), in which the transmitted microwave signals are mixed with the signals reflected from the sea surface to determine the phase shift between the two waves and thereby the range. Some FMCW radar gauges are capable of sampling sea levels as rapidly as subsurface pressure gauges can. In another method involving Fourier analysis, any target within the microwave beam shows up in the output spectrum. This includes the end of the antenna that is transmitting the microwave beam. The microprocessor compensates for the effect of propagation delay through the wave-guide on distance measurements by using the feed-horn echo as a reference signal. By dynamically compensating for errors arising from variations in the length of the wave-guide due to temperature changes or other factors contributing to length variations, the gauge is expected to provide superior performance even under adverse environmental conditions. The water level y with reference to a preferred datum (e.g., CD or harbor datum) is estimated from the known vertical height b of the sensing plane of the sensor head from the chosen datum. Wind-induced high-frequency oscillations on the sea-level are filtered out with the online application of an integrated software filter. Two commercially available microwave

(a)

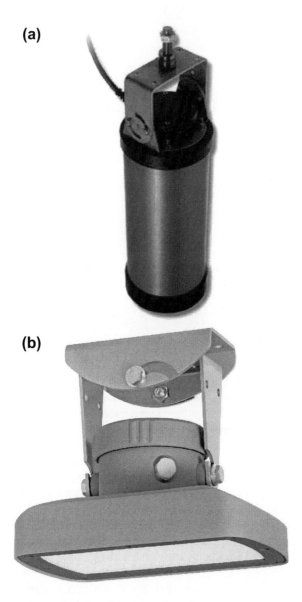

(b)

FIGURE 16.14 (a) OTT Kalesto micro-wave radar sensor; (b) OTT RLS microwave radar sensor. *Source: Courtesy of Helmut Hohenstein, OTT Hydro-met Gmbh; Reproduced with kind permission of OTT Hydromet GmbH, P.O. Box: 2140 / 87411, Kempten / Germany.*

radar gauges are shown in Figure 16.14a, b. The microwave radar gauges can be installed from coastal jetties, bridges, harbor piers, or offshore platforms with relatively fewer logistical problems (Figures 16.15). The measurements are transferred to a data logger by means of an RS-485 interface over distances of up to 1 km.

The main advantage of the microwave radar gauge over the air-acoustic gauge is that the electromagnetic transmission/reception signal used in the

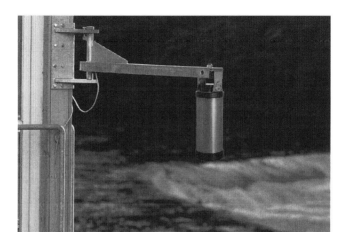

FIGURE 16.15 Typical site installations of microwave radar sensors on an offshore platform. *Source: Courtesy of Helmut Hohenstein, OTT Hydromet Gmbh; Reproduced with kind permission of OTT Hydromet GmbH, P.O. Box: 2140 / 87411, Kempten / Germany.*

former is independent of variations in site conditions such as air temperature, humidity, and rainfall. The accuracy of sea-level measurements using the microwave radar principle is claimed to be within ± 1 cm over the complete measuring range. Improved microwave signal processing methodologies are still evolving. Competing manufacturers continue to contribute significantly toward such improvements. For example, one of the unique features of Saab Marine Electronics' technology is claimed to be the incorporation of fast hardware adaptive signal technology (FHAST) method, whereby the microwave signal is passed through a digital hardware filter that removes irrelevant signals and background noise from the sea-level signal (Terdre, 1995). Another important benefit is the excellent stability of the sensor that exhibits virtually no drift; in other words, there is negligible variation in the signal for the same known distance in the case of varying environmental conditions such as temperature change and over the long term as the sensor ages.

From an operational point of view, an advantage with the noncontact measuring principle used in the microwave radar gauge is that problems such as disruption of measuring operation caused by high water, silt accumulation, debris, plant growth, and so forth, as well as time-consuming maintenance, are reduced considerably. In the event of any maintenance requirement, the difficulty involved is relatively less because of land-based operations. Because of its compact design and the noncontact measuring principle, the gauge can be installed easily and inconspicuously, without the use of cumbersome stilling-wells/protective-wells. Further, the device is particularly suitable for areas where conventional measuring systems cannot be used or where a station needs to be set up quickly and inexpensively. Because the gauge does not come into direct physical contact with water, it is particularly useful for measurement from locations where the water contains a large amount of suspended matter (e.g., Hugli Estuary, India).

As of now, microwave radar gauges are too new for any long-term data to exist. Comprehensive field-evaluation reports are also yet to emerge for a fuller understanding of its long-term performance. However, initial assessments by the user-communities appear to be encouraging. For example, Woodworth and Smith (2003) reported the experience in Liverpool waters (northwest England), where they made a one-year comparison of a microwave radar gauge (OTT Hydrometry Ltd.) with a conventional bubbler gauge. Liverpool is a challenging location for a test, with a tidal range of about 10 meters at spring tide phase (i.e., approximately during full-moon and new-moon). For intercomparison purposes, 15-minute averaging was used for both systems. The RMS difference between the two measurements was 1.28 cm (excluding outliers of $>\pm 5$ cm) for 15-minute values and 1.15 cm for hourly values. However, there were larger differences (spikes) of up to ± 10 cm between the two measurements during storms. Based on a one-year test at Liverpool over a wide range of weather conditions and tides, Woodworth and Smith (2003) made the following observations on the microwave radar gauge:

- Relatively low cost and ease of installation and maintenance (no need for divers or stilling-wells, etc.) make them attractive.
- Fully digital (so can be "real time").
- Accuracy approximately 1 cm as claimed.

This accuracy is consistent with the accuracy required for sea-level measurements for global networks.

Presently, field evaluation tests of radar gauges are underway at many locations in South Africa (e.g., Simon's Town, Richards Bay, Port Nolloth, and Cape Town) under the supervision of the South African Navy Hydrographic Office. An OTT microwave radar gauge was tested in Simon's Town in the beginning of 2002, and the results indicated its superior performance (Farre, 2004).

Shirman (2003) reported microwave radar gauge experiences in Israel. In the experiments conducted in Israel waters, comparison between the microwave radar gauge measurements and float-operated gauge measurements showed agreements in sea-level measurements within 1 cm accuracy. The observed differences were largely of a random nature.

Radar gauges encompass a new technology. Although they are widely manufactured and readily available, not all of them were rigorously tested. It is hoped that such independent tests in several water bodies in different parts of the world by independent researchers would shed the much-needed information on this newly emerging technology. Nevertheless, experiences so far are generally favorable—at least with the OTT Kalesto radar gauge that has undergone performance evaluation tests in several waters. Because radar technology has advantages over other types of gauges in ease of installation and maintenance, their findings suggest that radar has to be given strong consideration in future applications, especially at locations where variations in water density preclude the effective use of pressure systems.

Perhaps, a meritorious feature that can be attributed to radar gauges is their suitability for measuring large sea-level oscillations such as those occurring during tsunamis. However, this feature can be achieved only if the gauge is suitably installed, taking this requirement into consideration. For example, the measurement range of the Kalesto radar is 1.5 to 30 m. However, the footprint of the beam increases as the distance of the radar from the sea surface increases. Thus, the greater the distance between the radar sensor and the sea surface, the greater the safety distance between the center axis of the radar beam and the wall of the jetty where the gauge is erected. The recommended safety distance for the Kalesto radar (radiation angle $= \pm 5°$) to achieve the full range of 30 m is 3 m.

The National Institute of Oceanography (NIO), Goa, India, has designed and developed microwave radar gauges incorporating Kalesto radar sensors and established several of them on the Indian coasts, with plans to establish more on the Indian coasts and islands. Figure 16.16 illustrates the scheme of microwave radar gauge installations at Ratnagiri (OTT Kalesto radar sensor) on the west coast of India and at Kavaratti Island (OTT RLS radar sensor) in the Lakshadweep Archipelago in the southeastern Arabian Sea. Kalesto is the original version of the radar sensor and RLS is the modified version. The main advantages of RLS sensor over Kalesto sensor are in terms of weight/size and power consumption (weights of Kalesto and RLS are 8 kg and 2.1 kg, respectively; power consumption of Kalesto and RLS are 550 mA @12 volts DC and 12 mA @12 volts DC, respectively). The RLS covers a measurement range of up to 35 m. It is specifically designed for use in open air locations that have no requirement for mains power supply. The special flat antenna design construction and its minimal energy consumption make the RLS an economical, practical, and reliable alternative to conventional level gauges. The experiences of NIO with microwave radar gauges are encouraging so far. Its insensitivity to changes in water quality parameters, such as turbidity and suspended sediments, and hydrodynamic parameters, such as flows and waves, is particularly useful when water-level measurements must be taken at locations that are prone to sudden water-level changes arising from opening and closing of hydraulic barrages across estuaries. Figure 16.17 shows a real-time graphical report (obtained using a microwave radar-based NIO sea-level gauge installed at Reliance Jetty, Yanam, located on the Godaveri estuary on the central east coast of India) of a steady rise in water level at this location subsequent to the opening of a hydraulic barrage across the upstream segment of Godaveri estuary (about 50 km upstream of the gauge location) and the subsequent drop in water level due to continuous draining of water into the Indian Ocean.

FIGURE 16.16 Typical installation schemes of microwave radar-based NIO sea-level gauges. (a) at Finolex Port, Ranpar, Ratnagiri, on the west coast of India (Kalesto radar sensor) and (b) at Kavaratti Island (Lakshadweep Archipelago) in the south-eastern Arabian Sea (RLS radar sensor).

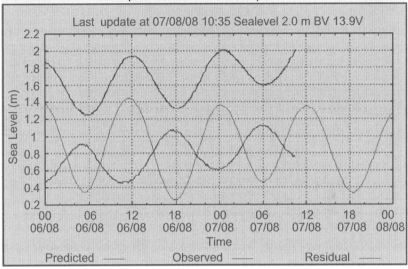

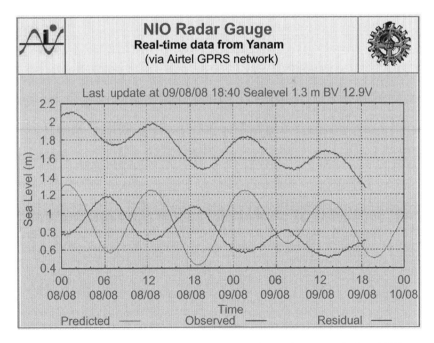

FIGURE 16.17 Real-time graphical report obtained using a microwave radar-based NIO sea-level gauge installed at Reliance jetty, Yanam, located on the Godaveri estuary on the central east coast of India. (a) Steady rise of water level at this location subsequent to the opening of a hydraulic barrage across the upstream segment of Godaveri estuary (about 50 km upstream of the gauge location). (b) Subsequent fall of water level due to continuous draining of water into the Indian Ocean. Influence of intense freshwater influx on tidal range underestimation (residual appearing in the form of an inverted tide) is clearly noticeable.

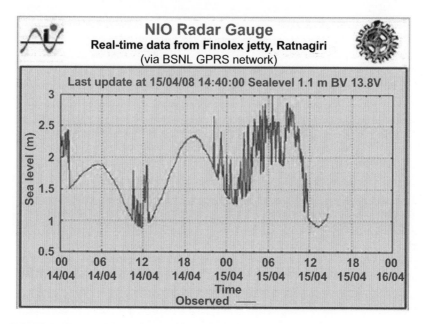

FIGURE 16.18   Artificial contamination introduced in sea-level measurements by the presence of a pontoon below the radar gauge in Ratnagiri, west coast of India.

Measurements using microwave radar gauges can, however, get severely contaminated by man-made lapses such as anchoring of floating objects below the radar sensor. Figure 16.18 shows artificial contamination introduced in sea-level measurements by the presence of a pontoon in the vicinity of the radar gauge (see Figure 16.16) in Ratnagiri. A similar problem has been reported from a Spanish harbor where the sea-level measurements used to be contaminated severely almost every day at about 12 noon. The culprit was found to be a fishing boat that was anchored below the radar sensor every day during the boatman's lunch hour (Private communication, Thomas Stepke, International Sales Director, OTT MESSTECHNIK GmbH & Co., April 2008). These incidents indicate the sensitivity of microwave radar gauges to floating objects and the importance of protecting the footprint area of the microwave radar from contamination by such objects if high-quality data are to be gathered.

16.6. SUBSURFACE PRESSURE GAUGE SYSTEMS

Although the recently introduced microwave radar gauges have advantages such as relative ease of installation and insensitivity to ambient temperature changes (Woodworth and Smith, 2003), they suffer from an inherent height limit in the case of tsunami measurements. Based on this, the IOC has recommended the use of subsurface pressure sensors as backups to the microwave radar gauges because of their ability to record sea-level changes that exceed the limited height of the radar sensor. Also, sea-level measurements by subsurface pressure sensors are not deteriorated by objects floating on the sea surface. Most importantly, subsurface pressure sensors have advantages such as the ability to be sampled at a higher rate, linear response to large-amplitude waves such as tsunamis and storm surges, and reasonably good accuracy, except in heavily suspended sediment-laden water bodies (Joseph et al., 1999c). Consequently, the subsurface pressure sensors are regarded as the main "tsunami sensors" and suitable for tsunami network applications (Woodworth et al., 2007a). Thus, delivery of scientific inferences on tsunami signals derived from subsurface pressure measurements is expected to be more realistic than those inferred from conventional stilling-well-based gauges, such as float-driven and guided air-acoustic gauges.

The approach used for estimating sea-level elevation using subsurface pressure sensors is to measure the hydrostatic pressure, P_d, at depth d below the sea surface and to convert this pressure to the water level above the pressure inlet of the transducer by using the relationship:

$$P_d = P_a + (\rho \times g \times d) \qquad (16.4)$$

where P_a is the atmospheric pressure, ρ is the mean density of the overlying column of water, and g is the local acceleration due to gravity. The measured pressure varies as the sea level varies. In coastal measurements, the subsurface pressure may be transmitted to a shore-based recorder through a narrow tube.

Local dynamic effects such as flows, waves, and a combination of both that operate in the vicinity of the pressure inlet influence the performance of a pressure-based sea-level gauge. However, these effects can be significantly reduced by the use of an appropriate hydromechanical front-end at the pressure inlet (Joseph et al., 2000). The response of a pressure transducer to a steady laminar flow field for different hydromechanical front-ends is shown in Figure 16.19. The response of a pressure transducer to various mainstream flows for different relative orientations of its housing and a cylindrical vertical piling is shown in Figure 16.20. This figure provides an indication of the response of a pressure transducer–based sea-level gauge when it is mounted on a cylindrical structure such as the pillar of a bridge/offshore platform. In these figures, D represents the diameter of the pressure transducer housing. Experiments conducted by Joseph and colleagues (2000) showed that the transducer's performance for steady laminar flow can be significantly improved if the pressure inlet is located at the center of a flat, thin, and smooth horizontal circular plate whose diameter is three times or more that of the pressure housing. This improvement is due to the isolation of the pressure inlet from the separated flows and from the vortices generated by the transducer housing. For turbulent flows, the addition of a similar plate parallel to the first and separated by a distance equal to the diameter of the housing leads to a much improved horizontal azimuthal response up to approximately 100 cm/sec. Beyond this

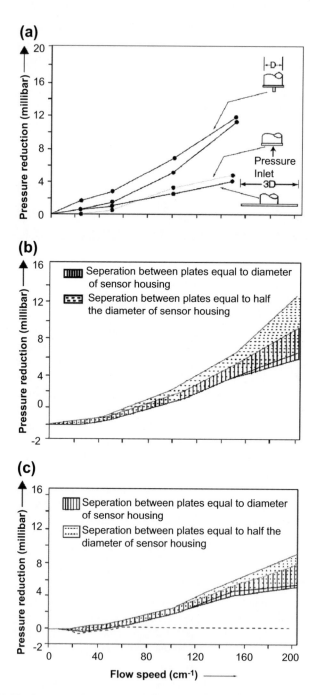

FIGURE 16.19 Response of pressure transducer to a steady laminar flow field for different hydromechanical front ends. *Source: Joseph, A., J.A.E. Desa, P. Foden, K. Taylor, J. McKeown, and E. Desa (2000), Evaluation and performance enhancement of a pressure transducer under flows, waves, and a combination of flows and waves, J. Atmos. Oceanic Technol., 17(3), 357-365; ©American Meteorological Society; Reprinted with permission.*

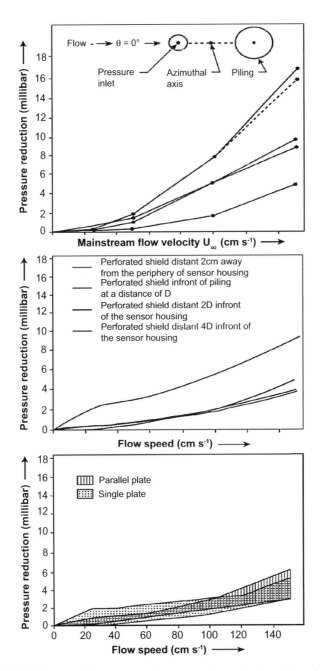

FIGURE 16.20 Response of pressure transducer to various mainstream flows for different relative orientations of its housing and a cylindrical vertical piling. *Source: Joseph, A., J.A.E. Desa, P. Foden, K. Taylor, J. McKeown, and E. Desa (2000), Evaluation and performance enhancement of a pressure transducer under flows, waves, and a combination of flows and waves, J. Atmos. Oceanic Technol., 17(3), 357-365; ©American Meteorological Society; Reprinted with permission.*

speed, the transducer's horizontal azimuthal response deteriorates faster, probably as a result of the loss of stability of Poiseuille flow at very large Reynolds numbers. Such a horizontal parallel-plate front-end was found to be useful also in the case of waves riding on currents. This result is of special significance to sea-level measurements from coastal waters in which waves usually propagate on tidal currents. Joseph and colleagues (1999c, 2004) also noticed that care must be taken in the use of pressure transducers for sea-level measurements from heavily sediment-laden or turbulent waters. An interesting observation made by these investigators is that such natural waters suffer a reduction in in situ density as a result of a variety of dynamically induced local factors such as a dynamic uplift force acting on the suspended sediment particles, turbulence, the presence of microbubbles in the upper layers of the water body, the tendency of suspended particles to accumulate air bubbles, and so forth. Inaccuracies in sea-level measurements arising from these factors can, however, be significantly reduced by the use of two pressure transducers located at two different but exactly known levels in a vertical plane or by in situ calibration based on tide staff readings.

16.6.1. Pressure Transducers

Pressure measured by a pressure transducer is always relative to a reference pressure. This reference pressure may be sealed within the pressure sensor body or accessed externally through a pressure reference tube, port, or vent hole that is usually located at the lead exit side (rear) of the sensor body. Accordingly, three types of pressure transducers are commonly used: *absolute*, *differential*, and *gauge pressure* transducers (Filloux, 1970; Baker et al., 1973; Peshwe et al., 1980; Joseph and Desa, 1984; Cox et al., 1984; Spencer and Foden, 1996; Sidor, 1983; Joseph et al., 1999a, b). In a pressure trans-ducer—based sea-level measurement system, frequent pressure samples are averaged over a duration of a few minutes to reveal only the low-frequency changes in the sea level. This rapid sampling and averaging technique removes wave-induced fluctuations from the sea-level data without the need for a stilling-well.

In an "absolute" pressure transducer, pressure is measured with respect to a vacuum reference (i.e., zero pressure) chamber. In this case, the reference pressure cavity is sealed (usually with electron beam welding) at the time of manufacture. Such a pressure transducer, which is used for sea-level measurement, has only one pressure port and senses the total pressure (i.e., atmospheric pressure plus pressure due to the water column present over the submerged transducer). The absolute sensor has an advantage that a vent tube is not necessary. The disadvantage, however, is that the atmospheric pressure must be independently and accurately monitored and it must be subtracted from the absolute pressure measurement to compensate for the error in water pres-sure measurement due to temporal changes in atmospheric pressure.

A "differential" pressure transducer detects the difference between the pressures applied at its two ports. Two classes of differential transducers include "sealed" and "vented" types. A sealed reference is generally used in applications where the reference port cannot be vented to a desired external pressure reference or where the external environment is not suitable. In this case, the reference pressure cavity of the transducer is sealed at a desired pressure at the time of manufacture. For example, sealed differential pressure transducers have been successfully employed in seafloor-mounted gauges for deep-ocean measurements of low-frequency sea-level oscillations. This arrangement allows amplification of the transducer output in a differential mode, suppressing the large offset-pressure signal arising from the depth of the ocean. "Gauge" pressure is a form of differential pressure measurement in which atmospheric pressure is used as the reference. A gauge sensor has one face of the pressure diaphragm exposed to the measuring fluid pressure and the other face exposed to atmospheric air pressure. Accordingly, in a gauge device the (−ve) port is vented to the external atmospheric pressure, and the (+ve) port is exposed to the total pressure (i.e., water pressure + atmospheric pressure). The output of the gauge pressure transducer provides a signal representing the pressure difference at the two ports. Because the differential characteristic of the gauge transducer cancels out the contributions from the atmospheric pressure, it automatically removes measurement errors arising from changing atmospheric pressure conditions. In all cases, the referenced pressure source must be clean, dry, nonconductive, and noncorrosive.

A number of different types of pressure transducers have been used for sea-level measurements. These include strain gauges and capacitance devices (inherently analogue) and quartz piezoelectric resonators (inherently digital). Because the inherent accuracy of a transducer is best maintained in a digital system, an inherently analog device such as a strain gauge is sometimes incorporated in an oscillator to provide a digital output (Sidor, 1983). All these transducers are sensitive to the ambient temperature variations (Rae, 1976). However, pressure transducers presently used for sea-level measurement are usually temperature compensated.

Because the sensing element in a transducer is critical to the overall performance, only the highest-grade materials are used for fabrication of the sensor. These materials include metals, alloys, semiconductors, ceramic, quartz, and so forth. Further, computer-controlled machine tools, which are capable of reproducing the desired parameters without deviation, are usually utilized for fabrication of the sensors. After being machined and inspected, the sensing elements are usually transferred to an environmentally controlled assembly department for testing, assembly, temperature-compensation, and calibration. The calibration is performed to full scale, preferably in the ascending and descending order of input pressure using calibration standards that are traceable to the National Institute of Standards and Technology (NIST). Computerized equipment is usually used for testing the pressure transducers. The computerized test facilities perform detailed testing of each transducer's performance indicators, such as accuracy, hysteresis and repeatability, linearity, temperature sensitivity, and so forth, over a specified temperature range and provide a record of the results.

16.6.1.1. Hydraulic Coupling of Pressure Ports

In the case of an in-water pressure-based sea-level gauge, a suitable hydraulic coupling device (HCD) is usually attached to the (+ve) port of the transducer in order to prevent corrosive destruction of the pressure port as a result of its physical contact with the saline seawater and to hydraulically couple the pressure port to the seawater without affecting its sensitivity to the subsurface pressure variations. The HCD consists of some kind of a jacket that is filled with a viscous fluid such as silicone oil whose viscosity is about 500 centi-Stokes.

Different types of HCDs have been incorporated with in-water pressure transducers. For example, Filloux (1970) described a sea-level gauge where the seawater pressure intake front end consists of a flexible membrane stretched across the pressure case of the instrument. A compartment located behind the membrane is filled with oil, which connects to the pressure inlet of the transducer. The membrane is exposed to seawater pressure through an insulating fluid that is devoid of air bubbles or other hollow, closed spaces. The HCD of the Harvard deep-sea gauge consists of an oil-filled bag in a separate enclosure. The oil in the bag is routed through a capillary tube to the pressure port of the transducer (Dietrich et al., 1980). The HCD incorporated with Paroscientific pressure transducers consists of a silicone oil–filled flexible plastic capillary tube of about 1 mm internal diameter and about 200 mm long, with one end attached to the pressure port of the transducer and the other end connected to the end-cap of the transducer housing. The HCD designed by Desa and colleagues (1999) consists of an oil-filled multiple-U-shaped capillary tube sandwiched between two halves of a plug. One end of this capillary terminates on a copper disc fixed on the end-cap of the transducer housing, and the other end dips to the pressure port of the transducer. These components, together with the viscous oil deposited in the front end of the pressure transducer, form the HCD. The oil medium hydraulically connects the pressure transducer to the surrounding saline water medium. The copper material of the HCD, because of its ability to repel marine growth (Huguenin and Ansuini, 1975), arrests the possibility of closure of the pressure inlet by biofouling during its prolonged exposure to the euphotic saline water medium. The viscous oil in the HCD inhibits physical contact of the pressure-sensing element with the seawater, while at the same time transferring the seawater pressure to the transducer (i.e., hydraulically coupling the transducer to the water pressure).

16.6.1.2. Venting Gauge Pressure Transducers

The gauge pressure transducer used for sea-level measurements measures the difference between the total pressures applied at its (+ve) input and the atmospheric air pressure applied at its (−ve) input. Atmospheric pressure detection requires an unobstructed passage that vents the (−ve) port of the transducer to the atmosphere (i.e., open environment). For this reason, gauge sensors utilize a vent tube to the surface. Vent tubes are generally encased in the sensor cable that connects the pressure sensor to the measuring instrument located above the sea surface. Using vent tubes is a very practical way to balance atmospheric pressure changes, but care must be taken with their deployment. When vented pressure transducers are deployed, there is usually a difference between the atmospheric temperature and the temperature of the transducer, which is deployed at a certain depth below the sea surface. This temperature difference might cause moist air from the surface to be drawn into the transducer through the vent line. For reliable operation, it is essential to prevent the entry of moisture into the sensor. This is because moisture causes material deterioration and shorting, and it also accelerates the effects of corrosion, thereby ultimately degrading the sensor performance or permanently damaging the sensor. Moisture can also condense on sensitive electronic components. Different manufacturers have reported various designs. For example, Entran has suggested backfilling the reference cavity of the pressure transducer with silicone oil or other similar instrument-oils to act as a fluid barrier to the external medium (*Entran Sensors & Electronics* catalogue, p. D3, April 1999). Honeywell has reported the design of a second level package that incorporates a filter that prevents environmental moisture from entering the sensor. The filter material is poly-tetra-fluoro-ethylene (PTFE) that exhibits outstanding chemical inertness. The material has a wide operating temperature range and excellent hydrophobic properties (i.e., it allows the passage of gases but repels water). In the Honeywell system, the PTFE "patch" is thermally bonded onto a plastic base through a process of thermal welding. This patch covers the exit hole of a passage that vents the interior of the second level package to the atmosphere. A vented transducer manufactured by Greenspan Company (Australia) incorporates silicon desiccant crystals in a venting system that is sealed against exposure to the atmosphere, still transmitting atmospheric pressure to the (−ve) port of the transducer. Silicon desiccant crystals easily absorb moisture, thereby maintaining dry air in the closed vent tube. For reliable operation of the desiccant system, it is necessary to partially inflate a breather bag, which is located inside the closed venting system, to enable it to expand and contract with atmospheric pressure changes. Another system, which is somewhat similar in principle, has been recommended by Aanderaa Instruments as well.

16.6.2. Commonly Used Pressure Transducers

Several different types of pressure transducers are used today. In the following sections, we address pressure transducers that are used for sea-level measurements.

16.6.2.1. Strain Gauge Pressure Transducers

Fluid pressure can be detected with a strain gauge that is mechanically coupled to an elastic wall (such as a diaphragm) on which the pressure to be measured is exerted. When pressure is applied to the diaphragm, its distributed load produces diaphragm flexing, which in turn creates strain in the strain gauges. The strain is proportional to the applied pressure. The deformation of the strain gauge caused by the exerted pressure affects the electrical resistance of the gauge. If the relationship between deformation and pressure is known, the unknown pressure can be estimated by measuring the change in electrical resistance.

For measuring the deformation, or *flexure,* of the diaphragm—and thus the fluid pressure—the diaphragm contains four small strain gauges, which together form a full Wheatstone bridge network. Because the strain gauges are located in areas of both tensile and compressive strain, the applied pressure produces a bridge imbalance. The imbalance of the bridge is a sensitive measure of the fluid pressure, provided that the strain gauges are properly located to take the best advantage of the mechanical strains in the diaphragm. With a bridge excitation voltage applied, the imbalance of the bridge produces a voltage change at the bridge output that is proportional to the pressure acting on the diaphragm. Two types of strain gauges are in common use: metal-alloy strain gauges and silicon strain gauges.

The change (Δr) in resistance in a deformed strain gauge is composed of two terms and is given by the expression (Gieles and Somers, 1973):

$$\Delta r = s \times \Delta \left(\frac{l}{a} \right) - \left(\frac{l}{a} \right) \Delta s \qquad (16.5)$$

where s, l, and a represent the specific resistance of the material of the conductor, its length, and cross-sectional area, respectively. The term Δs represents the change in resistivity caused by mechanical stress. The first term on the right-hand side of this expression describes the change of electrical resistance directly caused by the ordinary deformation effect of the conductor—that is, by the change of length and cross-sectional area under strain. It was Lord Kelvin who first reported in 1856 that metallic conductors subjected to mechanical strain exhibit a change in their electrical resistance. This phenomenon was first put to practical use in the 1930s. The second term is the contribution of the piezo-resistance effect, since the deformation is accompanied by a mechanical stress that changes the resistivity. In the mid-1950s, scientists at Bell Laboratories discovered the piezo-resistive characteristics of semiconductor materials (i.e., change in the resistivity of the materials caused by a mechanical stress). In metal-alloy strain gauges, the first term predominates and the second term is negligibly small for all practical purposes. However, in silicon strain gauges, the second term predominates and the first term is negligibly small.

16.6.2.2. Metal-Alloy Strain Gauge Pressure Transducers

Constantan and nickel-chromium, which are metal alloys of high resistivity (i.e., specific resistance) and that possess good elongation capability, are generally used in the construction of strain gauges. The elongation capability of constantan is approximately 3%. However, this is usually limited to 0.2% for precision transducers. Foil-type strain gauges of excellent linearity and strain characteristics are employed in the construction of pressure transducers. In this, four segments of photo-etched matrix of electrical resistance foil, with a copper soldering pad on each gauge (Figure 16.21), are carefully bonded on one side of a thin metallic diaphragm, under controlled conditions, using special epoxy adhesives. The resistors are supplied with copper soldering pads because of the difficulty of directly soldering to K-alloy. Additional resistors are usually incorporated into the Wheatstone bridge network for slope correction and to compensate for zero-imbalance and temperature-induced variations in the output. For example, in Figure 16.22, the two extending arms, *b* (known as *slope compensating bobbin resistors*), are nickel wires wound on a bobbin and are not subject to pressure-induced strain. Similarly, the extending arm, *a* (known as a *zero imbalance resistor*), is also not subject to pressure-induced strain. The *slope compensating resistors*, having positive temperature coefficients, are used to provide a self-correction against an effective negative temperature coefficient of the gauge factor. Such an

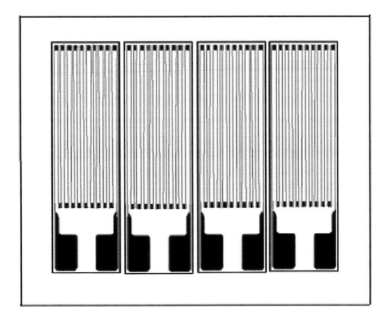

FIGURE 16.21 Typical metal foil strain gauges that form the four arms of a Wheatstone bridge.

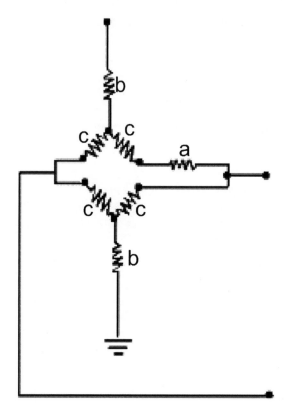

FIGURE 16.22 Gauge resistance foils connected in the form of a Wheatstone bridge network, together with (a) a zero imbalance resistor, and (b) slope compensating bobbin resistors.

effective negative temperature coefficient to the transducer arises from the temperature-induced changes in the geometrical and mechanical properties of the transducer diaphragm (usually stainless steel) on which the gauge resistance foils are bonded. Figure 16.23 shows a cross-sectional view of a typical metal-alloy strain gauge pressure transducer. Commercially available metal-foil strain-gauge pressure transducer manufactured by Macurex Sensors Pvt. Ltd are shown in Figure 16.24.

The electrical resistance metal-alloy strain gauge is a stable device when properly installed and used. However, improper bonding of the gauge and its leads may lead to transducer creep under load when the gauge is strained. For this reason, adhesive bonding is done under controlled conditions using special procedures to avoid airborne contamination, followed by postcuring under specified operating environments to achieve stable performance of the transducer. Chemical attack from residual solder flux or moisture intrusion into the protective coating can produce permanent changes in gauge resistance with time. To avoid such problems, the strain-sensitive elements are always sealed. Accuracy and long-term stability are maintained by protecting the gauges from the environment.

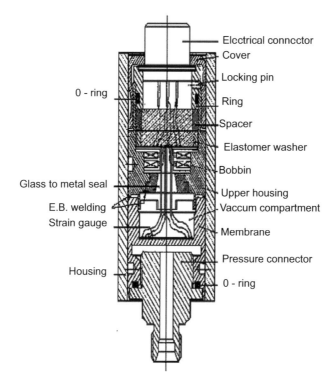

FIGURE 16.23 Cross-sectional view of a typical metal film strain gauge pressure transducer. *Source: Courtesy of Macurex Sensors (P) Ltd. Bangalore, India.*

Gauge creep (at high strains), thermal output (at both high and low temperatures), and poor electrical connections are common sources of "zero-shift" in strain gauge transducers. Drift—a gradual change in gauge zero-resistance with time—is often a function of the internal state of stress in the strain-sensitive gauge resistor. Properly manufactured modern strain gauges are capable of showing resistance drift of only a few parts per million per year when operated under ideal conditions such as complete protection against corrosive agents and operations only at temperatures within the stable range of the sensing alloy in the gauge. Because creep and drift are time-dependent effects with long time-constants, they present much less of a problem in purely dynamic environments.

A major limitation of metal-alloy strain gauge pressure transducers is hysteresis error, which is usually expressed as a combination of mechanical hysteresis and temperature hysteresis. Mechanical hysteresis is the output deviation at a certain input pressure or force when that input is approached first with increasing pressure or force and then with decreasing pressure or force. In other words, hysteresis error refers to differences between an upscale-sequence

FIGURE 16.24 Metal-foil strain gauge pressure transducer manufactured by Macurex Sensors Pvt. Ltd. *Source: Courtesy of Macurex Sensors (P) Ltd, Bangalore, India.*

calibration and a downscale-sequence calibration. Temperature hysteresis is the output deviation at a certain input before and after a temperature cycle. Because metal strain gauges are bonded to pressure-sensing members of dissimilar material, they tend to suffer from thermoelastic strain and complex fabrication processes. These gauges suffer also from instability problems arising from degradation of the bond, as well as temperature sensitivity and hysteresis caused by the thermoelastic strain.

16.6.2.3. Silicon Thin-Diaphragm Strain Gauge Pressure Transducers

Although semiconductor materials such as germanium and silicon exhibit substantial temperature-sensitivity, they possess pressure-sensitivities several times that of metallic wire or foil strain gauges. Silicon wafers are also more elastic than metallic ones (i.e., after being strained, they return more readily to their original shapes). Use of single-crystal silicon thin-diaphragm strain gauges began gaining prominence over the alloy metals because of these inherent superior qualities and the advantage of monolithic construction (i.e., formed from a single crystal). Piezo-resistive pressure sensing elements such as silicon wafers (that exhibit change in their resistivity when subjected to a mechanical stress) are fabricated using silicon-processing techniques commonly used in the semiconductor industry. For this reason, they have taken on some of the semiconductor terminology. Other names for piezo-resistive sensors are IC sensors, solid-state sensors, monolithic sensors, and silicon sensors.

In silicon strain gauge pressure transducers, the four resistors of a full Wheatstone bridge network are directly integrated on a thin diaphragm (wafer) of single-crystal silicon material. While the diaphragm used in the metal foil strain gauge transducer is usually stainless steel and the resistors are bonded to the diaphragm using adhesive, the diaphragm used in the piezo-resistive strain gauge transducer is made of silicon and the resistors are diffused into the silicon during the manufacturing process. This direct integration, with the absence of traditional bonding agents, ensures excellent long-term stability and repeatability. While the two resistors in the Wheatstone bridge configuration increase with positive pressure, the other two decrease in resistance. When pressure is applied to the sensor, the resistors in the arms of the bridge change by an amount ΔR. The alignment of the resistor on the silicon crystal determines if the resistor will increase or decrease with the applied pressure.

Since 1954 it has been known that the piezo-resistance effect is very much larger than the ordinary deformation effect in single-crystal silicon. In the late 1950s, Honeywell's Corporate Technology Center completed basic research on the piezo-resistive properties of silicon-diffused layers. In this material, the same stress causes a change in resistance that is much greater than that in the alloy metals, thereby resulting in a proportionately larger output signal from a transducer using such a strain gauge (Gieles and Somers, 1973). Piezo-resistance of a semiconductor can be described as the change in resistance caused by an applied strain of the diaphragm. Thus, solid-state resistors can be used as pressure sensors, much like wire strain gauges but with several important differences and advantages.

The high sensitivity, or gauge factor, is perhaps 100 times that of wire strain gauges. Piezo-resistors are implanted into a homogeneous single-crystalline silicon medium. The implanted resistors are thus integrated into the silicon pressure-sensing member. Diffused semiconductor strain gauges represent a great improvement in strain gauge technology because they eliminate the need for bonding agents. By eliminating bonding agents, errors due to creep are eliminated.

Silicon is an ideal material for receiving the applied force, primarily because it is a perfect crystal and does not become permanently stretched. After being strained, it returns to the original shape. Silicon wafers are better than metal for pressure-sensing diaphragms because silicon has extremely good elasticity within its operating range.

Distribution of mechanical stress in a thin (i.e., thickness much smaller than the diameter) silicon circular diaphragm that is clamped at the edge and flexed by the application of pressure is very nearly circularly symmetrical in the plane of the diaphragm. An interesting result that can be inferred from the radial (σ_r) and tangential (σ_t) components of the stress distribution curves is that the strain gauges located near the edge and near the center are subjected to stresses of opposite signs, which can increase the output signal of a bridge network and thus increase the sensitivity of the pressure transducer. The changes in resistance that occur are equal in magnitude but have opposite signs. In this arrangement, the strain gauge elements are made by the planar technology used for ICs and, therefore, form an integral part of the diaphragm. In contrast to metal-alloy strain gauges, the scheme incorporated in the silicon strain gauges dispenses with the need for a separate connecting layer, thereby avoiding the troublesome creep effects that would impair the stability of the pressure transducer.

The measuring range of the pressure transducers can be varied extensively by selecting the ratio between the thickness and diameter of the diaphragms. The optimum location of the gauges to obtain desirable features such as high pressure-sensitivity and low temperature-sensitivity is found from knowledge of the orientation of the crystallographic axes with respect to the diaphragm plane and stress distribution in clamped diaphragms. The pressure range and sensitivity are influenced most strongly by diaphragm thickness, with alignment an important secondary consideration. The differential output of the raw sensor is not precise in terms of calibration and temperature effects. These problems are resolved through signal conditioning, compensation networks, and temperature-compensation techniques.

An inherent advantage of silicon strain gauges is the miniaturization of pressure transducers, which improves its frequency characteristics. This is because the frequency response of a pressure transducer improves with decreasing mass and dimensions. Further, the single-crystal material is superior in ruggedness and strength and is free from creep, thereby providing enhanced reproducibility. Another advantage is that it is chemically not very reactive. Perhaps, the most attractive feature of diffused semiconductor pressure gauge is that it is small, inexpensive, repeatable, and generates a strong output signal.

Different manufacturers have employed different techniques for producing piezo-resistive pressure transducers. For example, the sensing element of Honeywell's MICRO SWITCH solid-state pressure sensor consists of four nearly identical piezo-resistors buried, by an ion-implanting method, in the surface of a thin, circular silicon diaphragm to form the Wheatstone bridge. When the silicon wafer is polished, the crystalline structure is open. The ion implanter forces the ions (which are resistive) down into the silicon to form a resistor in the crystalline structure (Honeywell Technical Note). Most, if not all, liquid molecules such as water molecules are too large to penetrate the crystalline structure. Usually a very thin layer of silicon oxide/silicon nitrite is placed on the surface of the sensor to protect the Wheatstone bridge.

The thin silicon diaphragm of the sensor is carved out from a single solid piece of silicon by chemically etching a square cavity into the surface opposite the piezo-resistors. The unetched portion of the silicon slice provides a rigid boundary constraint for the diaphragm and a surface mounting to some other member. A pressure or force causes the thin diaphragm to flex, thereby inducing a stress or strain in the diaphragm and the buried resistors. The resistor values will change depending on the amount of strain they undergo, which depends on the amount of pressure or force applied to the diaphragm.

The known limitations of uncompensated piezo-resistive pressure transducers are temperature sensitivity, poor long-term stability, and zero-point drift (through temperature, aging, humidity, and condensation). They are also not free from hysteresis effects (Vijaykumar et al., 2005). However, effective compensation methodologies have been invented against some of these undesirable features.

Temperature compensation can be accomplished by incorporating discrete components, compensating resistive-elements or two bridge circuits (one for pressure and one for temperature), and using look-up tables. Among these, using two bridge circuits together with look-up table(s) is considered to be the most efficient method. While trying to produce an accurate transducer over a large temperature range, there is no single equation that describes the relationship between pressure bridge output, temperature bridge output, and the corrected pressure. Each sensor is different due to lot-to-lot processing differences, packaging influences, pressure range differences, and so on. The trick used in the industry is to find an equation that best fits each sensor. In principle, the microprocessor that forms part of the transducer can be used to directly evaluate the correction equation in real time. However, unless the microprocessor is very fast and powerful, the direct evaluation method can negatively impact the transducer's frequency response. The more common method is therefore to use a look-up table generated from the correction equation. Table(s) can be "sparse" or "dense," depending on the transducer-accuracy requirements and/or the available microprocessor memory resources. Interpolation is performed for those values that fall between the table values.

Interpolation between points can be linear or nonlinear. Different manufacturers incorporate different methods for temperature compensation. For example, in a Honeywell pressure sensor, the differential output from each piezo-resistive bridge (integrated on to a single die) is amplified through an operational amplifier circuitry and then presented to an analog-to-digital converter. A built-in microprocessor then reads the digital values for each bridge. While in the digital domain, the pressure value is corrected for errors due to pressure nonlinearity and thermal effects. The Honeywell pressure transducer products generally use a sparse look-up table method with a proprietary interpolation method. In this way, temperature compensation has been incorporated into the Honeywell transducer to minimize its overall temperature sensitivity (Private communication, Glen Monzo, Applications/ Technical Support and Pressure Products Development, Honeywell, 2002).

Null and span errors can be reduced to a certain extent by laser trimming of the sensing devices but cannot be eliminated completely. Additional corrective circuitry is sometimes necessary for applications requiring extremely tight tolerances. Output drift with time, trimming tolerances, and changes in ambient temperature all contribute to a constant offset error, known as common-mode error. Changes in ambient temperature also add another deviation, known as sensitivity drift, that changes the slope of the pressure versus voltage curve of the sensor characteristics. A family of techniques known as autoreferencing provides a powerful tool to compensate for these errors. System design engineers find this method attractive because implementation costs are minor compared to ultrastable pressure sensors. Also, device accuracy is substantially increased. Either analog or digital autoreferencing is possible. However, the latter is the most cost-effective and easiest-to-use method. The overall performance characteristics are also influenced to a certain extent by the fundamental construction of the sensor and its package. For example, Honeywell uses an "elastomeric" mounting technology to package the silicon piezo-resistive sensor. In this method, the sensor die IC is sandwiched between two elastomeric seals. One of the seals is conductive, and when this sandwich is placed on the substrate and locked into place under compression, an electrical contact is made between the conductive pads on the substrate and the sensor die IC through the conductive elastomer. On the other side of the die is a flouro-silicone seal, which physically separates the sensor die from the external medium, at the same time coupling it to the pressure of the external medium.

The monolithic silicon-chip sensing diaphragm makes the silicon-based pressure transducer tiny in size, light in weight, and with excellent frequency response. Microminiaturization utilizing the techniques in silicon diffusion, ion implantation, and micromachining to fabricate the silicon sensor allows it to maintain excellent quality standards. When used for sea-level measurements, it is desirable to shield the silicon diaphragm from direct contact with the seawater with the use of an oil-filled protective diaphragm made of a suitable material. Weak tsunami signals at Takoradi Harbor, Ghana (Africa), on the east coast of the Atlantic Ocean, subsequent to the December 26, 2004, Sumatra earthquake that were detected using the Honeywell piezo-resistive pressure transducer is shown in Figure 16.25. Prabhudesai and colleagues (2008) reported the use of Honeywell piezo-resistive pressure transducers for clear detection of another weak tsunami at Goa, west coast of India, and at Kavaratti Island in the Lakshadweep archipelago due to the $M_w = 8.4$ earthquake in Sumatra on September 12, 2007.

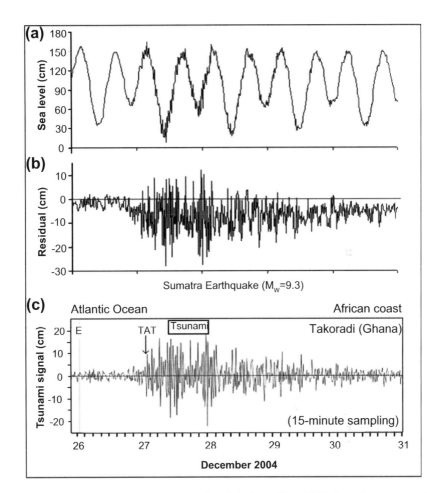

FIGURE 16.25 Weak tsunami signal at Takoradi harbour, Ghana (Africa), on the east coast of the Atlantic Ocean, subsequent to the December 26, 2004, Sumatra earthquake, detected using Honeywell piezo-resistive pressure transducer incorporated with NIO-make sea-level gauge. (a) Time-series sea-level record. (b) Detided sea-level record. (c) High-pass filtered detided record. (E) Earthquake time, TAT: Tsunami Arrival Time. *Source: Joseph et al., 2006; African Journal of Marine Sciences; Taylor & Francis first published the content; Reproduced with kind permission of Taylor & Francis.*

16.6.2.4. Variable-Capacitance Pressure Transducers

Variable-capacitance pressure transducers find a place in some commercially available devices used for sea-level measurement. Electrical capacitance is a function of the effective area of the conductors (plates) of the capacitor, separation between the conductors, and the dielectric strength of the material present within the separation between the plates. Any one of these three can be made variable, thereby causing a change in capacitance.

The electrical capacitance, C, between parallel plates is given by the expression:

$$C = \frac{KA}{D}$$

(16.6)

where;

K: dielectric constant
A: area of the conductor (plate)
D: separation between the conductors

The variable-capacitance pressure transducer depends on the change in separation between the two parallel plates that form a capacitor. In this transducer, two parallel conductors (one usually in the form of a block and the other in the form of a diaphragm) are used. The block is rigidly fixed and the diaphragm is subject to flexure under applied pressure. As the applied pressure varies, the diaphragm flexes, thereby changing the separation distance, D, and therefore its electrical capacitance, C. Thus, the capacitance change in the transducer results from the movement of a diaphragm element in response to pressure variations. The materials that are popularly used for construction of the diaphragm include stainless steel, metal-coated quartz, gold-plated ceramic, silver, and so forth. However, tantalum or high-nickel steel alloys such as Inconel or Hastelloy are used as diaphragm materials for pressure transducers that are specifically designed for long-term deployments in seawater because of their better performance in the corrosive saline water environment. A frequency response up to 500 kHz is usually possible. In operation, one side of the diaphragm is exposed to the water pressure, and the other side is exposed to a reference pressure (in the case of absolute-pressure transducer the reference pressure is vacuum, and in the case of gauge-pressure transducer the reference pressure is atmospheric pressure). In a two-plate capacitor sensor design, the movement of the diaphragm between the plates is detected as an indication of the changes in the pressure input. The change in capacitance can be measured by an ac bridge, a tuned circuit, or by alternate techniques.

The operation of a variable-capacitance pressure transducer is often somewhat similar to that of a strain gauge pressure transducer in the sense that in some designs a full-Wheatstone bridge circuit is formed out of four capacitors (Figure 16.26). In this type of a pressure transducer, a high-frequency oscillator is normally used to excite the sensing electrode elements. The deflection of the diaphragm causes a change in capacitance. The bridge circuit detects this change. The change in capacitance is amplified and observed by several different methods. A simple FM system using a tuned resonant circuit (Cerni and Foster, 1970) is said to give the most satisfactory performance. The ratio between the output voltage and the excitation voltage provides a measure of the input pressure. A process of calibration can indicate the exact relationship between the input pressure and the output signal. The signal is linear over

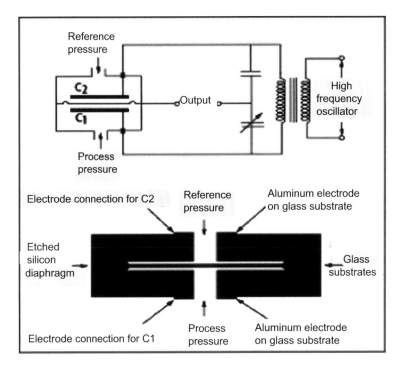

FIGURE 16.26 Constructional features of a variable capacitance pressure transducer. *Source: © 1998 Putman Publishing Company and OMEGA Press, LLC. Reproduced by courtesy of Omega Engineering, Inc., Stamford, CT 06907 USA www.omega.com.*

a fixed range. However, when large ranges of pressure variations are to be measured, a polynomial relationship is often used. For example, for its ceramic-based capacitive element pressure transducers, the Greenspan employs a fourth-order polynomial-fit curve.

Single-plate capacitor designs are also in common use. In this design, the plate is located behind the diaphragm. The variable capacitance is then a function of the deflection of the diaphragm. The detected capacitance is an indication of the applied pressure. The change in capacitance is converted into either a direct current or voltage signal that can be processed by a microcomputer-based data logger. Some designs employ two diaphragms. In this case, one diaphragm is a dummy that is used to minimize the acceleration sensitivity of the transducer (Dobson, 1980).

The main advantages of variable-capacitance pressure transducers are small size, excellent high-frequency response, adaptability for high-temperature operation, low drift in comparison to strain gauge transducers, good linearity and resolution, and the ability to measure both static and dynamic pressure input (Cerni and Foster, 1970). The great advantage of a capacitance gauge is its ability to detect extremely small diaphragm movements. If small size and weight are unimportant, the diaphragm of the transducer may be made larger, resulting in a higher sensitivity. Another advantage is their excellent stability and repeatability. Capacitance-based pressure transducers are capable of providing excellent accuracies (up to 0.01% of full scale). Because of these qualities, capacitance-type pressure transducers are sometimes used as secondary standards, especially in low-differential and low-absolute pressure applications.

The major shortcomings of variable-capacitance pressure transducers are sensitivity to temperature, high-impedance output, and the need for them to be reactively as well as resistively balanced. Because of these constraints, the electronic circuitries associated with variable-capacitance pressure transducers are relatively complex. Long lead lengths and moving leads might cause stray pickup and variation in capacitance, and these must be avoided. In fact, in the usual configurations of measuring the changes in capacitance due to changes in pressure, the capacitance changes are relatively small and can be readily overshadowed by the instabilities in the cable capacitance (Rangan et al., 1983). Because of stray pickup problems, it usually becomes necessary to locate a preamplifier very close to the transducer. An excellent method for isolating the cable capacitance effects is the use of a diode quad-bridge for signal detection, where the bridge is housed within the transducer-case itself (Hanson and Dioneff, 1973). However, newer capacitance-based pressure transducers are less sensitive to stray capacitance and vibration effects that used to cause "jitters" in older designs.

A variable-capacitance transducer-based sea-level gauge that was developed by the University of Washington consists of a 6-mm rod that is placed coaxially in a 12-mm tube. The capacitance technique uses the differences between the dielectric properties of air and water. As the water level rises in the concentric rod-tube pair (Rod), the capacitance of the Rod increases linearly. The water level is determined from the time it takes to charge the capacitor to a fixed voltage. It is claimed that the technology allows sea level to be measured with high precision and with a high resistance to clogging or drift. Because the Rod capacitance depends only on water level, atmospheric pressure measurement is not required. The sensor is compensated for air temperature changes. The Aquarod is designed to make field deployment easier. The device operates under the supervision of a microcontroller and is therefore programmable. It consists of multiple units of fully submersible sensors, each unit capable of measuring a level change of 2 m. Because the sensors can be staggered, the device can cover larger ranges of sea-level variations.

16.6.2.5. Resonant-Wire Pressure Transducers

The resonant-wire pressure transducer was introduced in the late 1970s. In this design (Figure 16.27), a fine wire is gripped under some tension by a static

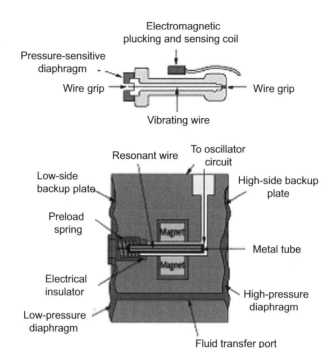

FIGURE 16.27 The constructional features of a resonant-wire pressure transducer. *Source: © 1998 Putman Publishing Company and OMEGA Press, LLC. Reproduced by courtesy of Omega Engineering, Inc., Stamford, CT 06907 USA www.omega.com.*

member at one end and the sensing diaphragm at the other. The wire is driven to oscillate by an oscillator circuit at its resonant frequency in a magnetic field. The frequency of oscillation is a function of the tension in the wire. A change in applied pressure changes the wire tension, which in turn changes the resonant frequency of the wire. The output signal is a pulse-stream, whose frequency can be converted to a digital value by counting the oscillations over a specific time interval. Because this change in frequency can be detected quite precisely, this type of transducer can be used for low differential pressure applications as well as to detect absolute and gauge pressures.

The most significant advantage of the resonant wire pressure transducer is that it generates an inherently digital signal and, therefore, can be sent directly to a digital circuitry or a microprocessor. Limitations of the resonant wire pressure transducer include sensitivity to temperature variations (for example, Vibrotron pressure transducers have been notorious for temperature-compensation requirements), nonlinear output signal, and some level of sensitivity to shock and vibration. Using a microprocessor to compensate for nonlinearity and temperature sensitivity minimizes these limitations. Typical accuracy is 0.1% of calibrated span.

16.6.2.6. Quartz Crystal–Based Pressure Transducers

An inherently analog pressure transducer can be converted to a digital pressure transducer by connecting its output to an analog-to-digital converter or a voltage-to-frequency converter, thereby transforming the analog signal to an equivalent digital signal. Such a transducer, whose output is inherently analog but can be transformed to digital signal, is known as a convertibly digital open-loop transducer. However, some pressure transducers are capable of producing digital output directly. Such transducers are known as inherently digital closed-loop transducers. After a decade of research on digital force sensors, Jerome Paros and Walter Kistler supervised the creation of a new quartz crystal technology in the year 1972, heralding the birth of a new type of quartz-based pressure transducer/instrumentation manufacturing company known as Paroscientific, Inc. (Paroscientific Digi-News E-Letter, Winter 2002 issue).

A familiar property of some crystals, such as quartz, is that they develop electric charges when they are pressed or stretched along some of their axes. This quality, which arises because the positively and negatively charged atoms in the crystal are not symmetrically placed and move to separate sides under pressure or tension, is called the piezoelectric property. The development of charges is very sensitive and quick. The opposite phenomenon is that if the crystal is exposed to a charge, it stretches or shrinks so application of an alternating charge makes it vibrate. A solid crystal of a given dimension possesses a very exact natural vibrating frequency under constant ambient temperature. However, applying tensile or compressive force along its crystalline axis alters its natural vibrating frequency in a nonlinear manner. Thus, the alteration of natural vibrating frequency of a piezoelectric crystal can be used to estimate the applied force/pressure under constant temperature. This is the principle used in the design of precision-pressure transducers using quartz crystals. Thus, an inherently digital closed-loop pressure transducer designed using a quartz crystal resonator possesses a high Q-factor, excellent repeatability, and negligible hysteresis, in contrast to a convertibly digital open-loop pressure transducer (Paros, 1976). The value of the Q-factor is proportional to the ratio of energy stored to energy lost per cycle in the vibrating system. The high Q-factor of the resonator means that a very small source of external energy needs to be supplied to achieve and maintain the oscillations. In such a transducer, the power dissipation is only a few milliwatts and, therefore, is most suited to oceanographic applications where power conservation is most important. The widespread use of and continuing trend toward digital information, together with the need for more accurate pressure instrumentation, have prompted the development of quartz crystal–based pressure transducers that are capable of precision measurements in the ranges from a fraction of an atmosphere to many atmospheres.

The essential sensing element in a quartz pressure transducer (QPT) is a quartz crystal beam whose resonant frequency of vibration varies with applied pressure-induced stress. The preferred choice of quartz crystal as the material for the manufacture of pressure transducers relies on its excellent elastic properties, long-term stability characteristics, and ease of vibrational excitation. Quartz oscillators are the most commonly used frequency standards because of these properties.

QPTs that are widely used at present for sea-level measurements are those manufactured by Paroscientific, Inc. (popularly known as Digiquartz pressure transducers). A *fixed-fixed* beam (i.e., both ends of the beam fixed) that vibrates in its first flexural mode is chosen as the resonant element in the Digiquartz pressure transducer because it can be made highly sensitive to force inputs. The entire resonator is fabricated from a single piece of quartz crystal to minimize energy losses due to joints. In order to stress the crystal, it is fastened to a structure that can transmit force to it. Its dimensions, composition, and stress load determine the resonant frequency of the vibrating beam. The mechanical isolation system that forms part of the vibrating beam has a much lower resonant frequency than that of the vibrating beam and, therefore, functions as a low-pass mechanical filter. This mechanical filter isolates the single-beam resonator from its mounting pads. This allows the single beam to vibrate in a high-frequency flexural mode without coupling energy to the pads. As a consequence, virtually no mechanical energy is lost, and this results in an extremely high operating Q-value, thereby contributing to the observed high accuracy and stability of the transducer.

A particular cut of crystalline quartz will strain under the application of an electrostatic field. Application of this electric field can be made through electrodes directly vapor-deposited on the beam and connected diagonally. With electrode pairs placed in opposition on the beam, the beam will bend upward for one sense of applied voltage and downward for the opposite voltage. Cyclical voltage changes will cause the beam to vibrate in a specific plane (Busse, 1978). The beam is forced into flexural vibration by an electronic oscillator circuit that tunes itself to the beam's resonant frequency. To achieve a significantly improved performance and stability, the quartz crystal resonator is located in an ultrahigh vacuum chamber, thereby eliminating air loading and damping effects. With this arrangement, the Q-factor increases from several thousand to over 40,000. Further, the crystal resonator is isolated from all external forces except the intended pressure-induced stress, thereby achieving better fidelity and repeatability.

The QPT operates on the principle that its natural resonant frequency is altered when subjected to strain. The resonant frequency of the vibrating beam increases under axial tension. Under compression, the frequency decreases. The external pressure is applied to the quartz crystal beam through a bellows or Bourdon tube. The bellows and Bourdon tubes function in different ways.

The bellows develop a force on the lever arm. This force is the product of the applied pressure and the bellows' effective area. The force develops a torque about the pivot that is counteracted by the rigid crystal. This imposed axially directed compressive stress in the resonator beam decreases the resonant frequency of oscillation, thereby providing a means of pressure measurement. Changing the size of the bellows and its position relative to the lever arm pivot allows the ratio of bellows force to crystal stress to be adjusted over a wide range. As a result, transducers of different pressure ranges can be fabricated with minimal changes in components (Busse, 1987).

In the design of pressure transducers meant for deep-water deployments, Bourdon tubes are often employed for conversion of pressure to load (Paros, 1984). In this case, a single-turn Bourdon tube constrains the quartz crystal beam. The Bourdon tube tends to uncoil under applied pressure, placing the resonator crystal beam under tension and thereby increasing the resonant frequency.

Both types of pressure mechanisms are balanced against acceleration by means of small, movable masses. These are adjusted so inertial torques and forces around the effective pivot or center of rotation are reduced to zero. As a result, the transducers are largely insensitive to linear acceleration.

Although the resonant frequency decreases with increasing applied pressure in the bellows mechanism and increases with increasing applied pressure in the Bourdon tube mechanism, the relative *change* in frequency follows the same physical principles in both cases and is, therefore, subject to the same algorithm when frequency is converted to pressure.

Electrodes, which are vacuum-deposited on the beam, are used with an electrical oscillator circuit to induce and maintain excitation and mechanical vibration. The quartz crystal resonator in the Digiquartz pressure transducer is essentially a precision oscillator similar to the crystal oscillators used in frequency standards and in clocks and watches, with long-term accuracies on the order of one part in 10 million (Busse, 1987). The oscillator piezoelectrically excites the crystal beam and detects the pressure that has induced the stress in the beam. The frequency of the electrical oscillator freely follows the crystal beam's resonant frequency as dictated by the applied axial load. Thus, on application of electrical power to the crystal oscillator, the crystal is piezoelectrically induced to vibrate in its resonant mode. A change in pressure correspondingly alters the axial load, and this results in a corresponding change in the resonant vibrational frequency of the crystal beam and, therefore, the oscillator frequency. Differential pressure transducers use a second bellows in an opposing, push-pull mode such that only a pressure difference between the inputs yields a resultant force on the crystal. With pressure P_1 applied to the bottom bellows and pressure P_2 applied to the top bellows, two coaxial and counteracting forces will be applied to the lever arm. In this manner, only the difference force will be applied to the quartz beam.

In the case of Paroscientific QPTs, typical pressure frequency varies from approximately 30 kHz to 42 kHz, so the excursion under full-scale load is approximately 12 kHz. The mathematical model that characterizes the pressure output (P) of a Digiquartz pressure transducer is given by the expression (Paros, 1976):

$$P = A\left[1 - \frac{T_0}{T}\right] - B\left[1 - \frac{T_0}{T}\right]^2 \qquad (16.7)$$

In this expression:

T: period output at applied pressure
T_0: period output at zero pressure input
A, B: curve fit coefficients

The output signal is measured with a period-averaging technique. In actual practice, it is inconvenient to measure T_0 because of the long pumping time required to reach low pressures. A more convenient method is to measure the transducer output at pressures across the full-scale range and derive all three calibration coefficients (A, B, and T_0) by curve-fitting methods.

Repetitive temperature cycling can generally induce mechanical strain relief in quartz-based pressure transducers. Usually three or four temperature cycles decrease strain effects by as much as 95%. In general, it takes a number of temperature and pressure cycles and weeks of continuous operation of the transducer in order to attain a fully relaxed condition. Stability of better than 0.01% has been demonstrated following this "burn-in" period.

The long-term stability of QPTs output is related to the aging of the quartz resonator, mechanical relaxation of machining and manufacturing stresses, and stabilization of the internal vacuum. Aging techniques such as continuous crystal operation (i.e., allowing the crystal to undergo mechanical flexing cycles—that is, vibrations—continuously) enhances the long-term stabilization of the quartz-based pressure transducers. In fact, crystal aging is the prime contributor to transducer drift. Wearn and Larson (1982) discussed the sensitivity and drift characteristics of QPT. Hewlett-Packard (HP) quartz sensor exhibited the lowest drift. The inherently lower drift of this sensor has been attributed to the absence of bellows. It is believed that elastic creep of the bellows under sustained stress causes the observed small drift in the Digiquartz sensor. Studies by Wearn and Larson (1982) have shown that the Digiquartz sensors are a factor of 10 lower in drift than the best existing strain gauges but are higher in drift than the HP quartz sensors. Based on the long-term stability test results for three Digiquartz pressure transducers for a period of 11 years, the median drift rate of the units tested was −0.007 mbar per year, and the largest rate observed for any of the units was −0.010 mbar per year over the 11-year test period. However, the author of this book has noticed a sudden shift in the zero-point offset of one of the Digiquartz transducers while used under laboratory environments.

The QPT is designed in such a manner that the pressure applied to the input port is transmitted directly to the quartz crystal beam. However, the mechanical coupling can transmit vibration and shock in the same manner. To avoid this scenario, the transducer is normally housed in a foam silicone rubber boot that isolates the device from vibration and shock. This isolator also provides thermal insulation, thereby reducing thermal effects. The present author has noticed that moderate-intensity acoustical noise might lead to a total failure of the Digiquartz transducer if it is not housed in the recommended silicone rubber boot that isolates the device from vibration and shock.

Properties of a quartz crystal are sensitive to thermal changes. Thermal sensitivity of quartz crystal arises from its pyroelectric property. This is a property of some materials to generate electric charges when they are heated or cooled. This property is closely linked to piezoelectric properties and appears in the same class of crystals, including quartz. In fact, all piezoelectric crystals are pyroelectric and vice versa. As in the case of piezoelectric crystals, the pyroelectric property appears in asymmetric crystals and is believed to arise because the positively and negatively charged atoms move to different sides when the crystals are heated. The presence of an electric field can affect the way the atoms move and the sense of the charge developed. The pyroelectric phenomenon is very sensitive, and a very small change in temperature can lead to a measurable charge difference.

The observed temperature effects that exist with quartz-based pressure transducers are related to its physical and geometrical parameters. Fortunately, these parameters are consistent throughout an instrument class, and, therefore, thermal characterization becomes possible by means of a simple mathematical expression.

Temperature effects on the quartz-based pressure transducers are the result of several distinct physical phenomena. The first is the inherent frequency-sensitivity of particular cuts of crystalline quartz related to the physical and mechanical properties. By cutting quartz in various directions relative to the principal crystalline axes, different temperature effects can be obtained. Judicious choice of the quartz crystal orientation can result in null temperature sensitivity that is a minimum at any desired temperature.

A second thermal effect is the change of coefficients A and B. These parameters generally relate to the linear thermal expansion coefficient of the bellows (larger effective area and greater force with temperature). The thermal expansion coefficient of the bellows results in greater transducer sensitivity to pressure at higher temperatures or correspondingly decreasing values of A and B. Empirical evaluation of the pressure transducer calibration coefficients A, B, and T_0 have demonstrated that all of these coefficients can be characterized by a simple parabolic expression (Busse, 1978).

Seawater temperature undergoes significant daily and seasonal variations, especially in coastal waters. For this reason, it is necessary that quartz-based transducers used for sea-level measurements need to be compensated for any sort of temperature sensitivity in order to achieve better accuracy in measurements.

The quartz crystal used for pressure measurement is designed to operate in a predominantly pressure-sensitive mode of operation by the proper choice of crystallographic orientation and force-producing structure to minimize thermal effects. However, in practice, it will have some degree of sensitivity to the ambient temperature. Digiquartz pressure transducers currently provide a second resonant quartz crystal beam for thermal compensation, which is designed to operate in a temperature-sensitive mode of vibration. A separate electrical oscillator is used to drive the temperature resonator. Thus, self-contained electronics have provided dual-frequency outputs, one for pressure and the other for thermal compensation. Thus, Digiquartz Intelligent Transmitters use separate pressure and temperature signal measurements to estimate temperature-compensated pressure values. The resolution of a pressure measurement is dependent on the length of time that the pressure and temperature signals are sampled. Longer sampling times result in better pressure resolution.

Because of the very high Q-factor of quartz crystal, both of the beams are excited by very low-power oscillator circuits. The temperature signal has a nominal frequency of 172 kHz, and it changes by about 50 ppm/°C. Typical temperature frequency varies from 168 kHz to 172 kHz. The two frequency outputs from the transducer represent an applied pressure signal that is contaminated with some temperature effects and a temperature signal that is contaminated with some pressure effects. The two signals contain all the information necessary to correct for temperature-induced inaccuracies in pressure measurements for any combination of pressure and temperature that occur in an oceanic environment.

Specific details of temperature-compensation methodology employed in Digiquartz pressure transducers have been reported by Yilmaz (2004). For a given temperature, an applied pressure will generate a specific crystal period, which is in some way related to the applied pressure. If the relationship between the crystal period and the applied pressure is known, the measured crystal period will give an indication of the applied pressure. This relationship is expressed in terms of three coefficients: C, D, and τ. The values of these coefficients are different for each transducer. For each temperature, these coefficients are expressed as polynomial expansions with coefficients C_1, C_2, C_3, D_1, D_2, T_1, and so on. Typically, it requires between 7 and 10 coefficients to fully describe the relationship between crystal output, temperature, and pressure. Determination of all the coefficients is carried out at the Paroscientific calibration laboratory. Digiquartz transducers are calibrated over the specified full temperature and pressure range so the indicated pressure as calculated by the thermal model will agree with the true applied pressure with a typical absolute accuracy of ±0.01% of the specified full scale of the transducer at all temperatures and pressures within the specified limits of a given transducer. Once the temperature frequency and pressure frequency outputs have been measured, temperature (°C) is calculated according to the equation:

$$T = (Y_1 + Y_2 + Y_3)U \tag{16.8}$$

where

T: temperature (°C)
U_o: temperature period (μs) at 25°C
U: temperature period (μs) $- U_o$ (μs)

The temperature coefficients U_o, Y_1, Y_2, and Y_3 are provided to the user. Equation (16.8) provides the temperature reading from the internal temperature sensor to compensate for the sensor's temperature.

Equation (16.9) expresses the temperature-compensated pressure output obtained from fixed-fixed vibrating beams (Busse, 1987):

$$P = C[1 - (T_0^2/\tau^2)]\{1 - D[1 - (T_0^2/\tau^2)]\} \tag{16.9}$$

where

P: pressure
τ: pressure period (μs)
$T_0 = T_1 + T_2 U + T_3 U^2 + T_4 U^3 + T_5 U^4$
U: temperature period (μs) $- U_o$ (μs)
U_o: temperature period (μs) at 25°C
$C = C_1 + C_2 U + C_3 U^2$
$D = D_1 + D_2 U$

In the preceding expressions, C_1, C_2, C_3, D_1, D_2, T_1, T_2, T_3, T_4, and T_5 are pressure coefficients. It may be noted that the temperature itself does not enter into the compensation except as the driving effect that changes both resonators simultaneously. Having estimated pressure and temperature values from period and coefficient data, the final pressure output is calculated using the expression:

$$P_{\text{Output}} = PM[(\text{units multiplier}) \times P + PA] \tag{16.10}$$

where

P_{Output}: final output pressure in psi
P: raw pressure value
PM: pressure multiplier
PA: pressure adder
Units multiplier: value used to convert psi to another pressure unit

Using this temperature compensation allows an accuracy of 0.01% of full scale to be achieved over the entire operating range.

Using a built-in microprocessor enables an online solution of the preceding expressions for pressure and temperature to allow for full plug-in interchangeability. An RS-232/485 link allows direct interface to a computer. The RS-232/485 interface allows complete remote configuration and control of operations such as resolution, sampling rate, integration time, and baud rate. However, all Paroscientific products need to be handled with extreme care. It is recommended that these products always be stored and transported in static-dissipative packaging with at least 3 inches of mechanical shock-absorbing anti-static material around each unit. A transducer packed in clear, non-dissipative bubble wrap or Styrofoam pellets can develop electrostatic voltages of up to 20,000 volts of ESD potential that can damage electrical components (Paroscientific Digi-News E-Letter, Winter 2002 issue). Some typical QPTs are shown in Figure 16.28.

Temperature-compensated quartz-based ocean bottom-pressure gauges provide high-quality data to detect offshore tsunamis before the arrival at the coastline with a peak-to-peak amplitude as small as several millimeters to centimeters (e.g., Gonzalez et al., 1991; Okada, 1995; Ritsema et al., 1995; Hino et al., 2001; Hirata et al., 2003; Baba et al., 2004; Titov et al., 2005a, b). Despite incorporating the best practically possible design mechanism for temperature compensation, some gaps have been reported in the case of some design variants of quartz-based pressure sensors. For example, in Japan, six cabled observatories (depths ~2500 m) with ocean bottom-pressure gauges (HP models 2813B or 2813E based on "quartz resonator pressure transducer"; Karrer and Leach, 1969) have been operated.

The 2813 series of the pressure sensors has two quartz resonator pressure transducers (Takahashi, 1981). One is the QPT, which is hydraulically connected to the ambient seawater through silicon oil in the buffer tube attached to the pressure inlet and thus affected by both pressure and temperature. The other is a quartz temperature compensator (QTC) that is not affected by the hydraulic pressure, but the ambient water temperature is conducted through the thermal conducting materials of the sensor. According to Takahashi (1981), in the temperature range from 0 to 10°C, the resonant frequency of QPT and QTC are given, respectively, by:

$$F_P = 4,992,000 + 19.8\,T_P - 2.0D \qquad (16.11)$$

and

$$F_C = 5,000,000 + 20.0T_C \qquad (16.12)$$

where F is the frequency, T is the temperature, D is the pressure of the unit in meters, and subscripts p and c denote the quantity concerning QPT and QTC, respectively. The coefficients of T_P and T_C are designed to become equal but have slightly different values for practical reason (Takahashi, 1981). The 2813-series produce the output resonant frequency that equals the difference between F_P and F_C as:

$$F = F_C - F_P = 8000 + 20.0(T_C - 0.99T_P) + 2.0D \qquad (16.13)$$

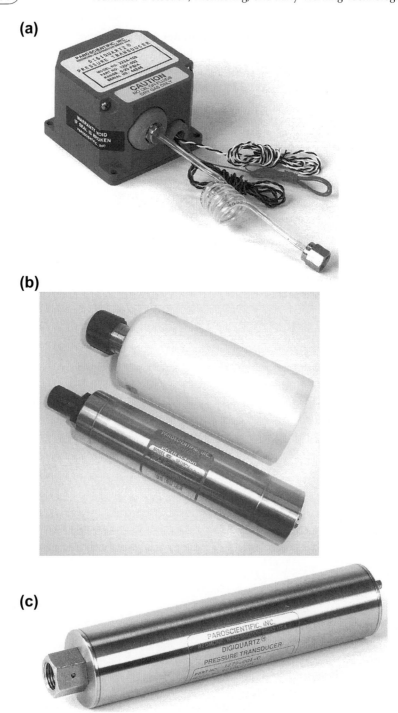

FIGURE 16.28 Some typical quartz-based pressure transducers. (a) Series 2000; (b) Series 8000; (c) Series 9000. *Source: Courtesy of Steve Smith, Sales & Applications Engineer, Paros-cientific, Inc., Reproduced with kind permission from Paroscientific, Inc.*

In the thermal equilibrium condition where the relation:

$$T = T_P - T_C \tag{16.14}$$

is satisfied, Eq. (16.13) results in:

$$F = 8000 + 0.2T + 2.0D \tag{16.15}$$

where T is the ambient temperature. The electronic implementation of Eq. (16.15) is accomplished by inputting the QPT and the QTC signals (digital frequencies) to a mixer circuit and further inputting the mixer output to a low-pass filter. The low-pass filter output is essentially the pressure gauge output. By this mechanism, the pressure gauge output is temperature-compensated under static thermal equilibrium condition of Eq. (16.14). In the pressure gauges deployed by Hirata and Baba (2006), preprocessed, static temperature correction as mentioned previously was applied, for which the manufacturer provided two-dimensional (pressure and temperature) polynomial functions with 16 calibrated coefficients. Such a correlation was discussed by Takahashi (1981, 1983). It has been found that regardless of the static temperature-compensation mechanism, numerous observations of offshore ocean-bottom pressures have deviated from ocean tidal motion in the absence of tsunamis. Amplitudes of such deviations were equivalent to 10 to 20 cm in water height, and these periods span from about 20 minutes to over an hour. Such sharp deviations have been found to highly correlate with sudden temperature changes (probably due to irregular deep ocean currents) around the pressure gauges. Hirata and Baba (2006) ascribed this deviation to the transient thermal response of the pressure gauges, which produce apparent pressure fluctuations when the ambient temperature changes quickly so the temperature equilibrium condition of the two quartz transducers (which are integral parts of the pressure sensor) is no more valid. It has been pointed out that the bottom-pressure measurement errors arising from such rapid ambient temperature changes may have to be corrected for a robust tsunami warning based on offshore tsunami observations with pressure gauges. Hirata and Baba (2006) put forward an empirical method to estimate the transient thermal response correctly. This empirical method evaluates the effect of sudden temperature changes on pressure measurements by estimating an impulse thermal response of ocean-bottom pressure gauges so the noises other than those originated from the temperature change are considerably suppressed.

16.6.3. Subsurface Pressure Transducer Installation Schemes

One way to install subsurface pressure transducer-based sea-level gauges in a harbor environment, where concrete walls or other purpose-built structures are readily available or easy to erect, is rigidly mounting the transducer on a structure. The output signal from the transducer can be brought to a nearby cabin and recorded on some suitable electronic media for future use or displayed/transmitted in real time for port-control or other operational applications such as navigation, hydrographic survey, and so on. Such an installation methodology is shown in Figure 16.29. Figure 16.30a indicates a widely used method wherein the sea-level gauge is deployed in a conventional tide-well. The gauge can be hung from a cable or rope and weighed down with adequate weights to keep it taut. The transducer can also be rigidly mounted on a ladder deployed within the tide-well. This method is suitable if the currents in the site are not large (more than 50 cm/s). If large currents are present, it is advisable not to mount the gauge in this manner because the Bernoulli effect might adversely affect the performance of the gauge. Modification of the orifice of the well with the addition of a pair of parallel plates, as recommended by Shih and Baer (1991), would circumvent this problem. Providing a few holes drilled radially on the submerged portion of the wall of the well will prevent trapping of lower-density water in the well. In the absence of a tide-well, an arrangement shown in Figure 16.30b can be conveniently used. This arrangement, which is successfully used in the Gulf of Kachchh (India) by Adani Port, employed commercially available standard pipes with flanges to erect a tide-well. The pressure transducer is conveniently hung from a cable, which carries the signal. Figure 16.30c illustrates another scheme in which the gauge is rigidly mounted on a ladder.

In situations where the gauge needs to be installed on the bank of an estuary, a long cable run may become necessary, since most sites have no structures or piers. Often, cables need to be laid across the mudflats. In this situation the cables need to be protected in a sturdy pipe and then buried below the riverbed so they will not be damaged by the breaking wave force. The data recording/ readout units may be located in some existing buildings or huts. In such installations, the sea unit that houses the pressure transducers must be kept submerged at a fixed level below the CD. In this situation, the sea unit usually remains in the photic zone and tends to attract marine growths of differing kinds.

When sea-level measurements need to be made from locations away from the coast, a widely used method is deployment of a pressure transducer-based gauge on a U-mooring line (Figure 16.31). In this arrangement, the sea-level gauge needs to be located as close to the seafloor as possible. This is because datum stability is of prime importance in sea-level measurements. Because the gauge is usually positioned close to an anchor weight (say, an anchor chain), it may be advisable to provide sufficient protection to the gauge so it will be protected from any direct impacts either due to its own impingement on the seafloor or to the anchor weight hitting it during deployment. When deployed in a water body that experiences strong flows (say, tidal currents), locating the gauge at a distance from the seafloor would cause serious contamination of the sea-level measurements. This occurs as a result of the flow-dependent inclination of the mooring line and the consequent vertical oscillation of the transducer during a tidal cycle. The contamination is usually

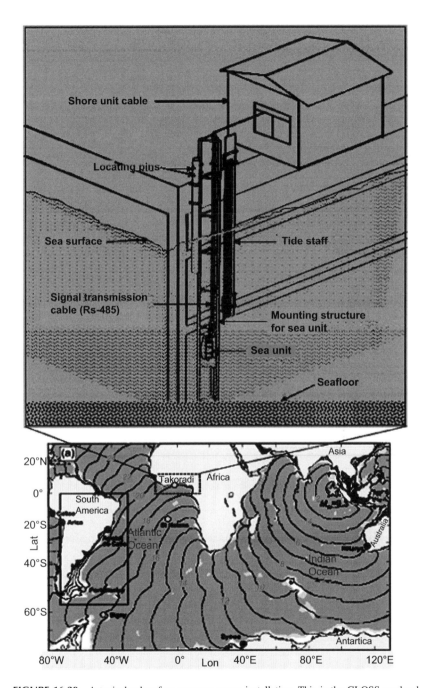

FIGURE 16.29 A typical subsurface pressure gauge installation. This is the GLOSS sea-level station at Takoradi harbor (Ghana, West Africa; See Joseph et al., 2006). The tide staff (graduated scale) seen close to the pressure gauge is a part of every tide station. The blue curved lines (isochromes) on the map represent the wave front of the December 2004 tsunami that originated in Sumatra, indicated by a star. *Source: Candella et al., 2008; Advances in Geosciences; Copernicus Publications on behalf of the European Geosciences Union.*

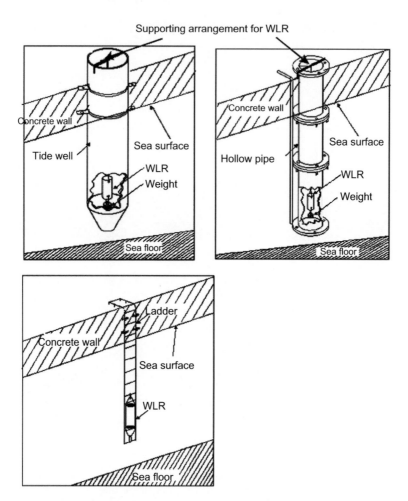

FIGURE 16.30 Some typical ways to install pressure transducer−based sea-level gauges in a harbor environment. (a) Deployment in a conventional tide-well, (b) Deployment in a well constructed from standard pipes with flanges, (c) Deployment on a ladder.

manifested in the form of stretching the tidal range. In extreme cases, the recorded change in water column over the pressure transducer is primarily due to the vertical oscillation of the mooring line rather than the tidal oscillations of the sea level. In this case the water-level record measured by the gauge was found to suffer from two types of errors: stretching of the tidal range and doubling of the peaks. Another problem that might occur with a pressure gauge, especially with a capillary-type pressure inlet, deployed in heavily suspended-sediment-laden waters or water bodies having heavy deposits of clay on the seafloor is occasional reduction in its pressure sensitivity as a result of entry of sediment particles in the pressure inlet. The

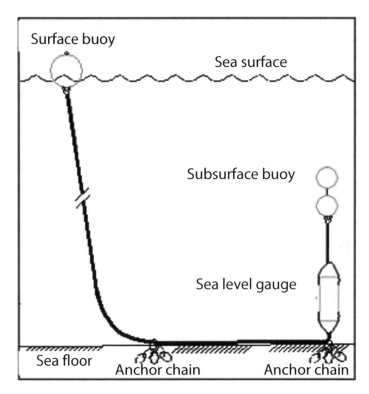

FIGURE 16.31 Deployment of a pressure transducer–based sea-level gauge on a U-mooring line.

reduction in sensitivity is manifested as a reduction in tidal range. However, the entire corrupted record can easily be corrected if one or two sample measurements of tidal range obtained from a tide staff in the vicinity of the pressure gauge are available. In areas of intense fishing activity, the pressure gauge could be dragged by the fishing nets, resulting in datum changes. Despite these limitations, subsurface pressure gauges that are installed on U-mooring lines with proper care and close to the seabed have provided valuable records of sea-level measurements.

Different manufacturers provide the pressure gauge outputs in different units. Ultimately, the pressure measurements need to be converted to water level measurements using the relationship: $p = h\rho g$, where p, h, ρ, and g represent water pressure, water height above the pressure port of the pressure sensor, average density of the water column above the pressure sensor, and acceleration due to gravity, respectively. If an absolute pressure sensor is used for pressure measurements, the local atmospheric pressure at the mean sea level (MSL) must be subtracted from the absolute pressure measurements to obtain the water pressure above the pressure port of the sensor. The following relationships can be used to convert the measured water pressure to cgs unit:

$1 \text{ bar} = 10^6 \text{ dynes/cm}^2$
$1 \text{ bar} = 14.504 \text{ psi}$
$1 \text{ bar} = 100 \text{ kPa}$

16.6.4. Gas-Purged Bubbler Gauges

A common problem encountered with coastal installations of subsurface pressure gauges is that the sea-unit usually remains in the photic zone and tends to attract marine growths of differing kinds. Figure 16.32 shows two such instances of luxuriant growth of barnacles and shells on the sea units made of two different materials. Overgrowth of this kind can affect the measurements. Periodic cleaning and servicing are required with such coastal installations. This problem is overcome by the use of bubbler gauges. A large majority of sea-level gauges used in the United Kingdom are of this type.

A bubbler gauge works on the principle that if compressed gas is bubbled freely into a liquid from a submerged fixed end of a tube, the pressure in the entire length of the tube, irrespective of the elevation of its other end, is

FIGURE 16.32   (a) The growth of barnacles on a stainless-steel sea unit of a subsurface pressure transducer–based sea-level gauge deployed at Zuari estuary, Goa, India. Copper and teflon components leading to the pressure port on the unit have no barnacles. (b) Growth of shells on PVC sea unit of a subsurface pressure transducer–based sea-level gauge. Brass plates on the unit have no shells.

approximately equal to the pressure head of the liquid column over the bubbler orifice (Pugh, 1972; Ling and Pao, 1994).

In operation, compressed air or nitrogen gas from a cylinder is reduced in pressure through one or two valves so there is a small steady flow down a connecting tube to escape through an orifice in an underwater canister called a bubbler (Figure 16.33). Depending on the sea-level elevation above the bubbler orifice, an air pressure equal to the hydrostatic pressure is established inside the measuring tube. This is also the pressure transmitted along the tube to the measuring and recording system, apart from a small correction for pressure gradients in the connecting tube. A pressure transducer attached to the land-ward end of the tube measures the gas pressure (which is equivalent to the fluid pressure above the bubbling point). The measured pressure is converted to sea-level elevation using the principle previously described in this chapter. The pressure transducer used can be either an absolute or a gauge type. However, the usage of an absolute pressure transducer would make the gauge completely unaffected by humidity and condensation.

The error due to pressure drop in the tube increases in proportion to (l/a), where l and a are the length and radius of the tube, respectively. For this reason, special care is necessary when designing for connecting tube lengths in excess of 200 m. Consequently, in locations having a broad beach and shallow topography, the use of the pneumatic pressure transmission principle must be made with some caution. Further, an unreasonably long pneumatic tube would introduce a delay in pressure transmission.

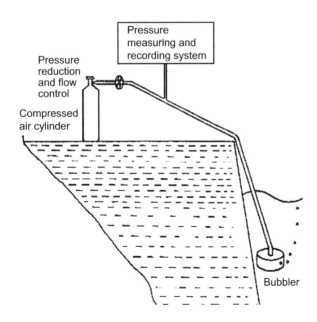

FIGURE 16.33 A bubbler gauge. *Source: Pugh, 1987; Tides, surges and mean sea-level: A handbook for engineers and scientists, 472 pp; Reproduced with kind permission from Wiley-Blackwell, UK.*

The advantages of pneumatic bubbler systems include the following:

- Easily installed where other gauges are difficult to install (e.g., rocky coasts)
- Easier to install in flat tidal areas
- Clearly defined datum
- Long-term stability
- Expendable nature of the vulnerable underwater parts
- Measurements possible even in contaminated waters
- Avoids the use of expensive pressure transducers or other hardware underwater

The technology associated with the bubble-in principle has improved substantially with the introduction of maintenance-free compressor and microcomputers. For example, unlike with conventional bubble-in gauges, the modern gauges do not require external compressors or nitrogen bottles. Instead, the compressed air is generated by a maintenance-free piston pump with integral valve function. In this arrangement, a piston pump inside the instrument enclosure generates compressed air, which flows through a dedicated line into the bubble chamber at programmable intervals, where it bubbles out uniformly into the water. The modern sea-level gauges that employ the bubble-in technology possess most of the following features:

- Intelligent sensor card with integrated data storage memory module
- Internal, maintenance-free air compressor
- Variable sampling and storage intervals
- RS-232 and SDI-12 interfaces
- In situ communication via palmtop/notebook computer
- Well-arranged user matrix for fast configuration of the data logger via integral touch-key path
- Compatibility with wireless communication equipment (radio, satellite)
- Infrared interface and LCD display for clear and easy-to-read indication of system and sensor parameters

However, because bubbler gauge systems are pressure-based, it is necessary to have a good idea of water density for translation of pressure to water level, especially if there are large density changes with seasons and the tidal phases. Large density changes occur in estuarine sites where there is considerable freshwater influence from the river that joins the estuary. Bubbler gauges in their present form are also biased by waves.

In analogy with stilling-well gauges having large well/orifice diameter ratio, bubbler gauges in which damped nitrogen or compressed air is allowed to bubble out from the submerged end of a long narrow tube is also known to have a nonlinear response to large-amplitude, short-period waves. For example, Kulikov and colleagues (1996) observed that because of the nonlinear response of bubbler gauges at short periods, the recorded November 3, 1994, landslide-generated 3-minute tsunami heights at

Skagway Harbor, Alaska, were considerably smaller than those of the actual tsunami.

16.7. RADIOWAVE INTERFEROMETRY

Interferometric techniques are widely used in optics for spectroscopy and metrology. This technique has subsequently been applied at radio frequencies as well and came to be known as *radiowave interferometry*. In the field of sea-level measurement, radiowave interferometry has been tried out because it is a nonintrusive remote sensing method. The method relies on the fact that water is a good reflector of radio-frequency (RF) energy. The radiowave interferometer measures sea level by sensing the path difference between direct RF signals and those reflected from the sea surface. The path difference, which is a function of the geometry of the system and, therefore, the sea level, generates an interference field at the receiving antenna. Thus, interference measurement at the receiving antenna enables sea-level measurements. The remote character of operation in this method is useful in estuarine and coastal environments. Both ground-based and satellite-borne RF transmitters have been attempted. In both cases, the RF—receiving antenna is located on the ground. The operations of these two systems are briefly addressed in the following sections.

16.7.1. Ground-Based Transmitter Systems

Remote radiowave interferometric techniques employ a bistatic scheme where an S-band or L-band radiowave transmitter sends a continuous wave (cw) RF signal across a body of water to a receiver located at a horizontal separation ranging from 400 to 2700 m (Glazman, 1981, 1982). The geometry of the method is given in Figure 16.34. The received signals are composed of two coherent parts, one direct from the transmitting antenna to the receiving antenna and the other by way of low-angle reflection from the sea surface. Sea-level information is contained in the path difference of signals arriving over the two paths. This path difference causes the two signals to interfere at the receiving antenna with a resultant that depends on the signal frequency and the geometry of the system, which incorporates the instantaneous sea level. Thus, with the rise and fall of the sea level, this geometry also undergoes a corresponding change. With the bistatic scheme as in Figure 16.34, it becomes possible to derive information about the relative sea-level height by monitoring the interference field along a vertical direction at the receiver. Using two receiving antennae spaced by a known vertical distance, it becomes ideally possible to obtain unambiguous information on sea-level elevation, independent of sea surface roughness caused by wind-waves. In practice, the accuracy of sea-level measurements using this method is in the range of 3 to 9 cm, depending on the sea state and tidal range.

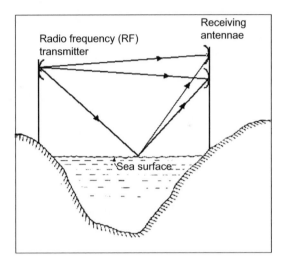

FIGURE 16.34 Radiowave interferometric method for measuring sea levels. *Source: in part from Glazman, 1982; IEEE J. Oceanic Engg; Reproduced with kind permission of IEEE.*

An advantage of this system is that it has a remote character of operation and operates around the clock regardless of weather conditions. A disadvantage is that it can be operated only on such locations where the transmitter and the receiver can be suitably located, with the horizontal distance between them not exceeding a few kilometers. For large distances (many tens of km), the Earth's curvature and atmospheric refractions can have a significant effect on the received signal.

16.7.2. Satellite-Borne Transmitter Systems

The concept of radiowave interferometric methods for measuring sea-level elevation using satellite-borne transmitters is shown in Figure 16.35. The method was first suggested and implemented by Anderson (2000). The satellites of the existing constellation of global positioning systems (GPS) are conveniently used as the radio frequency transmitters. The receiving antenna is located on top of a tall tower. For a given satellite that appears in the horizon, there are two paths for the signal from the GPS satellite to a ground-based receiver when there is a water surface between the receiver and the horizon. One is the direct path, or direct ray, and the other is the reflected path, or the reflected ray. As the satellite moves along its orbit, the difference in path length between the direct ray and the reflected ray changes many wavelengths. As a result, the signal at the receiver will appear as an interference pattern when plotted as a function of time or subsatellite range, which is the range from the receiver to the nadir point of the satellite on the Earth. For very low-elevation angles from the receiver to the satellite (less than a few degrees or so),

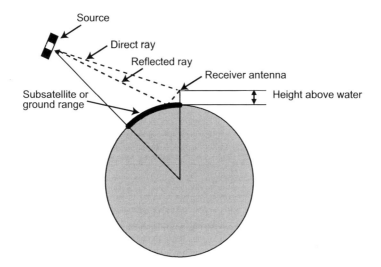

FIGURE 16.35 Radiowave interferometric method for measuring sea levels using a satellite-borne transmitter. *Source: Anderson, K.D. (2000), Determination of water level and tides using interferometric observations of GPS signals, J. Atmos. and Oceanic Technol., 17(8), 1118-1127;* © *American Meteorological Society; Reprinted with permission.*

atmospheric refractive effects strongly influence the spacing between peaks (or nulls) in the interference pattern. However, at higher-elevation angles (above a few degrees or so), refractive effects are minimal, and the spacing between peaks in the interference pattern is almost entirely due to the height of the receiving antenna above the reflecting surface (i.e., sea-level). Because of this, it is possible to deduce the height of the receiver antenna above the sea surface by careful measurements of the interference pattern at the receiver. As the receiver antenna is fixed and the sea level is changing, the change in sea level with reference to a given chosen datum (say, CD) can be estimated. Thus, the combination of the orbiting GPS satellites, together with the ground-based receiver module, would function as a practical remote sensing sea-level gauge.

The accuracy that can be achieved in sea-level measurements depends largely on efficient signal processing algorithms and careful location of the receiver antenna system. This is because the received reflected signal is usually contaminated from various sources during its propagation between the GPS satellite and the receiver. For example, wind ruffling of the sea surface scatters the reflected ray. This is a major source of noise for estimating the height of the receiver antenna. Further, spatial and temporal variations in the refractivity—predominantly in the first few kilometers above the sea surface—cause the rays to wander. However, this is considered to be a minor problem because the rays traverse a relatively short path. In the ionosphere, the direct and the reflected rays travel on nearly the same path, and as a result, the path length difference is not affected by variations in ionospheric refractivity. Another source of noise, which can be significant, is additional reflections into the receiver antenna from nearby objects (i.e., multipath). However, proper antenna placement can circumvent this problem. Anderson (2000) employed two receiver antennas separated by vertical distance of 10 cm for extraction of sea-level data from GPS-based interferometric methods. Both antennas were tilted a few degrees from the zenith toward the sea surface to improve reception of the reflected ray. A correlation signal processing scheme was employed for extracting useful information from the received signal.

Site selection has been found to be an important consideration in obtaining realistic sea-level information using interferometric methods. Figure 16.36 shows a comparison of a time-series of sea-level measurements obtained from GPS-based interferometric method and an in situ sea-level gauge in the vicinity. The light gray solid curve represents the in situ tide gauge data, and the vertical gray lines represent the standard deviation of the samples about the mean. The standard error in the comparison of GPS-interferometric measurements to a well-calibrated in situ tide gauge measurements was less than 12 cm. However, a receiving antenna that was incorrectly located was found to severely contaminate the estimated sea-level data. For example, comparison of sea-level measurements obtained from a GPS-based interferometric sea-level gauge located near a wet to damp shallow beach and an in situ sea-level gauge showed large errors during low tide when the beach was uncovered (Figure 16.37). On a shallow beach, the GPS signal might intercept a large portion of the beach during low tide, and the reflected ray might originate from

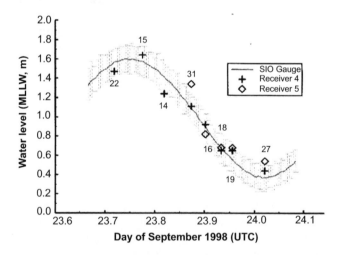

FIGURE 16.36 Comparison of time-series of sea-level measurements obtained from the GPS-based interferometric method and an in situ sea-level gauge. *Source: Anderson, K.D. (2000), Determination of water level and tides using interferometric observations of GPS signals, J. Atmos. and Oceanic Technol., 17(8), 1118-1127; © American Meteorological Society; Reprinted with permission.*

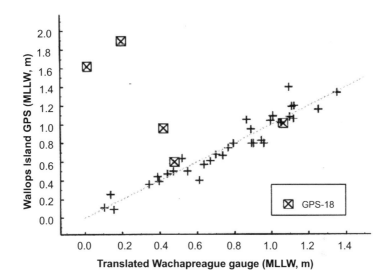

FIGURE 16.37 Comparison of sea-level measurements obtained from a GPS-based interfero-metric sea-level gauge located near a wet to damp shallow beach and an in situ sea-level gauge, showing large errors in sea-level measurements during low tide when the beach was uncovered. *Source: Anderson, K.D. (2000), Determination of water level and tides using interferometric observations of GPS signals, J. Atmos. and Oceanic Technol., 17(8), 1118-1127; © American Meteorological Society; Reprinted with permission.*

the beach rather than from the water surface. This possibility occurring at an incorrectly sited location gives rise to significant errors in GPS-based inter-ferometric sea-level measurements. Results of long-term measurements need to be analyzed to fully understand the merits and limitations of interferometric methods of sea-level measurements.

16.8. DIFFERENTIAL GLOBAL POSITIONING SYSTEMS ON FLOATING BUOYS

Measurement of sea levels using the differential global positioning systems (DGPS) based sea-level gauge relies on a method called *kinematic differential positioning* based on the highly stable carrier phase signals transmitted by NAVSTAR global positioning satellites. By observing the signals from two receivers simultaneously (one placed on a precisely known stable position known as *reference station*), the relative elevation of the moving receiver (i.e., *gauge station*) can be derived in a geocentric reference frame or with respect to that of the reference station. The technique of using DGPS receivers in floating buoys allows reliable sea-level measurements to be made as far as 20 km from the coast and requires only simple anchoring of buoys containing a GPS receiving antenna, tilt transducer, radio unit, and the required power supply housed within a buoy and a solar panel to charge the battery (Figure 16.38). The

FIGURE 16.38 Picture of the GPS-buoy deployed in the Japan Sea. *Source: Courtesy of the Ports and Harbours Bureau of Ministry of Land, Infrastructure, Transport and Tourism of Japan.*

GPS receiver at the reference station determines its three-dimensional position using signals received from many GPS satellites in the horizon (at least three at a time). The measured GPS data from the gauge station along with the tilt transducer data are time-tagged and transmitted to the reference station for processing. Analysis of the phase observations in this differential mode provides elevation of the buoy antenna relative to that on the land-based reference station. Analysis of kinematic GPS observations requires that the initial phase ambiguities be determined during motion, since the buoy is never in a static position. An initialization is also necessary after periods of loss of phase lock to the satellites and/or when switching the GPS receiver on or off for observing only over certain time periods. For longer separations (i.e., more than 20 km) between the reference station and the buoy antenna, the position accuracy in the z-axis (i.e., sea-level elevation) will gradually degrade as a function of atmospheric conditions. Thus, separations less than 20 km are preferred in GPS-based sea-level measurements. Figure 16.39 shows the concept diagram of DGPS-technique used for sea-level measurement.

The method of using two GPS receivers provides accuracies considerably better than what would be achievable with a single receiver. This is due to the fact that measurements made by receivers that are separated by 20 km or less have very nearly the same systematic errors, and these get canceled out because of the "differential" nature of processing. The systematic errors relate to satellite ephemeris, satellite clock, propagation, and so forth. The particular

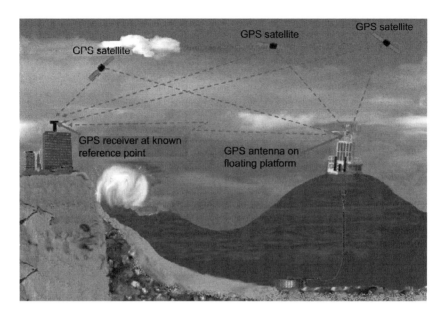

FIGURE 16.39   Concept diagram of DGPS-technique used for sea-level measurement.

processing technique used in the GPS receiver for estimation of a three-dimensional position significantly affects the level of achievable accuracy. In DGPS systems, differential processing techniques are used where corrections are provided to receivers, ideally reducing all bias errors to zero and leaving only the contributions of the random error sources. Applying Kalman filtering significantly reduces the influence of random errors. Improvements in the receiver technology have enabled reduction in receiver noise. Some natural sources of errors are removed by modeling algorithms incorporated into the receiver's microcomputer. This allows the sea surface to be determined with an accuracy of 1 to 2 cm in the reference system. While most of the common errors associated with GPS measurements get canceled in a DGPS system, some errors are specific to the case of sea-level measurements. Such an error is the dynamics of the receiver antenna resulting from the wave-induced motion of the buoy. Receiver dynamics can impair the measurement accuracy by introducing noise in the phase-lock-tracking loop. The noise performance of these loops can be improved by decreasing the loop bandwidth to as small a value as possible. However, beyond a certain limit this method causes a serious degradation in the dynamic tracking performance of the loop. To take care of wave-induced motions of the sea surface, observations can be carried out once per second over a few minutes and then averaged. Another source of inaccuracy in the GPS measurements of sea levels is that tidal and other ocean current forces pulls the buoy deeper into the water when the buoy arrives at the extreme position allowed by the length of the anchor chain. Additionally, other sources of errors are ionospheric, tropospheric, and orbit-related factors. However, these errors cancel for a short baseline. A unique advantage of DGPS systems is that they can complement the satellite altimeters as crossover points.

GPS buoys are widely used in tsunami monitoring. For example, the JMA makes use of data received from such buoys in the near-shore regions of Japan. In Japan, the Ports and Harbours Bureau of Ministry of Land, Infrastructure, Transport and Tourism (MLIT) is the deployer of the offshore GPS buoy system network to observe deep-sea waves and tsunamis. Figure 16.40 shows the location map of the network of DGPS-buoy system based sea-level gauges around Japan as of September 2010. The GPS buoys are deployed farther offshore but within 20 km of the coast (Kato et al., 2005). Because the GPS buoys can be installed in deeper and more offshore areas, this method facilitates earlier tsunami detection with appropriate filtering techniques (Nagai et al., 2008). In addition, they can observe deep-sea waves and swells. These buoys can detect tsunamis with amplitudes of several centimeters. In the present JMA procedure, offshore tsunami observations, which are monitored by operators on a real-time basis, help determine quickly if a tsunami was actually generated. For a GPS buoy located approximately 10 km offshore from the Cape Muroto, Shikoku Island, the tsunami from the 2004 Kii-Hanto-Oki earthquake could be detected about 10 minutes earlier than at the nearest tide gauge (Nagai et al., 2004, 2005; Hirata, 2009).

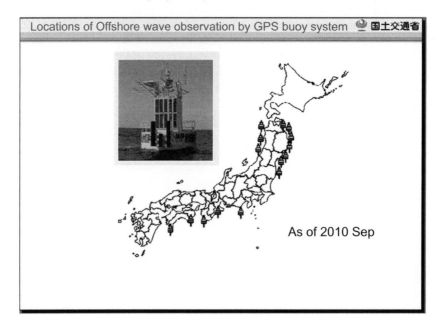

Locations of Offshore wave observation by GPS buoy system　国土交通省

As of 2010 Sep

FIGURE 16.40 Location map of the network of DGPS-buoy system based sea-level gauges around Japan. Picture of the buoy is shown in the inset. *Source: Courtesy of the Ports and Harbours Bureau of Ministry of Land, Infrastructure, Transport and Tourism of Japan.*

16.9. APPLYING A TIDE STAFF FOR DATUM CONTROL IN COASTAL SEA-LEVEL MEASUREMENTS

The tide staff is the oldest device used for sea-level measurements with reference to a recognized datum level called CD. The tide staff consists primarily of a single graduated staff that is either driven into the seabed or erected vertically on the side of a jetty wall. The tide staff reading can be understood by the average person, and it provides visual information from a close distance regarding the instantaneous sea level at a given coastal/ estuarine site. Usually, the zero reading of the tide staff coincides with the CD. However, there are a few instances in which this tradition has been found to have been violated. Therefore, the connection between the tide staff's zero-reading and the CD must be ascertained before arriving at conclusions pertaining to important scientific issues, such as the interpretation of MSL variability at a given location. Although the tide staff may look old-fashioned, it is found in conjunction with every modern sea-level gauge, primarily for dynamic calibration of electronic tide gauges and even satellite altimeter measurements of sea levels. Dynamic calibration is usually accomplished by establishing a correlation between the instrument outputs during a spring tide cycle and the corresponding tide staff readings. The tide staff forms an important component of all coastal sea-level stations and is often used for the purpose of detecting any datum level shift in the autonomous recording device. Precise vertical referencing is especially important for identifying small, elusive changes in long-term sea-level records.

The tide staff is usually used for datum adjustment of a newly installed gauge. The datum of the new gauge is initially adjusted to give the expected tide height as indicated on a local tide staff. It is also used for datum control of other gauges. Tide gauge datum control is an essential issue for any installation. Consequently, even if a tidal station is equipped with the most modern tide gauge, it is essential to provide confirmation of the datum from time to time by means of an inexpensive tide staff (IOC Manuals and Guides No. 14, Vol. 3, 2002). Individual measurements by tide staffs may not be especially accurate, but they can be used to guard against gross errors in the datum of sophisticated sea-level instrumentation. When the tide staff readings are intended for this application, however, it is important to be aware that tide staff observations must be made when the sea is calm so the readings obtained are free from wave-induced uncertainties.

Traditionally, in the ports spread all over the world, experienced human observers make visually averaged sea-level readings from tide staffs at specified sampling intervals, usually 30 minutes. However, in regions where the tidal height changes rapidly, tide staff readings need to be recorded at much closer time intervals—say, 5 minutes.

Tide staffs have been modified from time to time. Guth (1981) reported a photo tide staff, which is an autonomous device and thus eliminates the need for a tide observer. This modified tide staff records sea-level oscillations using a vertical free-floating graduated rod staff and time-tagging the water-level readings with a chronometer. The photo tide staff comprises a vertically oriented graduated rod staff that is attached to a float to give it buoyancy. The rod staff is floated within the confines of a vertical cylindrical tube, which functions as a tide-well. A T-shaped junction box attached to the tide-well near its upper end at a point above the highest expected tide level houses a chronometer, an index mark, a time-lapse camera, and a light flash that is synchronized with the camera. A plastic tube that is mounted vertically above the T-shaped junction box functions as a vertical guide for the rod staff as it rises and falls with the tidal oscillations. The tide-well has an orifice near its bottom end to allow the free flow of water in and out of it as the tide rises and falls, respectively. Further, it dampens the high-frequency sea surface wave action within the tide-well. The length and diameter of the tide-well, rod staff, and guide tube, and the height of the index mark and the camera above the orifice are variable and are designed to suit the tidal range (i.e., the extent of vertical swing in sea level) at the given place. The elevation of the rod staff is considered to be relative to the fixed index mark. This index mark is located at a fixed height above the orifice, and its level with reference to a local benchmark can be ascertained from conventional survey techniques. The fixed index mark enables comparing the rod staff graduations against a stable reference level as the staff rises and falls with the tide. An automatic time-lapse camera located opposite the index mark enables recording the rod staff readings and simultaneous time readings from a chronometer, which is located near the index mark. This arrangement enables relating the sea-level elevation record to specific times of the day, during which the rod staff readings are made. An advantage of this system is that by photorecording the staff elevation from a fixed camera platform relative to a fixed reference level, it eliminates the need for the presence of a human observer for direct reading of the sea-level oscillations from the staff. Another addition to the tide staff family is an electric step gauge, which was described earlier.

The use of benchmark-leveled tide staff data sets would be especially important in deriving CD-referenced sea-level records from seafloor-mounted gauges deployed in near-shore regions. The adjustment calculated from the linear regression between the tide staff readings and the corresponding measurements from the autonomous device can be used to reduce the sea-level measurements from the latter to tide staff zero. An important aspect that needs to be considered in the use of regression analysis or any type of intercomparison measurements is that sea-level readings from both devices must have the same time reference (i.e., readings have been made simultaneously). Any difference in the times would be interpreted by the two systems as a difference in the measurements because of the continually changing character of the sea level.

Sea Level Measurements From Deep-Sea Regions

If sea-level data from only coastal tide gauges are used, tsunami warning messages will not be helpful to the coasts located closest to the earthquake source. Thus, coastal sea-level measurements are unsuitable to forecast the impact along the coasts that are near the epicenter, which are hit too rapidly. Small distances from potential seismic sources to the impacted regions imply very short travel times and therefore require rapid detection and warning processes. Taking into account the seismotectonic setting and the processing capacities, the sea-level network that would most properly detect and measure tsunamis needs to be examined. The objective is to most efficiently notify the public of an impending event but avoid excessive warnings.

The main justification for extensive research on tsunamis and for seismic and sea-level stations networks is to develop further the ability to issue an *early, accurate, and reliable* warning for endangered areas. An early warning, in principle, can be issued for tsunamis originating at a point distant from the affected area. Sea-level measurements from the open ocean can be categorized under this class of events, although such measurements have several applications other than tsunami detection. Ideally, an accurate warning should contain the following data (Zielinski and Saxena, 1983):

- Time of tsunami arrival
- Direction of arrival
- Tsunami amplitude or degree of inundation
- Character of tsunami (i.e., depression, flooding, single or multiple waves)
- Expected duration of event

Tsunamis: Detection, Monitoring, and Early-Warning Technologies. DOI: 10.1016/B978-0-12-385053-9.10017-1

Tsunami generation, propagation, and coastal transformations are very complex processes that can be handled only after several simplifying assumptions are made. An earthquake-generated tsunami disturbance propagates over complex ocean-bottom topography and is subjected to spatial and time domain evolution as it progresses. Coastal areas, which are quite wide compared to tsunami wavelengths, drastically change a tsunami's character and amplitude. The considerable reduction of propagation speed, and thus of wave length in shoaling water along distant shorelines, results in a considerable magnification of the tsunami's amplitude and often in extensive flooding (Miller, 1964; Van Dorn, 1968). Local topography and its attendant refractive distortion of wavefronts compound the problem by focusing tsunami energy onto preferential coastal features (Filloux, 1982).

Tsunami deep-water energy decay can deviate from simple geometric spreading, depending on the earthquake's parameters (Kajiura, 1963). Far-field tsunami radiation patterns may exhibit remarkable directivity with amplitude variations of up to 14:1, depending on the angular position (Ben-Menahem and Rosenman, 1972). A midocean tsunami signature evolves from nondispersive to dispersive in character (Kajiura, 1970). An ideal method of monitoring this complex process would be to obtain distortion-free midocean observations from a sufficiently dense grid of measuring stations. Present monitoring networks consist largely of coastal tide gauges, both on continents and on some islands. Nonuniform distribution of suitable islands, however, causes nonuniform distribution of the stations and leaves a large ocean area unmonitored. The complexity of tsunami propagation, combined with an insufficiently dense sampling network, leaves tsunami prediction problems to the knowledge, experience, skill, and intuition of an operator of the tsunami warning center. The subjective element perhaps will long remain a factor in decisions to issue warnings, but additional data can still contribute greatly to the reliability and accuracy of such warnings (Zielinski and Saxena, 1983).

Midocean tsunami measurements can be of great value for prediction and warning purposes. Undistorted by coastal effects, a tsunami signature provides an objective measure of tsunami energy and character at the measurement location. Several studies using ocean-bottom instruments (e.g., Gonzalez et al., 1991) have shown that tsunami waveform data that are free from complex distortions can be retrieved from offshore sea-level observations. Such distortion-free offshore tsunami data have enabled Hino and colleagues (2001) to estimate reliable tsunami source parameters. The following objectives can be accomplished by such measurements (Zielinski and Saxena, 1983):

1. *Tsunami warnings*: Real-time tsunami measurements will confirm for warning purposes the existence of tsunamis in midocean areas.
2. *Tsunami spatial coherence*: It has been speculated (Kajiura, 1972) and confirmed from several subsequent measurements that multiple reflections from midocean ridges and a coast near the tsunami origin may decrease

spatial coherence in the midocean area. Tsunami signatures from two or more sufficiently separated locations, therefore, can be used to investigate this aspect. For coherent tsunamis, these signatures can be used to determine tsunami directivity, velocity, rate of decay, and direction of propagation.

3. *Small island response*: It is believed that tsunami measurements taken off small islands are relatively distortion free (Van Dorn, 1965). If this can be demonstrated with available midocean data, then measurements from a small island station can be substituted for a true midocean signature.

4. *Onshore response*: It has been observed that local onshore response to very different tsunamis, although different in amplitude, has remarkably similar spectral character (Miller, 1972). It is quite conceivable that the magnitude of local response is related to some gross features of the midocean signature. The midocean signature in combination with suitable onshore tsunami measurements can be used to investigate such a relationship.

Some of the preceding objectives will require permanent midocean stations with real-time measurement and communication capabilities.

Many techniques have been developed and successfully used for sea-level measurements from offshore regions. Offshore instrumented buoys, such as deep-ocean data capsules and satellite altimeters are important devices that can be used for detection of tsunami events far away from land. Among these, satellite altimeters hold a prime place in coverage and practical utility. These are discussed in the following sections.

17.1. SEAFLOOR PRESSURE SENSOR CAPSULES

High-quality bottom pressure recorder measurements in the deep ocean contribute to a fundamental understanding of oceanographic processes over a wide range of time scales. The information contained in the sea-bottom pressure recordings may contribute to the investigation of an extended range of phenomena of ocean geophysics (Filloux, 1980). These vary from long-period fluctuations induced by planetary waves, oceanic tides, and meteorological forcing events to relatively short-period phenomena such as long surface gravity waves, microseisms, and tsunamis. Further, the spectrum of pressure fluctuations at the seafloor ranges from the very low frequencies associated with geological changes in the ocean depths to very high-frequency sound waves. By the application of appropriate digital filters on the measured seafloor pressure data, the signal of interest can be recovered.

Although deep-sea tsunami signal amplitude is substantially small compared to that in coastal water bodies, measurement and real-time reporting of tsunami signals from offshore deep-sea regions are of considerable practical significance for tsunami research as well as tsunami warning applications. The importance of tsunami measurements from deep-sea regions stems from the fact that whereas the tsunami signals at coastal waters are generally contaminated due to a variety of shallow-water nonlinear effects, harbor resonance, background noise from wind-waves and swell, and so on, the deep-sea tsunami signal is usually free from such contamination (except the influence of strong seismic waves in near-field regions), and, therefore, deep-sea tsunami signals are useful for quantitative tsunami research and near-real time operational applications. Such precise tsunami data with a high signal-to-noise ratio can be used to detect tsunamis before they reach the coast (Gonzalez et al., 1991; Okada, 1995). Geophysical tsunami generation depends on earthquake source parameters (Okamoto, 1994). It is known that the characteristic period of tsunamis increases with source depth and depends also on other source parameters (Yamashita and Sato, 1974; Ward, 1980). In general, the tsunami generation is controlled by a combination of source parameters, including focal depth, fault size, amount of slip along the fault, and the focal mechanism. The size of the rupture area is usually defined by its length and width, assuming rectangular fault plane geometry.

It may perhaps appear to be an irony that instead of seismic waves, tsunami waves are used for estimation of seaquake parameters. It is true that since seismic waves propagate upward and downward as well as horizontally, seis-mological studies essentially provide the most reliable focal depth estimate in most regions of the world. However, the focal depth of subduction zone earthquakes may not always be well constrained by regional and global seismic network data because of a lack of seismic station coverage over the source region. The seismic wave propagation and, therefore, most seismological determinations of earthquake source parameters are influenced by the near-source seismic velocity structure (Okamoto, 1994). Because of no or little seismological station coverage for subduction zone earthquakes, there is a severe dearth of reliable information on the three-dimensional velocity structure with sufficiently high accuracy for many subduction zones.

In contrast to the meager knowledge about seismic velocity structures in the solid earth, we have a great deal of knowledge about bathymetry in most oceans and, therefore, a very detailed knowledge of tsunami propagation. Thus, there is a clear advantage of using microtsunami waveform analysis to determine the source parameters of subduction zone earthquakes. Moreover, microtsunami waveforms recorded in the near-field (though not necessarily above the source region) are sensitive to source depth as well as fault size. This indicates an advantage of using tsunami waveform analysis for determining the focal depth and other source parameters for subduction zone earthquakes, in particular for regions where there is a lack of seismic stations near the source region. Therefore, microtsunami waveform analysis may become an additional tool for determining focal depth as well as other source parameters of subduction zone earthquakes. Even seismic stress drop could be derived through microtsunami waveform modeling (Hirata et al., 2003). This is an important application because stress drop, focal depth, and focal mechanism are the essential parameters needed to distinguish between intraplate and interplate earthquakes in subduction zones. The focal depth of intraplate earthquakes is larger than that of interplate earthquakes in the same region because intraplate earthquakes occur below the interface between the subducting and overriding plates. In addition, the stress drop for intraplate events is considered to be larger than that for interplate earthquakes (Kanamori and Anderson, 1975).

For a prescribed depth and length, the average amount of slip on the fault is estimated by fitting the observed microtsunami waveform $O_i(t)$ and the calculated microtsunami waveform $C_i(t)$ at the ith station using the least-squares method. The degree of similarity between $O_i(t)$ and $C_i(t)$ is measured by a root mean square (RMS) variance reduction, which is defined as (Hirata et al., 2003):

$$\text{RMS variance reduction} = 1 - \left\{ \sum \left[O_i(t) - C_i(t) \right]^2 / \sum O_i(t)^2 \right\} \quad (17.1)$$

In Eq. (17.1), $C_i(t) = D_0 Gi(t)$, where $Gi(t)$ is the tsunami Green's function, which is generated from a unit amount of average slip on the fault, for the ith deep-sea tsunami recording station, and D_0 is the slip estimated by the least-squares method. The RMS variance reduction in Eq. (17.1) is 1.0 when the calculated microtsunami waveforms exactly match the observed ones. Studies by Hirata and colleagues (2003) suggest that single-record analysis may result in an inaccurate estimate for the source depth. Therefore, tsunami records from two or more deep-sea stations are preferred for source depth estimation. If there are n spatially separated deep-sea tsunami recording stations, the summation in Eq. (17.1) is performed over $i = 1$ to n. The tsunami Green's function $Gi(t)$ is numerically calculated using linear long-wave (shallow-water) theory (e.g., Satake, 1995) with a detailed bathymetry. Use of the *linear* theory is justifiable for deep-sea environments. Initial conditions of sea-level displacement, which is assumed to be the same as the vertical crustal displacement on the ocean bottom, are calculated using the analytical expressions of Okada (1985). In general, the characteristic period of a tsunami waveform also increases with fault length (hence fault size) as well as source depth. Indeed, the best-fit depth tends to be shallower when the prescribed fault length increases. Therefore, there is a trade-off between the depth and length estimates from the tsunami waveforms (Hirata et al., 2003).

An uncontaminated deep-sea tsunami signal is therefore useful in a reliable estimation of the source parameters. For example, a study of a microtsunami from an interplate earthquake off Sanriku, Japan, was done by Hino and colleagues (2001), who performed forward modeling to match the observed waveform. In an analysis by Hirata and colleagues (2003), the depth and size of the fault were varied in a prescribed grid space, and the amount of slip for a prescribed depth and size of the fault was determined by least-squares analysis. Thus, the sensitivity of the uncontaminated tsunami waveform to focal depth and fault size can be effectively used to accurately determine the fault parameters if the tsunami waveform can be measured with high precision.

A widely used method for measurement of tsunamis as well as low-frequency oscillations in the far offshore waters is the use of self-recording pressure sensor capsules that rest on the seafloor over long periods. The desired frequency band of oscillations can be recovered from the measured seafloor pressure by the application of suitable digital filters on the dataset. The requirements that must be satisfied by such a recording pressure gauge include (Cox et al., 1984) the following:

- It should have adequate sensitivity and such low noise that it can detect the pressure signals in the band of interest.
- It must be stable in the high ambient pressure of the seafloor and in the presence of the temperature fluctuations and accelerations to which it will be exposed.
- It must withstand the stresses of transportation on shipboard and to and from the seabed.
- The electrical power required to operate the gauge must be moderate so a battery supply can be used.
- It must be simple and easy to use.

As indicated earlier, the only demonstrated method for in situ measurement of midocean tsunamis is based on pressure fluctuations measured by bottom-pressure sensors. Because deep-ocean (~5000 m) tsunami amplitude is on the order of a few tens of cm, such measurements present a formidable task. Apart from instrumental and system noise, several other external noise sources and influences are present (Filloux, 1980). Fortunately, the noise spectrum in the tsunami frequency band is low, and preliminary studies by several investigators have indicated the possibility of reliable detection of tsunamis with amplitude on the order of 1 cm. The first recorded glance of a tsunami traveling in the open ocean, 1000 km away from the source region, took place in 1979. This was achieved by the use of open-ocean seafloor pressure measurements (Filloux, 1982). Thus, tsunami detectability in deep-ocean areas has thus been demonstrated to be well within the state of the art in the design of seafloor pressure transducers. The other problems to overcome are associated with the deployment of sensor.

Major earthquakes, even quite distant ones, produce conspicuous signatures on deep-seafloor records. Specifically, the vertical seafloor motions associated with the earliest seismic energy packet (Rayleigh seismic wave, R_1) are sufficiently intense to produce well-defined pressure signals, P_s, as they accelerate the mass of several km of seawater column, with $P_s = \rho h \, (d^2 a \, / \, dt^2)$, where P_s is the pressure generated by vertical seismic seafloor displacements, a, and ρ is the seawater density (Filloux, 1982). Thus, seismic waves are also recorded by ocean-bottom pressure gauges. There have been surprising instances of such recorded seismic events having been accompanied by low-amplitude tsunamis (Figure 17.1). It is interesting how the seafloor pressure transducers, which permit tsunami detection, are also very sensitive to the

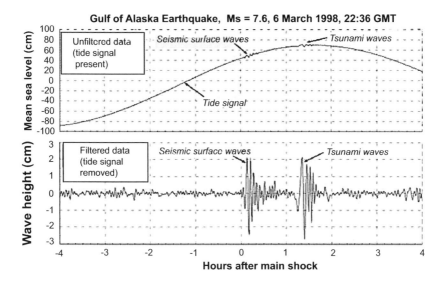

FIGURE 17.1 Earthquake-generated seismic impulses that occur during the tsunami generation phase, as well as the relatively slowly propagating tsunami signals detected by deep-ocean seafloor pressure transducers. *Source: Courtesy of PMEL, United States.*

seismic impulses that occur during the tsunami generation phase. During ocean-bottom pressure measurements from a location off northeastern Japan, Hino and colleagues (2001) noticed that the pressure change due to seismic waves was very large and high-frequency vibrations were clearly evident. However, the tsunami signals in the form of a single sinusoidal wavelet could easily be identified.

In the use of ocean-bottom pressure sensors for tsunami measurements, some not-so-well-known effects have been reported by some investigators. For example, Hirata and colleagues (2003) observed pressure changes that are suspected to be from fluctuations of deep-water currents and atmospheric pressure changes at the period range of tsunamis (e.g., Wunsch and Stammer, 1997) and wind-generated waves (e.g., Mitsubayashi and Honda, 1974) or due to very long seismic surface waves (e.g., Filloux, 1982). These effects cannot be removed by simple signal processing routines. Evaluating such effects on ocean-bottom pressure is difficult, but such an exercise is important for completely retrieving the microtsunami signal from moderate-sized subduction zone earthquakes using ocean-bottom pressure gauges.

17.1.1. Permanent Installation

Real-time operation needed for tsunami prediction requires permanent mid-ocean installations. Saxena and Zielinski (1981) suggested a buoy-based system with two-way communication capabilities via Geostationary

Operational Environmental Satellite (GOES). Bottom-pressure data can be transmitted via an acoustic telemetry link to the buoy. In subsequent years, the suggested methodology began to be implemented in the form of DART buoys, discussed earlier. Alternatively, a submarine cable can be used to link surface and bottom units. A main mooring line can be used to guide the electric cable. Such a solution will allow for a power supply from the surface, thereby reducing maintenance costs. Utilization of existing and planned data buoys as platforms for tsunami systems can further reduce the cost of the system. Another practical solution for transmission of bottom-pressure data to the shore is the use of submarine cables laid along the seafloor. This method has been successfully tried out by the Russians (Dykhan et al., 1983) and the Japanese (Momma et al., 1998; Kawaguchi et al., 2002). In fact, as detailed earlier, the existing tsunami warning system (TWS) in Japan is based on this technology. Canada is also in the process of implementing cable-mounted TWS systems as part of its multipurpose NEPTUNE Program (Mofjeld, 2008).

Installation of a reliable midocean tsunami station poses a difficult and costly task even with the present technology. Devastating tsunamis occur relatively infrequently in a given location, and general awareness (including availability of funds) decreases rapidly with time since the last disaster. Tsunamis, however, are much more frequent on a global scale, and the cost of installation can be shared among affected countries. This started happening after the December 2004 global tsunami.

17.1.2. Seafloor Pressure Measurement Techniques

Various techniques have been used to capture the water pressure variations on the seafloor.

17.1.2.1. Sensors

Several types of transducers have been incorporated into pressure sensor units. Quasi-absolute pressure measurements have been made with diaphragm gauges in which sensing of the motion of the diaphragm has been carried out by vibrating strings (Nowroozi et al., 1966; Snodgrass, 1968), capacitive detectors, and so forth; strain-gauge devices (Hicks et al., 1965; Wunsch and Dahlen, 1974); specially shaped quartz crystals (Irish and Snodgrass, 1972); Bourdon tubes with optical readouts (Filloux, 1970, 1971; Baker et al., 1973); and bellows sensors with quartz crystals as deformation-to-frequency converters (Hayes, 1979). The strain gauge was developed and extensively used by researchers at the Institute of Oceanographic Sciences, United Kingdom. Many designs have been used, most of them incorporating a Wheatstone bridge to increase the sensitivity and to minimize temperature effects. Gwilliam and Collar (1974) and Collar and Cartwright (1972) reported on the successful uses of strain gauge sensors to measure sea levels. They are not, however, well suited for long-term deep deployments where a large dynamic range, high sensitivity, and low drift are required. In order to detect changes of water head of 0.5 cm in 5 km depth, an instrumental stability and resolution of 10^{-6} is necessary.

An early Hewlett-Packard sensor utilized a piezoelectric (quartz) resonator pressure transducer characterized by increased pressure sensitivity. This design represented an improvement over previous designs incorporating the vibrating wire, capacitance plate, and strain gauges (Gwilliam and Collar, 1974). Most of these devices suffer from at least one basic deficiency, such as temperature sensitivity, limited sensitivity to pressure variations, restricted frequency response, long-term instability, or excessive power requirements. More recent quartz crystal pressure transducers such as those currently marketed by Paroscientific, Inc., display much improved long-term stability and significantly greater accuracy than their predecessors. However, these transducers are not entirely free from drift when used under extreme pressure and temperature conditions, such as those occurring in the abyssal depths of the oceans. There are many other reasons as well for the observed long-term instrumental drift. These include slow sinking of the pressure capsule into the soft ocean-bottom material, time-base drift, transducer crystal aging, degradation of the vacuum in the crystal chamber, and creep of either the metal of the Bourdon tube itself or the joint material used to connect the crystal to the tube. Laboratory simulation and measurement of the drift characteristics of the pressure transducers meant for deployments on the deep-ocean floor is impractical because long-term maintenance of a stable high-pressure and low-temperature environment characteristic of the deep-ocean bottom is very difficult. Instrumental drift is a severe limitation with respect to investigations of low-frequency phenomena because the separation of instrumental drift and the environmental signal is difficult to be accomplished. Therefore, the most desired quality of an open-ocean seafloor pressure transducer is considered to be its long-term stability and repeatability.

Hicks and colleagues (1965) first reported the application of seafloor-mounted pressure transducers for measurement of sea-level oscillations from offshore regions. Subsequently, Eyries (1968), Snodgrass (1968), and Nowroozi and colleagues (1968) reported further studies in this area. Eyries (1968), Snodgrass (1968), and Filloux (1969) are recognized as the early pioneers in the development and deployment of self-contained pressure recording devices.

Filloux (1970) reported the deployment of a pressure data capsule on the seafloor to record sea-level oscillations offshore. In this system, an adequately heat-treated Bourdon tube with a sensitive optical detector measured sea-level oscillations with ample resolution, negligible temperature sensitivity, and predictable creep characteristics. The instrument was housed in a heavy-wall aluminium case. In this instrument, the upper end of a freely hanging multiturn (42 turns) Bourdon tube helix that was linked to an oil chamber functioned as the pressure inlet. The Bourdon tube was made of an optimum, properly heat-treated ferro-nickel alloy. The lower end of the Bourdon helix was located in a cup of oil to dampen its natural oscillations. The rotations generated by the changes of water head were sensed by a frictionless, grid-type optical detector and an electrooptical feedback system (Figure 17.2). A mirror attached to the Bourdon tube deflected a beam of light that was generated by a miniature lamp at intervals controlled by a clock. The light focused on the mirror was reflected back through the objective to form an image of the lower-grid ruling on the upper-grid ruling. A change in seafloor pressure caused a corresponding deflection of the mirror that was attached to the Bourdon tube. The deflection of the mirror caused the image of the ruling to move with respect to the real ruling, thereby modulating the transmitted light. The modulated light beam was intercepted by a linear array of optical detectors, thereby converting the deflection of the mirror into electrical signals.

Because diffraction reduces the sharpness of the image, an ingenious use of a feedback technique achieved the required linearization. In this technique, the amplified signal was fed back to a coil that surrounded a permanent magnet that was attached to the rear portion of the mirror. The coil functioned as an elec-tromagnet. The magnetic field produced by this coil, as a result of the feedback electrical signal, interacted with the magnetic field of the permanent magnet, resulting in a torque that twisted the Bourdon tube. The resulting rotation of the mirror modified the signal until equilibrium was reached. The nulling current provided an indication of the pressure change, and, therefore, measurement of these currents and use of appropriate calibration relationship yielded the pressure values corresponding to the sea-level oscillations.

Pressure-signal calibration is frozen in quantities that are inherently unchangeable, such as focal length of a lens, elastic constant of Bourdon tube,

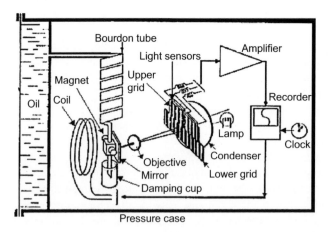

FIGURE 17.2 A deep-sea capsule for recording sea-level oscillations. *Source: Filloux, (1970); Deep-sea tide gauge with optical readout of bourdon tube rotation, Nature, Reproduced with kind permission from Nature Publishing Group, The Macmillan Building, 4-6 Crinan Street, London N1 9XW.*

magnetic moment of a small magnet, and number of turns of a coil. Recording was on Philips cassettes, with a capacity of 4.2×10^6 bits equivalent to 4×10^5 data points per cassette. Thus, the endurance or maximum uninterrupted duration of observation was over 4 months at 128 data per hour. The smoothness of the seafloor pressure record obtained by this method was reported to be considerably better than the sea-level record obtained from coastal gauges (Filloux, 1970). An advantage of this type of pressure sensor is its relative insensitivity to temperature, as well as to temperature transients, and that it can be operated with extremely low power. Another feature of great importance is its inherently fast response (Filloux, 1980). Bourdon helixes, with electrooptical feedback, could be used with reasonable success in long-term studies, even in the presence of a considerable inherent drift. The reason for this drift is that to preserve a high sensitivity, the sensor must be heavily strained and therefore is subject to a finite, though very smooth and gradual, plastic deformation.

The behavior of this anelastic transducer strain, however, is sufficiently well understood to allow for a high degree of correction of the resulting drift. Furthermore, the ratio of anelastic to elastic deformation can be minimized by appropriate optimization of a variety of design parameters such as geometry, fabrication methods, heat treatment, and postfabrication stabilization. The behavior of Bourdon helixes, particularly when made of properly heat-treated Ni Span C nickel-iron alloy, leads to some of the most desirable properties of deep-sea-floor pressure sensors (Filloux, 1969). Temperature sensitivity results principally from the changes in shape due to the thermoexpansion and thermoelastic constants of the transducer material. For Ni Span C alloy 902, thermoelasticity is adjustable by heat treatment in the neighborhood of zero, allowing both positive and negative values. Because the effect of the already low thermoexpansion coefficient can be further compensated by proper action on the thermoelastic coefficient, the overall temperature sensitivity of Bourdon helixes can be negated for a given pressure and then remain small in the area of this pressure. The remarkably low sensitivity to temperature transients generally brought about by temperature gradients through the sensor results from the inherent impossibility of sustaining appreciable temperature differences between the two ends of the Bourdon tube because of its excellent heat conduction (Filloux, 1983).

Baker and colleagues (1973) reported deployment of a deep-sea pressure gauge at a depth of 5400 m in the Sargasso Sea and the collection of deep-ocean pressure records corresponding to the offshore sea-level oscillations. Measurements were made using a new pressure sensor designed and constructed at Harvard University, which is suitable for measuring pressure and temperature at any depth in the ocean. The device was a self-contained package that measured pressure with a resolution of 0.1 mbar and could record up to 75,000 samples of time-tagged measurements on a digital cassette tape recorder at a sampling rate of up to 1 per second. The battery capacity allowed operation for 6 months at the seafloor. The pressure-sensing element was a helical Bourdon tube, with a sensitivity of 0.27 degrees of rotation per mbar, made of fused quartz for high stability. It operated in a differential mode so only those pressure variations from the initial reference pressure at the sea floor were measured. This design was an improvement over a successful deep-sea tide gauge, reported by Filloux (1970), which used a Bourdon tube in the absolute mode. The differential design of Baker and colleagues (1973) allowed a greater control over factors causing long-term drift. The Bourdon tube and a solid-state optical servomechanism system that followed the rotations of a mirror attached to the tube end were mounted in a case filled with silicon oil. A polyvinyl-chloride (PVC) membrane was exposed to the external pressure. The inner volume of the tube was filled with air and connected by a stainless-steel valve to an external flexible PVC container filled with air. After the gauge capsule is lowered to the seafloor, the valve closes the inner part of the quartz Bourdon tube, and the sensor reacts on pressure variations in its surroundings. In other words, the valve was programmed to remain open until the gauge came into temperature equilibrium at the seafloor and to remain closed subsequently in order to establish a constant reference volume of air inside the Bourdon tube. Subsequent changes in ocean pressure caused the Bourdon tube to rotate by an amount proportional to the difference between the external pressure and the reference pressure. The resulting turning of the mirror was recorded via photocells and then converted by a servo system into capacitance changes of a capacitor. This capacitor forms part of an oscillator and was used for the generation of an electrical alternating voltage signal with variable frequency. Because the reference pressure was a function of temperature at the depth of deployment, temperature measurements were also made to correct for thermal effects on the reference pressure. The variable frequency of the oscillator output and a corresponding signal from a quartz thermometer were measured electronically and stored in digital format on a magnetic tape, together with time data from a quartz clock. The Harvard deep-sea gauge could be deployed at more than 6000 m water depths up to 6 months, and it showed a stability of the pressure registration corresponding to a few millimeters of the water column.

Taira and colleagues (1985) reported long-term measurements of ocean-bottom pressure in the Kuroshio region by using Anderaa tide gauges incorporating quartz sensors manufactured by Paroscientific, Inc. Since 1978, a pressure gauge with a quartz sensor has been kept at 2200 m depth off Omaezaki for warning of tsunamis.

17.1.2.2. Deployment Methods

A pressure sensor can be installed on the bottom of the ocean by several different methods. The mooring technique is important because the pressure gauge must be mounted rigidly on the seafloor. In 1976, Taira and colleagues (1977) made moorings with surface radio-buoys at 40 m depth in Otsuchi Bay and at 190 m depth on the western slope of the Izu Ridge. However, this simple mooring method cannot be applied for the long-term measurements or to the deep seas. In 1977, they made a bottom mooring in the northern part of Sagami Bay by the use of a rope canister (110 m rope coiled inside the canister) and subsurface buoys and obtained a 40-day record. However, this method has limitations for deep-sea measurements. In 1983, they invented a bottom mooring device in which two pressure cases for data loggers were deployed on a frame that sits on the seafloor. A vertical array of subsurface buoys provided the requisite buoyancy to keep the pressure gauge capsule in a vertical orientation. The capsules can be recovered by ship. This method can be adopted both for deep seas and for long-term measurements. The bottom moorings of this method have been recovered without any failures.

Another successful deep-sea capsule extensively used for offshore sea-level measurements is the Multi-Year Return Tidal Level Equipment (MYRTLE) developed by the Proudman Oceanographic Laboratory in the United Kingdom (Spencer et al., 1994). This self-contained instrument package records sea levels using precision quartz pressure transducers mounted on the seabed. The recorded data are transferred via infrared link to buoyant glass data capsules that can be periodically released (Figure 17.3). The capsules can be recovered by ship, or the data can be transmitted via satellite.

The seabed instrument consists of a circular frame of tubular aluminum containing a number of pressure cases for data loggers, acoustic systems, and buoyant releasable data capsules (RDCs). The RDC is a glass sphere that contains a data logger and an acoustic transponder. The sphere also has a recovery strap held captive in a cup and lid and released with the capsule. The frame is attached to disposable, steel ballasts by means of a titanium release system that can be activated by one of two acoustic release systems. Glass spheres provide the buoyancy and also house the acoustics and the instrumentation required for the four RDCs. A data logger stores pressure and temperature on the seabed and also transfers the data to the four RDCs using a microcontroller and infrared links. This ingenious data transfer solution enables the RDCs to be released from the seabed without the need for server cables. A flashing light and radio beacon are used to aid recovery. Obviously, operation of these gauges requires a high level of technical skill for deployment and recovery of the equipment. The ability of this system to achieve time-series sea-level measurements from the deep ocean is unique.

An advantage of MYRTLE is that the datum of the measurements is retained, since the pressure transducer is left undisturbed while the data capsules are released. These measurements can also be used to validate satellite altimetry measurements from the deep-ocean regions where other tools for altimetry validation are not available. However, a source of uncertainty can be the possibility of slow settling of the instrument platform.

Using pressure transducers in the deep ocean has some disadvantages. They drift when operating at high pressures. The drift usually diminishes with time

(a)

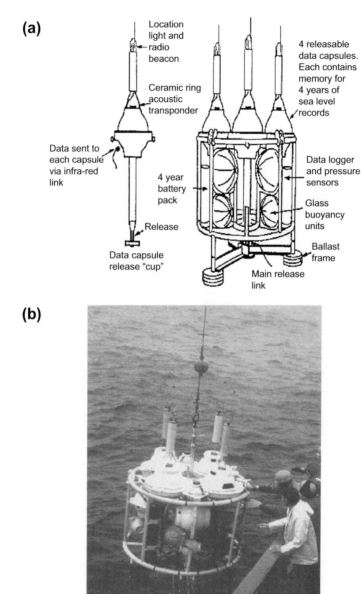

Location light and radio beacon

Ceramic ring acoustic transponder

4 releasable data capsules. Each contains memory for 4 years of sea level records

Data sent to each capsule via infra-red link

4 year battery pack

Data logger and pressure sensors

Glass buoyancy units

Release

Ballast frame

Data capsule release "cup"

Main release link

(b)

FIGURE 17.3 (a) Structural details of the deep-ocean data capsule MYRTLE. (b) Deployment of MYRTLE. *Sources: (a) Spencer and Foden, (1996); Reproduced with kind permission from Sea Technology magazine; www.sea-technology.com*; (b) Courtesy of Joe Rae, Proudman Oceanographic Laboratory, United Kingdom.

but can be reintroduced with subsequent recovery and redeployment cycles (Spencer and Foden, 1996). Another problem with a quartz pressure transducer is its sensitivity to ambient temperature. Although temperature-compensated quartz pressure transducers in the range of $-54°C$ to $107°C$ are available, temperature corrections to the pressure output are adequate only when the ambient temperature changes slowly enough that the transducer can be considered to be in thermal equilibrium (Kusters, 1976). Quartz transducers generally have time constants on the order of half an hour when adjusting to changes in temperature. In situations where field temperatures vary in time scales of this order, pressure errors arise that cannot be ignored (Chiswell, 1991). Factory calibration of quartz pressure transducers allows for static temperature corrections, assuming that the temperature changes slowly enough that the transducer is always in thermal equilibrium.

When the transducer is deployed in some shelf regions where the ambient water temperature changes quickly with changing tidal heights, the preceding assumption of thermal equilibrium may not hold. When this is the case, static temperature correction is not sufficient, and pressure errors arise that are induced by the dynamic response of the transducer to temperature changes. In such cases, the pressure signal has been found to be related to the time derivative of temperature, and a temperature correction to the recorded pressure data is thus possible (Boss and Gonzalez, 1995). Researchers from Proudman Oceanographic Laboratory in the United Kingdom also observed similar results (Private communication, J.M. Vassie, 1995). Deployment of subsurface pressure transducers at ice-laden waters has always been a difficult endeavor. Some typical installation schemes are shown in Figure 17.4.

17.2. SATELLITE RADAR ALTIMETRY

Little sea-level data are available from the vast offshore regions. To circumvent this problem it was necessary to have a spatially dense set of observations made over the entire globe, once every few days. Satellite altimetry provides a technique for collecting just such a data set (Parker et al., 1992). Satellite altimeter is a "sea-level gauge in space" and is considered to be a very important device for measuring offshore sea-level variability. A satellite altimeter measures the distance of the sea surface below an orbiting satellite in the same way a shore-based air-acoustic gauge and microwave radar gauge measure coastal sea-level changes. A polar orbiting satellite orbits the Earth many times in a single day, covering a large area very quickly.

Measuring the sea surface topography with a satellite-borne microwave radar altimeter system is fairly simple in concept. It involves (principally) two measurements. The first is a precise microwave radar measurement of the distance between the satellite and the sea surface. The second measurement involves determining, from independent tracking data, the height of the satellite with respect to a reference ellipsoid. The difference between these two

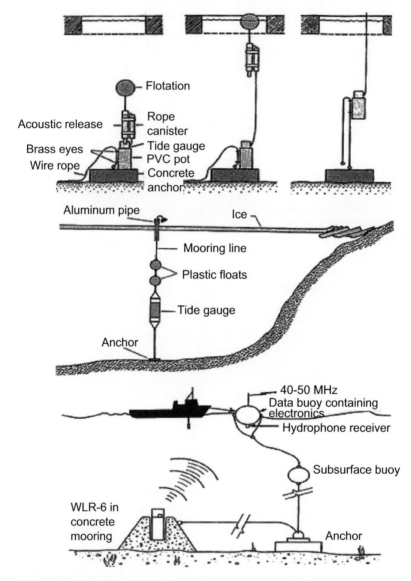

FIGURE 17.4 Deployment of pressure transducer—based sea-level gauges in ice-laden waters.
Source: Reproduced from IOC manual.

measurements gives the height of the sea surface (spatial average over the footprint of the altimeter) relative to the reference ellipsoid. Whereas the ground coverage repeats at regular intervals, any differences in the levels measured over the same area by repeated orbits must be due to sea surface variability.

The radar altimeter measures the time delay between the transmission and return reception of a very short (nanosecond) compressed radar pulse, which, along with precision satellite orbit determinations, is used to construct the large-scale topography of the sea surface as a function of the distance along the satellite's suborbital path. From a sequence of these measurements, information could be deduced on the shape of the global marine geoid, the underlying gravimetric features, tides, and other distortions/swellings of the ocean surface. Though the principle of sea-level measurement using an altimeter may appear to be simple, much ancillary information, as well as sophisticated mathematical algorithms, is needed to arrive at error-free measurements.

The orbit of the satellite (800 to 1300 km above Earth's surface) does not remain precisely stable due to many atmospheric, astronomical, and geophysical forces that act to disturb the dynamics of the satellite. Other small orbit distortions are introduced by changes of the satellite's mass after maneuvers, by gravitational attractions of the Moon and Sun, and even by changes in gravity due to the water mass movements of the ocean tides. Because of these uncertainties, the satellite positions and velocities must be tracked at intervals relative to fixed ground stations.

Telemetry of Sea-Level Data

Tsunami and storm-surge monitoring systems require that the data be reported to the disaster management authorities as soon as possible. There are various communication technologies that could be utilized for real/near-real-time reporting of sea-level data. A variety of real-time communication options are in use today, and newer options are being examined. These include submarine cable communication, acoustic communication, wired telephone connection, VHF/UHF transceivers, satellite transmit terminals, and cellular connectivity.

18.1. SUBMARINE CABLE COMMUNICATION

Submarine cables have a history of being used for data communication. In the past, most common instrument systems and unmanned vehicles were tethered, and the tethering cable could also serve as an electrical communication link. Transoceanic telecommunication cable is another example of the use of submarine cable for communication. In the past, experiments have been proposed for examining the use of decommissioned telecommunication cables and conceptual development or designs for multidisciplinary sea floor observatories, both in the United States and Japan. In the past, several technical workshops have been conducted, with the intention of promoting better mutual understanding between ocean scientists and engineers from multiple disciplines and a broader perspective on future submarine observations (Kasahara et al., 2003). Submarine cables laid along the seafloor have been used for

Tsunamis: Detection, Monitoring, and Early-Warning Technologies. DOI: 10.1016/B978-0-12-385053-9.10018-3

transmission of bottom-pressure data to the shore for the purpose of real-time monitoring of tsunami signals from the offshore regions for early tsunami warning applications. In the history of tsunami monitoring, this method has been successfully tried out by the Russians (Dykhan et al., 1983) and the Japanese. In fact, the existing tsunami warning system (TWS) in Japan is based on this technology. Canada is also in the process of implementing cable-mounted TWS systems as part of its multipurpose NEPTUNE Program.

In addition to the use of submarine cables laid along the seafloor for data communication over long horizontal distances, they can also be used to electrically link the nearly vertically-oriented surface and bottom units of a deep-ocean moored buoy system. In this application, a main mooring line can be used to guide the electric cable. Such a solution will allow for a power supply from the surface, thereby reducing maintenance costs. However, this solution is hardly implemented in practice because of various difficulties, such as entanglement of electrical cables (e.g., around the foundations of offshore platforms, anchor chains, etc.). Such issues precluded the use of electrical cable for data communication between the surface and bottom units of deep-ocean moored buoy systems such as DART-type buoys used for tsunami warning purposes.

18.2. ACOUSTIC COMMUNICATION

Untethered in-water data communication systems would provide attractive operational and economic benefits, since they remove the requirement for cable handling equipment needed to work with cables. Also, mobility is enhanced in untethered systems as drag forces on tethering cables are eliminated. If an automated system located on the sea surface could rely on in-water acoustic channels to interrogate cable-free systems deployed on the sea bottom and to receive transmitted data from them, it would provide the basis for a new generation of tether-free instrumentation, with a greatly enhanced range of useful in-water communication applications. Such "sonar modems" are presently being used to effectively transmit environmental data from seafloor-mounted data acquisition systems to seasurface-mounted receiving systems, thus eliminating the need for submarine cables.

The use of acoustic signals to transmit data through seawater has been well known since the advent of sound ranging (sonar) during World War II. However, one of the major difficulties of transmitting data at high rates in an ocean medium is absorption of sound energy in the ocean, which prohibits long-range propagation at the high frequencies required for high data rates. In addition, inhomogeneities in the sea create multiple communication paths and reverberation. However, near-vertical relatively short-range transmissions (a few km) are not as vulnerable to the reverberation phenomenon as horizontal signaling. An acoustic in-water communication system (Figure 18.1) consists of (1) a data source, encoding and modulation hardware, and a transmitting hardware located in the ocean in a remote self-contained and self-powered

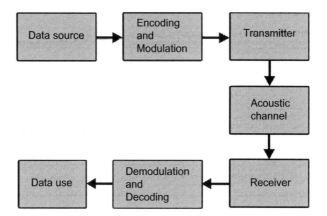

FIGURE 18.1 Components of an acoustic communication system for deep-sea applications. *Source: Doelling, 1980. A new underwater communication system; MITSG Report #80-6, Reproduced with kind permission from MIT Sea Grant.*

instrument package; (2) an acoustic communication channel (i.e., the ocean medium linking the transmitter and the receiver); and (3) an acoustic receiver, a demodulation and decoding device, and a data reception unit located on the sea surface. To minimize energy storage requirements on the remote self-powered systems, the general design philosophy is to incorporate low-power-consuming encoding and transmitting systems that minimize computational complexity. On the other hand, power and space requirements for the sea-surface unit are less restrictive, and, therefore, the computational sophistication need not be limited.

With the advancement of microprocessor technology, it has become feasible to develop such systems. Efforts in the development of an underwater acoustic communication system were initiated primarily by the Massachusetts Institute of Technology and the Woods Hole Oceanographic Institution in 1980 (Doelling, 1980). The system was designed for use in near vertical paths and to operate at maximum data rates consistent with the limitations of the ocean environment. The design used algorithms developed in modern communication theory using microprocessor implementations and was an adaptation to the ocean environment of modern techniques that have been applied for the communication of data over hard wires or through the troposphere to satellites and other forms of electromagnetic wave communication. An important feature of this acoustic communication system is that because of the microprocessor implementation, it can be adapted through programmable software changes to satisfy different trade-offs among the key parameters of data rate, range, and error rates.

Data rates for the application of readouts of stored data from a remote instrument package such as an ocean bottom-pressure recorder can be relatively low—say, 1000 bits per second. In general, the lower the data rate, the lower the error rate that can be achieved for any given communication system. The maximum data rate is determined by the range over which the information needs to be sent, the acoustical characteristics of the particular ocean environment, and the error rate that can be tolerated.

Since digital data cannot be transmitted directly, it is first encoded and then modulated for acoustic transmission through seawater (the ocean medium). For implementing the desired encoding and modulation procedures in the self-powered instrument package, incorporation of a microprocessor with its low power consumption is essential. Literature on information theory abounds in extensive studies of coding methods, and a large number of algorithms have been developed. However, until the advent of microcomputers, these techniques could not even be considered for remote in-water use because of the large size of the computer needed and the attendant electric power requirements.

For certain acoustic environments, it may be useful or necessary to use frequency hopping techniques through which successive messages are encoded on slightly different frequencies. In this case, the frequency hopping code is included in the signal sent out so the receiving system understands the encoding technique.

An efficient method used for reducing power requirements at the seafloor-mounted transmitter system, and at the same time decreasing multipath effects, is using a directional data transmission system. Since the surface buoy will move about with respect to the seafloor-mounted transmitter system, it is essential to be able to control the direction of maximum transmission. Borrowing from radar and sonar technology, the MIT team reported the design of a steered array (Doelling, 1980). In this, an array (rectangular or circular shape) of small transmitter hydrophones is made directional by introducing prescribed signal time delays between the individual elements of the array. It was found that a 4 × 8 element array, with a beam width of about 15° × 7.5° at the 50 kHz center frequency, provides an engineering balance between signal gain and the possibility of pointing errors (i.e., missing the signal because it is very narrow and not pointed directly at the receiver). If beam steering is employed at the seafloor-mounted transmitter system, another transmitter system located on the buoy transmits a set of command signals to the seafloor-mounted system for pointing the array. These signals are received and tracked through filters and phase-locked loops for steering the beam of the seafloor-mounted transmitter system. The microprocessor on the seafloor-mounted transmitter system computes the required amount of steering delays for beam steering.

The receiver hydrophone is located on the surface buoy and can be an omnidirectional system. However, if multipath or other signal-to-noise ratio problems are severe, the buoy-mounted hydrophone could also be a steered array. The incoming encoded data signals, the Doppler signals arising from the relative motion between the transmitter and the receiver, and the synchronization signals are separated by filtering techniques. Demodulation and decoding of the encoded signal are carried out under microprocessor control. The signal spectrum is obtained from a complex fast Fourier transform (FFT) and then decoded by using an inverse of the previous encoding algorithm. Depending on the encoding algorithm, various error correction or error detection schemes may be incorporated in the decoding process.

18.3. VHF/UHF TRANSCEIVERS AND WIRED TELEPHONE CONNECTIONS

In the past, the method of data communication depended largely on the distance over which the data had to be transmitted. Thus, for short links (e.g., harbor operations), a VHF/UHF radio link was often convenient. More than one communications medium can be used within a system, and data can be transferred from remote to central location and vice versa using a high-integrity data-communication protocol. This ensures that the probability of an undetected error is kept to a minimum. A simple scheme such as a VHF/UHF telemetry system attached to a sea-level gauge provides greater flexibility of remote access to the real-time data. Hillier (1987) reported automated telemetry of sea-level data from 11 sea-level stations via a network of UHF radio links along the Thames River over a distance of over 125 km. The telemetry system is driven by a software package running on a personal computer located at the Thames Navigation Service (TNS). A communications interface relieves the host computer of the burden of actually prompting the remote site polling and processing of replies to requests for data. Each gauge in the network that feeds data to the TNS is polled every minute. An entry-point command station located at the TNS, where data validation software continually screens the incoming observed and predicted tidal heights and flags any errors, receives the telemetered data. Subsequently, the information is processed on a network of computers and observed, predicted, and surge sea-level heights are displayed simultaneously on color monitors in the TNS complex. An archival computer continually logs data to a large memory module. Butler (1989) also reported successful execution of VHF-telemetry of sea-level data.

Valeport Ltd Company in the United Kingdom reported the development of a telemetry buoy that is used for wireless transmission of oceanographic data, including sea-level (Figure 18.2). The fully bidirectional UHF link not only transmits data to a PC-based monitoring station in real time or on demand but also allows for changes to system setups. There is a choice of a rigid or an inflatable flotation unit, with all electronics and batteries inside the submerged stainless-steel pod. With a 500 mWatt antenna output, data transmission of up to 20 km (line of sight) is achieved. In practice, communication via VHF/UHF transceivers is limited by line-of-sight distance between transceivers and normally offer only point-to-point data transfer. Ideally, nationwide and even

FIGURE 18.2   Valeport Ltd (UK), Model 750 Telemetry Buoy used for wireless transmission of oceanographic data including sea-level. *Source: Brochure of Valeport Ltd, U.K.; Reproduced with kind permission from Kevin Edwards, Sales & Marketing Manager of Valeport Limited, UK.*

global scale links are necessary for storm surge and tsunami warning communication network systems.

For countrywide links, Subscriber Trunk Dialing (STD) or dedicated telephone lines of the Public Switched Telephone Network (PSTN) have been successfully used (e.g., tide gauge network of Proudman Oceanographic Laboratory (POL) in the United Kingdom). The telephone line networking system of POL, known as "Data Ring," enables a central monitoring station (CMS) located at POL to poll any sea-level station in the network. The general rule adopted by POL is to interrogate all stations in the weekend during the night hours when the national telephone network is not too busy. The telephone network accomplishes polling and data retrieval automatically under command from the CMS. The retrieved data is checked for data quality and then submitted to the databank of PSMSL (Permanent Service for Mean Sea Level). In addition, whenever a storm surge is expected at a given location (say, based on a weather forecast), the station is periodically monitored remotely using this network. Based on the observed sea-level and the prediction, the CMS can display the currently occurring sea-level anomaly and its growth or decay. Also, this scheme facilitates inspecting the condition of the gauge periodically, since any failure becomes apparent on reading the data received. However, a disadvantage of such a sea-level network is that the real-time data are not readily available to the public.

18.4. SATELLITE COMMUNICATION

The most significant advantage of satellite-based data transmission, in contrast to radio telemetry, is its unlimited range. Satellite communication via PTTs has wider coverage and, therefore, allows data reception from offshore platforms. For the last decade or more, sea-level installations in a few countries have used satellite systems (ARGOS, GOES, ORBCOMM, IRIDIUM, METEOSAT, GMS, and INMARSAT) for data reporting. Nevertheless, data transfer speeds of many satellite-based communication systems are limited to 9600 baud or less.

18.4.1. ARGOS

ARGOS operates worldwide using polar orbiting satellites with an orbital period of approximately 100 minutes. The number of accessible satellite passes per day is latitude-dependent, varying from about 7 passes at the equator to 28 passes at the poles. The ARGOS sea-level station is designed to interface with all existing types of sensors, regardless of the type of data output: analog (volts or frequency) or digital. The main features of the ARGOS station (Figure 18.3) are simplicity and independent operation. The PTT is housed in a watertight case. An electronics board interfaces with the sea-level gauge. The processing board encodes the data received and generates an ARGOS-compatible message. The board also triggers the start of the measurement cycle. It is designed to dialog with a PC-type portable field terminal. Through the terminal, the station can be configured on site, time can be checked and modified, and operation tests can be conducted. In the nominal operating mode, the ARGOS station triggers measurements at intervals preset by the user and stores them in memory. Subsequently, the station processes the data to be transmitted in real time and encodes it in an acceptable format that is compatible with the ARGOS message. The message is then sent to the PTT memory for transmission to the satellite. As soon as the messages are received at the ARGOS ground stations (GPC) in Toulouse (France) and Washington DC (United States), they are decoded and processed. After quality control, the data are available in the following three forms (ARGOS Technical News Bulletin):

- Online data (by interrogating the ARGOS centers via packet-switching networks)
- Transmitted to regional or world databanks
- Offline (listings, tapes, or floppy disks)

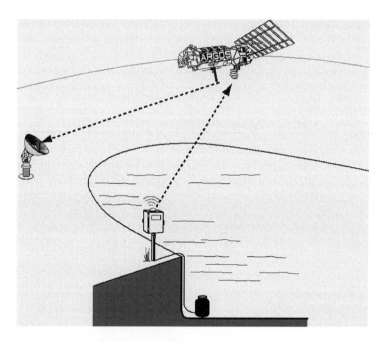

FIGURE 18.3   ARGOS sea-level reporting station. *Source: ARGOS Technical News Bulletin, CLS; Reproduced with kind permission from Yann BERNARD, CLS — Science Department, Geopositioning and Data Collection Systems Division.*

The stations established in the tropical regions are solar powered. The first station was installed at Lome (Togo) in early July 1989, and subsequent installations were carried out to establish an operational South Atlantic network of sea-level stations. In the Pacific Ocean, the Australian Bureau of Meteorology (BOM) and the Committee of Tide and Mean Sea Level, represented by Flinders University of South Australia, look after the implementation of such a network of transmitting sea-level stations. In the Indian Ocean, similar satellite-transmitting sea-level stations were planned for Kerguelen Island, Amsterdam Island, Crozet Island, and Dumont d'Urville (Antarctic) under the leadership of the French oceanographic institute IFREMER in cooperation with the IMG (Institut Mecanique de Grenoble), the TAAF (Terres Antarctiques et Australes Francaises), the CNRS (Centre National de Recherche et d'Etudes Scientifiques), and Service ARGOS. To use ARGOS for sea-level data reporting, a PTT with a data bandwidth capacity of 256 bits per pass is interfaced to the sea-level gauge. Data are received by the authorized users through the ARGOS Global Processing Centers at Toulouse (France) and Largo (United States). Depending on location, the delay in data reception by the user can be several hours. Thus, utility of the ARGOS satellite-communication system for tsunami and storm surge warning purpose is limited.

Many satellites (for example, GOES, INSAT) permit data transfer only at predefined time slots, thereby inhibiting continuous data access. To ensure that data transmission by the sea-level gauges takes place at these precise times, the gauge electronics should have their clocks set to a common "time source." This requirement has necessitated incorporating GPS receivers with each data collection platform (DCP). A network of geostationary satellites comprising GOES (United States), METEOSAT (Europe), and GMS (Japan) offer overlapping longitudinal coverage and appreciable latitude coverage (75°). However, the size of data that can be transmitted to a satellite during the specified time is limited to 649 bytes.

18.4.2. ORBCOMM

ORBCOMM consists of a space segment of 36 low earth orbit (LEO) satellites with ground segments known as Gateway Earth Stations (GES) and Gateway Control Centers (GCC). With ORBCOMM, data communications can be achieved in near real time in only those areas where the receiving satellite can simultaneously communicate with a GES and the subscriber. In regions where GESs are sparse (e.g., Indian Ocean and much of Africa), ORBCOMM operates in a special mode known as GLOBALGRAM. In this mode, data communication is routed through a specified Internet service provider. This would cause substantial delays (as long as an hour or more) in receiving the data. Thus, ORBCOMM has limited applications for monitoring and warning of anomalous sea-level changes.

18.4.3. IRIDIUM

Another option is a network of satellite mobile telecommunications known as IRIDIUM, which is designed to enable data reporting regardless of the user's location on land or sea. Subscribers can use their Iridium modem to communicate with any other telephone anywhere in the world. The Iridium space sector consists of a network of 66 satellites in orbit at 780 kilometers above the Earth. Unlike geostationary telecommunication satellites that are at an altitude of 36,000 kilometers, Iridium satellites cover the Earth with narrower beams as a result of their low orbit. This low orbit results in clear links with powerful signals. Iridium satellites are small and lightweight and electronically interconnected via a network to give worldwide coverage at all times. When an Iridium modem is activated, the closest satellite automatically determines the validity of the user's account and his or her position. The modem communicates with the satellite network directly. The call is then transferred from satellite to satellite as far as its final destination. The Iridium transit station makes the interconnection between the satellite network and the fixed or cordless land infrastructure anywhere in the world.

Experience reveals that IRIDIUM works very well. However, it is expensive to dial to an IRIDIUM terminal. Also, IRIDIUM has limitations of low baud rates (4800) from a landline. To download a reasonably large dataset (say, 1 megabyte) at this baud rate, it may take approximately 35 minutes, and therefore service charges can be quite high (Private communication, Peter Foden, Proudman Oceanographic Laboratory, 2005).

18.4.4. METEOSAT

METEOSAT is a geostationary earth observation satellite launched by the European Space Agency (ESA) and now operated by EUMETSAT. This satellite is placed on a geostationary orbit over the Gulf of Guinea, and its field of view is restricted to that obtained from its location 36,000 km above the intersection of the equator and the Greenwich Meridian. Its operational base station is located in Germany. At present, the Pacific Tsunami Warning System (PTWS) telemeters sea-level data in real time using the meteorological satellite (e.g. Meteosat or GOES) Data Collection Platform (DCP) system. However, in contrast to fast tsunami travel times (TTT) in the deep Pacific Ocean, the TTT are much shorter in other ocean basins, and the latency in DCP transmissions is less acceptable (Woodworth et al., 2007).

18.4.5. INMARSAT

INMARSAT provides better data communication features than IRIDIUM. The INMARSAT Standard-C also uses a network of geostationary satellites giving worldwide coverage except for latitudes above 75 degrees. This system allows two-way data communication in near real time at a rate of 600 bits per second (bps), with data messages up to about 8 Kbytes. The Australian Hydrographical Service has successfully used the INMARSAT satellite communication system for reporting sea-level data.

18.4.5.1. BGAN

Although several methods are available for reporting sea-level data for early tsunami warning purposes, a telephone or an Internet infrastructure (which are prone to failures and damages during natural calamities, especially in the event of a major disaster) is often a part of the overall communication network. A state-of-the-art form of telemetry, a satellite-based system independent of the vulnerable telephone or Internet infrastructure, is the recently introduced Broadband Global Area Network (BGAN) by INMARSAT. BGAN provides reliable and high-speed data communication and data file transfers at broadband speed. Its predecessor was the Regional Broadband Global Area Network (RBGAN) by INMARSAT. The RBGAN service began with the launch of F1 INMARSAT-4 (64°E) in March 2005. The satellite footprint covered up to 99 countries stretching across north, central, and western Africa, as well as

Europe, the Middle East, central Asia, and the entire Indian subcontinent. Subsequently, this satellite was joined by other satellites to give virtually entire world coverage. Introduction of this new service is considered as an emerging technology in the field of high-speed data communication. With the introduction of BGAN, the RBGAN service virtually disappeared. The advent of BGAN has made "always on" technology available almost anywhere on the Earth's surface.

BGAN is a satellite broadband system that employs highly portable terminals the size of a laptop. BGAN is based on Internet Protocol (IP) and General Packet Radio Service (GPRS) technology. Using the lightweight, portable satellite IP modem, BGAN offers reliable and cost-effective access to the Internet. Because the service is based on IP packet technology, users only pay for the amount of data they send and receive and not for the amount of time spent online. This enables them to stay "always connected" to the Internet. The BGAN units provide bidirectional IP data streams of up to 492 kbps as well as Voice Over IP (VOIP) telephony and Short Message Service (SMS) text messaging. A major advantage of BGAN from a tsunami warning point of view is that it does not have to depend on local telephone or Internet infrastructures (Holgate et al., 2008).

The INMARSAT BGAN broadband terminal is a drop-in replacement for landline broadband modems. It shares most of the advantages and disadvantages of conventional broadband, but it is capable of operating in remote areas and is optimized for low power operation (Holgate et al., 2007). Perhaps the biggest advantage of BGAN over fixed-line broadband is its independence from the local telephone infrastructure. This means that the communication network can be expected to continue operating during extreme weather conditions. INMARSAT satellites, which use the BGAN technology that provides "always on" connectivity from almost anywhere on Earth, are the currently preferred alternative satellite technology for real-time sea-level telemetry (Holgate et al., 2007).

A modern sea-level gauge consists essentially of a microcontroller-based embedded electronic system interfaced to a sea-level sensor and a communication system. POL in the United Kingdom has implemented real-time reporting of sea-level data through BGAN service (Holgate et al., 2008). Communication from the embedded system through BGAN can be made in several ways. In one method, the BGAN terminal in effect provides the system with an IP address through which communication across the Internet (either public Internet or dedicated line) with established protocols can be accomplished (Figure 18.4a). In this method, the BGAN technique is conceptually the same as conventional broadband telemetry, but it routes its information via a satellite rather than a landline. Using the IP connectivity, full communication is possible with the sea-level gauges despite their remote locations. This broadband connection provides the ability to diagnose any problems with the gauge remotely and also to reconfigure the system without costly visits to the field.

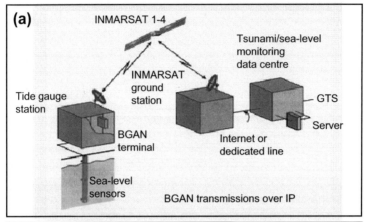

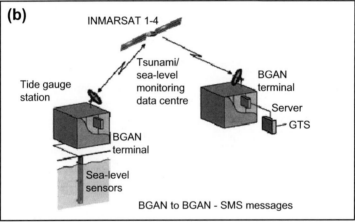

FIGURE 18.4 Schematic of (a) a sea-level gauge station controlled by an embedded system linked via a BGAN terminal to the Internet and tsunami warning center and onward data distribution via the GTS; (b) SMS data transfer between a BGAN terminal connected to an embedded system at a sea-level station and another BGAN terminal at a warning center. *Source: Holgate, S.J., P.L. Woodworth, P.R. Foden, and J. Pugh (2008), A study of delays in making tide gauge data available to tsunami warning centers; J. Atmos. Oceanic Technol., 25, 475-481; (©) American Meteorological Society; Reprinted with permission.*

The terminal can also function as a telephone with SMS capability, such that data can be sent via the satellite either to a telephone capable of receiving SMS messages or to one or more other BGAN terminals (Figure 18.4b). The latter option is particularly attractive for use in monitoring tsunamis and other disasters. This is because once a sea-level station and a warning center are both equipped with BGAN terminals, data can pass between them via the INMARSAT satellites irrespective of whether the telephone or Internet infrastructure has been damaged. A particularly important advantage of this option is data security. By keeping the SMS messages entirely within the private BGAN system (i.e., by sending messages from one BGAN terminal to another), the data never enters a public network. Furthermore, the delay in data exchange is only about 2 seconds because INMARSAT does not employ any message queuing, unlike public domain network operators. Also, its bandwidth (i.e., number of bytes allowed per SMS message) is ideal for sea-level data reporting.

18.5. CELLULAR MODEMS

While satellite communication is relatively expensive, the proliferation of wirelesses networking infrastructures and the ubiquity of cellular phones have together made cellular communication affordable. Low initial and recurring costs are an important advantage of cellular data communication.

Various services exist for data communication, such as SMS, Data Call, and GPRS. SMS is a common method of sending text messages of up to 140 bytes between cellular devices. This mode of transmission is probably the easiest to implement when the data size is small. However, the recurring cost can be substantial if data need to be reported frequently. Further, keeping track of lost SMS can easily override its simplicity. Data Call (similar in operation to the conventional STD call) is another alternative, which can also turn out to be expensive if data have to be frequently reported. Both SMS and Data Call services would require modems on the remote reporting device as well as on the receiving end. This adds to the hardware cost as well as software overheads on the receiving end to check the data integrity for transmission errors. The main benefit of cellular connectivity with GPRS technology is that it utilizes the radio resources only when there are data to send. In addition, GPRS offers improved quality of data services as measured in terms of reliability, response times, and features that are supported.

Another advantage of GPRS over other data communication services is that it provides an "always on" communication channel without incurring time-based costs. That is, there is no difference in costs whether data are collected once a minute or once a day. Further, GPRS data transmission speeds are of the order of three to four times that of the traditional cellular data connection. The GPRS allows data rates of 115 kilobits per second. Also, GPRS enables a constant TCP/IP connection with the Internet so data can be easily uploaded. Since data can be posted on the Internet server, there is neither a need for a modem on the receiving side nor the requirement to have special software at the server side to collect data from the remote site.

Given the popularity of the Internet on a global scale, providing state-of-the-art sea-level information accessibility at significantly low cost is a feasible idea (Joseph and Prabhudesai, 2005). Subsequent to the December 26, 2004, Sumatra tsunami, Prabhudesai and colleagues (2006) designed and developed in house a cellular-based real/near-real time reporting and Internet-accessible sea-level gauge and established sea-level monitoring stations at a few locations on the Indian coast and Kavaratti Island in the Indian Ocean. In this system, by using the existing cellular phone network, continuous real-time updates of coastal sea-level elevations are realized on a Web server. The data received at the Internet server are stored in its backend database and simultaneously presented in graphical format together with the predicted sea level and the residual. The residual, which is the measured sea level minus the predicted sea level, provides a clear indication and a quantitative estimate of the anomalous behavior of sea-level oscillation and can be monitored via the Internet from any part of the world. The driving force for such anomalous behavior could be atmospheric forcing (storm) or geophysical (earthquake). The cellular-based system established by the National Institute of Oceanography, India, at Goa was useful in the real-time monitoring of the weak September 2007 tsunami at Goa.

18.6. TELEMETRY FROM POLAR REGIONS

In some special situations, telemetry becomes important even over short distances. A classical example is the polar regions, where sea-level data collection has been historically very sparse. Isolation and the harsh polar environment present a multitude of technical challenges that must be overcome before reliable polar sea-level stations can be made operational and the data can be received in near real time. In the polar regions, due to the tremendous force of moving ice and the noise it creates, acoustic or direct cable links between shore-based stations and a submerged sea-level gauge are not practical. To alleviate this special problem that is peculiar to the polar regions, Wrathall (1991) reported the use of an electromagnetic communications system that made good use of the unique ability of audiofrequency electromagnetic signals to propagate through water, ice, air, and solid rock as a potential solution to the long-term problem of real-time telemetering from sea-level gauges that remain submerged below ice barriers. In this system, under microcomputer control, a DC-powered transmitter and associated electronics electromagnetically transmit the sea-level data through seawater and the floating ice cap to an electromagnetic receiver located inside a shore-based station. The receiver control logic converts the incoming electromagnetic signals to an RS-232 format and presents the data to an operating system for transmission via satellite.

The components of this electromagnetic marine communications system include a bracelet-shaped audiofrequency (630 to 2970 Hz) high "Q" ferrite loop transmitting antenna resonated to the carrier frequency, a microcomputer-controlled signal generating system to create an accurate data transfer between the transmitter and receiver, a shore-based receiving antenna, and a very sensitive and narrow bandwidth receiver. Using this technology, Wrathall (1991) was able to successfully transmit sea-level data from a submerged transmitter to a small DC-powered electromagnetic receiver located inside the second floor of a building.

Chapter 19

Evaluating and Assessing Tsunamis Technologies for Specific Situations

The 2004 Indian Ocean tsunami encouraged all of the countries on the ocean to prepare for a possible tremendous disaster. One of the most useful and powerful tools to be deployed to face this challenge and save lives, if not property, is tsunami early detection and warning systems. Since the year 2004, a number of tsunami monitoring systems have been deployed, and some are still being developed. Many countries are setting up or trying to enhance tsunami observing and reporting systems at the coastal and offshore regions. It is likely that a considerable number of new systems will be developed and deployed in several new locations in the near future. But researchers, design engineers, operational engineers for tsunami early warning systems, and governmental decision makers who have to relate to development or operational works of tsunami observing/data transmission systems are not always familiar with these kinds of technologies and methods.

Adequate knowledge on technical issues and methods for improving tsunami warning systems (TWSs) and effective data communication is essential in justifying augmented observational systems to government agencies responsible for public safety. Such knowledge is also to be acquired by tsunami warning center personnel, reviewers for government funding agencies, and consulting firms that hold, or bid for, contracts to install and maintain tsunami observing and reporting systems. Unfortunately, regardless of the recent worldwide concerns about tsunami monitoring or tsunami early warnings, no

Tsunamis: Detection, Monitoring, and Early-Warning Technologies. DOI: 10.1016/B978-0-12-385053-9.10019-5

books have been published that deal with all of the important aspects of tsunami detection, monitoring, and real-time reporting technologies that have been developed extensively during the last two or three decades.

Given the ever-increasing pace of technology development in the area of tsunami detection, monitoring, and telemetry, it is hard to keep up with the new technologies that have become available since the Indian Ocean tsunami of 2004. It is generally accepted that there is a dearth of literature on the technological aspects of tsunami detection, monitoring, and real-time reporting. Destructive tsunamis are infrequent and episodic in nature, and as a result, the technologies applied and reported on from the last major tsunami are lost on the new technologists who have not carefully canvassed or examined the technology literature. Various cost-effective data telemetry techniques required to retrieve information in real time are still emerging. Information is available, but it is still hard to find because the technical aspects of tsunami observational methods and data communication are scattered throughout the literature in the form of journal articles and conference proceedings.

Countries that are confronted with the possibility of tsunamis and storm surges must know which technologies are available and appropriate for their specific needs. Because many of the countries threatened by tsunamis are resource-limited, choosing the wrong or inappropriate technology may lead to a greater number of deaths during the next tsunami. Also, countries without expertise may fall victim to exploitive technologists and manufacturers who want to sell their products and make a quick profit. The availability of comprehensive literature that offered an unbiased comparative evaluation and assessment of the optimum technology suitable for specific situations would give nonexperts the required information. It would also enable the poorer countries to draft successful proposals to international bodies. Thus, relative comparison of technologies available to serve as a reference for personnel responsible for protecting their coastal populations is appropriate.

19.1. OPTIMAL OCEAN-BOTTOM PRESSURE RECORDERS

Due to the infrequent occurrence of large tsunamis, the midocean measurement program should be designed for the detection of much more probable, small tsunamis. This requirement calls for high resolution in the midocean measurements. The requisite good-quality observations are absolutely necessary for a thorough quantitative analysis of the tsunami wave properties and its propagation pattern. In general, our ability to detect tsunami waves in a sea-level gauge record depends strongly on the signal-to-noise ratio (i.e., the ratio between the tsunami signal and background noise in sea-level measurements). Unlike most of the coastal stations that are beset with the presence of a myriad of disturbing environments, the seafloor provides a relatively benign environment free from such disturbances. For this reason, the accuracy of tsunami measurements obtained from seafloor-mounted pressure stations is not negatively impacted by exterior influences such as those prevailing at the coastal regions. Thus, tsunami signal-to-noise ratios in deep-ocean measurements are considerably larger than those obtained from coastal measurements. This would mean that clear tsunami signals can be obtained from deep-ocean sea-level measurements if the sensing device possesses most of the ideal performance qualities, such as high sensitivity, freedom from hysteresis and electrical noise, and excellent resolution, accuracy, and linearity. To achieve these, apart from inherently superior qualities such as negligible hysteresis, the sensor should be microprocessor controlled to manage in situ temperature compensation, linearization, and so forth. If these criteria are achieved, even weak tsunami signals in the sea-level record will be clear and distinguishable. Thus, new digital sensors capable of measuring sea-level variations with high precision and fast sampling intervals are needed to detect and record even weak tsunamis.

Among the presently available ocean-bottom pressure sensors, those with the most sophisticated qualities are temperature-compensated quartz crystal −based sensors. Quartz crystal−based temperature-compensated pressure sensors for oceanographic measurements are manufactured by two well-known agencies: Hewlett-Packard Corporation and Paroscientific Corporation. Sea-level gauges incorporating these sensors are manufactured by several oceanographic instrument manufacturers such as NEC Corporation (incorporates the quartz pressure sensor manufactured by Hewlett-Packard), Aanderaa Instruments (incorporates the quartz pressure sensor manufactured by Paroscientific Corporation), and Sea-Bird Electronics, Inc. (incorporates the quartz pressure sensor manufactured by Paroscientific Corporation). Although quartz pressure sensors are the costliest among all the presently available pressure sensors, their use for ocean-bottom tsunami measurements is justifiable because the excellent properties provided by them is achievable in practice at the relatively interference-free environment prevailing at the deep-ocean seafloor so accurate measurement of even small tsunamis is possible. Considering the fact that deep-ocean early TWSs provides tsunami alerts based on an anomaly of ± 3 cm in deep-ocean sea-level measurements, the high level of precision in sea-level measurements offered by such sensors is worth the higher cost.

Because the time of arrival of tsunami waves at the measurement location is an important parameter required by tsunami analysts, it is necessary that the sea-level recording device have a drift-free real-time clock (RTC). An excellent time base is required for deep-ocean tsunami gauges because there is no way to correct the time after the deployment (submarine cable-mounted measurement devices are exceptions).

19.2. OPTIMAL DEVICES FOR MEASURING COASTAL TSUNAMIS

Irrespective of the measurement device's accuracy under laboratory conditions, the accuracy of sea-level measurements actually obtained from coastal environments is deteriorated as a result of a myriad of site-related disturbances such as wind-waves and swell, water currents, large suspended sediments, microbubbles, floating objects, temperature gradients, water density anomalies, and so on. This means that it would be illogical to deploy prohibitively costly instruments for coastal sea-level measurements for tsunami studies, especially when the concerned agencies or nations are resource limited. Despite the difficulty in achieving the desired accuracy from coastal waters, the best possible optimal accuracy can be achieved by proper selection of measuring devices and the use of optimal deployment methodologies.

An important technical requirement for realistic measurements of tsunamis is the linear response of the sea-level measurement device so the tsunami waveform is not unduly attenuated or distorted. Based on considerable field experiments and theoretical evaluation, it is generally accepted that stilling-well-based sea-level gauges are unsuitable for tsunami measurements, although they might be adequate for measuring slow changes in sea level. Just as stilling-well gauges with their large well-to-orifice-diameter ratios are known to suffer from nonlinear responses to large-amplitude short-period waves such as tsunamis, bubbler gauges incorporating long, narrow tubes are also subject to such nonlinearities. As a result, measurements obtained from these two devices are not considered to be useful for realistic tsunami analysis.

A relatively new device that is commercially available is "electric step gauge," which is an electromechanical variant of the age-old tide staff. Because this device does not require hydraulic damping devices for waves, it is free from the inherent nonlinearity that is associated with such wave-damping devices. Although this is a desirable feature of electric step gauges, sea-level measurements obtained from this device can be contaminated by direction-sensitive water piling at the gauge staff due to relatively large flows and waves associated with the coastal tsunamis.

While considering a suitable coastal sea-level gauge, weightage must be given to microwave radar gauges because of their immunity to hydrodynamic contamination of sea-level measurements. This immunity comes from the fact that the radar gauge does not come into direct physical contact with seawater. Radar sensors installed at sufficiently large heights matching their specified measurement ranges would be suitable for error-free measurements of even considerably large tsunami heights. However, as indicated earlier, sea-level measurements using microwave radar gauges can get severely contaminated by man-made lapses such as anchoring of floating objects below the radar sensor or by debris deposited below the sensor by the tsunami waves. Two variants of downward-looking aerial microwave radar sensors are now in use: OTT Kalesto and OTT RLS. Table 19.1 provides an intercomparison of their specifications.

Subsurface pressure sensors are widely used in sea-level measurements. They have advantages such as the ability to be sampled at a higher rate, linear response to large-amplitude waves such as tsunamis and storm surges, and a reasonably good accuracy, except in heavily suspended, sediment-laden water

TABLE 19.1 Intercomparison of Specifications of OTT Kalesto and OTT RLS Downward-Looking Aerial Microwave Radar Sensors

Specifications	Kalesto	RLS
Weight	~8 Kg	2.1 Kg
Size	~500 mm × 160 mm (L × Ø)	222 mm × 152 mm × 190 mm (L × W× D)
Power consumption	10.5−15 VDC Active mode: 550 mA Standby mode: 1 μA	9.6−28 V DC Active mode: 12 mA Sleep mode: 0.05 mA
Frequency	24.125 GHz	24 GHz
Beam angle	±5°	±12°
Measurement range	1.5−28.5 m	0.8−35 m

bodies. The usually observed quality deterioration of coastal sea-level measurements, arising from large water currents and waves, can be substantially improved by the use of a suitable hydraulic coupling device at the pressure inlet of the subsurface pressure sensor. Despite this, deployment of a highly sensitive and accurate pressure sensor (which is naturally much costlier) may not be of great advantage because onshore subsurface pressure measurements are usually associated with high site-related background noise and, therefore, the high price paid for the sensor may not be justifiable in the interests of economy.

A variety of relatively lower-performance, but substantially cheaper, subsurface pressure sensors (both absolute and differential types) are commercially available. In terms of performance, the next candidate is piezo-resistive pressure transducers, which are commercially available from a few manufacturers. The latest in the list is the PPTR transducer manufactured by Honeywell Inc. Like the Paroscientific quartz pressure transducer, Honeywell also provides temperature-compensated pressure output in RS-232 or RS-485 formats based on the user's needs. While the RS-232 format is useful for interfacing to a self-recording environment, the RS-485 format is useful for cable-driven applications. Piezo-resistive pressure transducers are available with other manufacturers for the user to choose from as well.

A relatively recent entry among subsurface pressure transducers is capacitance type. The salient features of this type of pressure sensor are its small size, excellent high-frequency response, good linearity and resolution, and the ability to measure both static and dynamic pressure input. Although the older types of capacitance sensors suffer from instabilities in the cable capacitance, newer capacitance-based pressure transducers are less sensitive to stray capacitance and vibration effects. An example of such a sea-level gauge is that originally manufactured by Sequoia Scientific, Inc., and presently taken over by Advanced Measurements & Controls, Inc. (http://www.advmnc.com/).

Another cheaper, lower-quality pressure sensor that can be used for applications in coastal waters is the metal alloy strain gauge sensor. Because of the configuration of the strain gauges in a Wheatstone bridge format, the sensor's temperature sensitivity is inherently minimal. However, hysteresis is substantial in this type of sensor, so errors can be quite considerable in large tidal regimes. Drift is another disadvantage.

Yet another subsurface pressure sensor used for sea-level measurements and extraction of tsunami signals is the resonant wire pressure transducer. This sensor generates an inherently digital signal and, therefore, can be sent directly to a digital circuitry or a microprocessor. Vibrotron pressure transducers come under this category and have been used in tsunami monitoring systems of Russia.

19.3. SUMMARY

Sea-level gauges of different types, sensing elements, data storage media, and communication facilities have evolved over a long period. Some oceanographic research laboratories and many private companies have played significant roles in this endeavor. A couple of nationally and internationally sponsored oceanographic research programs have supported the widespread use and proliferation of some particular types of sea-level gauges. Some of the gauges have been designed specifically to suit the requirements of water level measurements from bore wells and dams as well. World Bank funding for hydrological studies have promoted and accelerated the development and widespread use of such gauges. Some gauges have been designed primarily for navigational and hydrographic survey applications. Some others have been constructed for specific studies.

Table 19.2 summarizes the information available on most of the popular sea-level/water-level gauges. The choice of a particular type of gauge would largely depend on various factors such as cost, size, performance specifications, communication interface, and so forth.

Radiowave interferometer gauges, which are useful for sea-level measurements from estuarine environments and coastal environments, have been reported in the literature. However, commercial availability of these devices is yet to be realized.

DGPS-buoy-based sea-level measurements have emerged from research stage to operational stage. These devices have now become important tools for real-time detection and monitoring of offshore waves including tsunamis, approximately within 20 km from shore, and they are routinely used in the offshore sea-level network in Japan.

Two schemes have been popularly used for tsunami detection and measurements from the offshore regions: submarine cable-mounted

TABLE 19.2 Commercially Available Sea-Level Gauges

Manufacturer	Sensor Type	Gauge Type	Area of Application, Advantages, Limitations
AMASS Data Technologies Inc. Website:www.amassdata.com/ edaprod/pars.htm	Shaft encoder system	Motor-driven plumb bob with digital display	Applicable for sea-level measurement from tide-wells only. Useful for deployments from bore wells for groundwater monitoring and from stilling-wells for dam water monitoring. Free from drift.
Etrometa, P.O. Box 132, 8400 AC Corredijk, The Netherlands Fax: (+31) 5133 3112	Electric step gauge	Digital data on computer	Sea-level measurement from harbor environment. Free from drift.
Sonar Research & Development Ltd., Unit 1 B, Grove Hill Road Industrial Estate, Beverley, North Humberside HU 17 OJW, UK Tel : (0482) 869559 Fax : (0482) 872184	Noncontact acoustic sensor	Onboard data storage	Sea-level measurement from harbor environment. Useful for dam water monitoring. Measurements affected by air temperature gradient along the acoustic beam path.
Sutron Corporation, 21300 Ridgetop Circle, Sterling, VA 20166, U.S.A. Tel: (703) 406-2800 Fax: (703) 406-2801 E-mail: sales@sutron.com service@sutron.com	Guided air-acoustic sensor	Self-recording of digital data. Option for transmission via satellite.	Sea-level measurement from harbor environment. Useful for dam water monitoring. Self-calibration for sound velocity. Still, measurements affected to some extent by air temperature gradient along the acoustic beam path.

(Continued)

TABLE 19.2 Commercially Available Sea-Level Gauges—cont'd

Manufacturer	Sensor Type	Gauge Type	Area of Application, Advantages, Limitations
R.S. Aqua Ltd., Units 4-6 Hurst Barns, Privett, Alton, Hants GU34 3PL, U.K. Tel: +44 (0)1730828222 Fax: +44 (0)1730828128 E-Mail: c.deeley@rsaqua.co.uk Website: www.rsaqua.co.uk	Microwave radar gauge	Self-recording of digital data.	Trouble-free installation from offshore platforms, jetties, and harbors. Sea-level measurements not affected by local changes in water density, suspended sediments, water currents, etc. Sea-level measurements sensitive to floating objects below the radar sensor. Free from nonlinear response to high-frequency large-amplitude waves such as tsunami waves.
TSKA, Inc., 46208 SE 139th Place, North Bend, Washington 98045. Fax: 011 425888 3895 E-mail: ght@centurytel.net Website: www.tsk-jp.com	Microwave radar		Trouble-free installation from offshore platforms, jetties, and harbors. Sea-level measurements not affected by local changes in water density, suspended sediments, water currents, etc. Sea-level measurements sensitive to floating objects below the radar sensor. Free from nonlinear response to high-frequency large-amplitude waves such as tsunami waves.
OTT Hydromet GmbH P.O.Box: 2140/87411 Kempten/Germany	Float-driven gauge	Automatic plotting on chart recorder. Event controlled recording, 1 minute … 24	Trouble-free installation from offshore platforms, jetties, and harbors. Sea-level measurements not affected by local

Visitors: Ludwigstraße 16 87437 Kempten Tel: +49 831 5617 - 230 Fax: +49 831 5617 - 209 Website: http://www.ott.com	Two types of microwave radar sensors	hour storage interval etc. Data downloading or configuration can easily be done directly via IBM-compatible notebooks, palmtops or with OTT's rugged VOTA Multifunctional Field Unit. RS232 output/SDI12 interface are provided for remote control and data transmission via serial modem (land line), GSM (cellular), and so on. This allows communication worldwide.	changes in water density, suspended sediments, water currents, etc. Sea-level measurements sensitive to floating objects below the radar sensor. Free from nonlinear response to high-frequency large-amplitude waves such as tsunami waves.
Richard Brancker Research Limited., 27 Monk Street, Ottawa, Ontario K1S 3Y7, Canada. Tel: (613) 233-1621 Fax: (613) 233-4100 E-mail: info@brancker.com Website: www.rbr-global.com	Piezo-resistive strain gauge pressure transducer	RS-232	Sea-level measurement. Groundwater monitoring. Dam water monitoring. Measurements likely to be affected by large flows and suspended sediments. Measurements insensitive to floating objects.

(Continued)

TABLE 19.2 Commercially Available Sea-Level Gauges—cont'd

Manufacturer	Sensor Type	Gauge Type	Area of Application, Advantages, Limitations
Druck Limited, Fir Tree Lane, Groby, Leicester LE6 0FH, England. Tel: 44(0) 116 231 7100 Fax: 44(0) 116 231 7100 E-mail: Sales@druck.com Website: www.druck.com	Piezo-resistive silicon strain gauge bridge pressure transducer	Self-recording. RS-232 output.	Sea-level measurement. Groundwater monitoring. Dam water monitoring. Measurements likely to be affected by large flows and suspended sediments. Measurements insensitive to floating objects.
Glanford Electronics Ltd., Glanford House, Exmoor Avenue, Skippingdale Business Park, Scunthorpe, North Lincolnshire, DN15 8NJ, U.K. Tel: 017 24 289600 Fax: 017 24 289008 E-mail: Info@glanfordelectronics.com	Piezo-resistive silicon strain gauge bridge pressure transducer	Analog output.	Sea-level measurement. Groundwater monitoring. Dam water monitoring. Measurements likely to be affected by large flows and suspended sediments. Measurements insensitive to floating objects.
Viatran Corporation, 300 Industrial Drive, Grand Island, NY 14072. Fax: 716-773-2488	Pressure	4–20 mA signal output.	Suitable for coastal and offshore sea-level measurements.

Revlis Electronics Ltd., P.O. Box 146, Vernon, B.C. Canada V1T – 6m1; Tel: (604) 549-2762 Fax: (604) 545-3399	Quartz pressure sensor (Paroscientific) used with gas purge pneumatic system	Compensated for variations in atmospheric pressure and ambient temperature. RS-232, HPIL, and ARGOS interfaces.	Tidal or hydrographic surveys. Permanent or temporary field installations. Onsite tidal measurement to correct survey data. Measurements likely to be affected by large flows and suspended sediments. Measurements insensitive to floating objects.
Pacer Systems, Inc. 900 Technology Park Drive, Billerica, MA 01821. Tel: (508) 667- 8800 Fax: (508) 663-5155	Quartz Pressure sensor (Paroscientific)	RS-232 interface. data storage on battery—backed RAM.	Sea-level measurements from coastal and estuarine environments. Measurements likely to be affected by large flows and suspended sediments. Measurements insensitive to floating objects.
Aanderaa Instruments, Fanaveien 13B, 5050 Bergen, Norway. Tel: (05) 13 25 00 Fax: (47) 5 13 79 50	Quartz Pressure sensor (Paroscientific)	Adaptable to data communication via ARGOS satellite.	Sea-level measurements from coastal and estuarine environments. Measurements likely to be affected by large flows and suspended sediments. Measurements insensitive to floating objects.
SUBER Sainte-Anne du Portzic 29200 Brest, France. Tel: 9845 8510	Quartz pressure sensor (Paroscientific)	Data storage in battery packed CMOS memory RS-232 interface.	Seafloor deployment. Measurements likely to be affected by large flows and suspended sediments. Measurements insensitive to floating objects.

(Continued)

TABLE 19.2 Commercially Available Sea-Level Gauges—cont'd

Manufacturer	Sensor Type	Gauge Type	Area of Application, Advantages, Limitations
Sea Data, Inc., (A Pacer Systems Company), One Bridge Street, Newton, MA 02158 U.S.A. Tel: (617) 244-8191. Fax: (617) 244-3624	1. Strain gauge pressure sensor 2. Quartz pressure sensor	1. Transfers the data by cable to shore in real-time. 2. Transfers the data by cable to an RF transmitter and then via RF telemetry to an RF receiver that provides an RS-232 Interface.	To measure tides and seiches. Measurements likely to be affected by large flows and suspended sediments. Measurements insensitive to floating objects.
Sea-Bird Electronics, Inc., 13431 NE 20th Street Bellevue, Washington 98005 USA Tel: (+1) 425-643-9866 Fax: (+1) 425-643-9954 E-mail: seabird@seabird.com Website: http://www.seabird.com	1. Strain gauge pressure sensor 2. Quartz pressure sensor	Self-recording. RS-232 interface.	Sea-level measurements. Measurements likely to be affected by large flows and suspended sediments. Measurements insensitive to floating objects.
Ocean Data Equipment		Data transmission via RF telemetry to base station. Adaptable to satellite telemetry.	Remote monitoring of tide, stream, or river levels for dredge operations and hydrographic surveys.

AGROS Geopositioning and Data Collection Systems Division Tel: +33 (0)5 6139 3909 E-mail: ybernard@cls.fr Website: http://www.argos-system.org/	Any sensor (suitable/ interfacing required)	Transmission via ARGOS satellite.	(1) Implementation of operational monitoring networks in the tropical ocean to meet the objectives of the TOGA program. (2) Supply of mean sea-level data within a short time compatible with the calibration requirements of satelliteborne altimeters.
InterOcean Systems Inc., San Diego, U.S.A.	Pressure sensor	Self-recording, Two-way acoustic transmission.	Hydrographic survey. Measurements likely to be affected by large flows and suspended sediments. Measurements insensitive to floating objects.
Greenspan Technology, 22 Palmerin Street, Warwick 4370 Qld, Australia. Tel: 0746 601888 Fax: 0746 601800	Capacitance rod		Sea-level measurements. Bore well water-level measurements. Dam water-level measurements.

(Continued)

TABLE 19.2 Commercially Available Sea-Level Gauges—cont'd

Manufacturer	Sensor Type	Gauge Type	Area of Application, Advantages, Limitations
Richard Brancker Research Ltd., 27 Monk Street, Ottawa, Ontario K1S 3Y7, Canada. Tel: (613) 233-1621 Fax: (613) 233-4100 E-mail: info@brancker.com Website: www.rbr-global.com	Piezo-resistive semiconductor strain Gauge absolute pressure transducer	2 MB of flash memory (nonvolatile), upgradable to 4 MB. Plastic and titanium casings. Compact (23 cm length and 4 cm outer diameter). Communications: RS-232. Depth: 20 m to 4000 m.	Temporary and permanent field installations for tidal or hydrographic surveys. Capture of reservoir/dam levels. Well-level measurements. Measurements likely to be affected by large flows and suspended sediments. Measurements insensitive to floating objects.
Marimatech, Samsovej 30, DK-8382 Hinnerup, Denmark. Phone: +45 86 91 22 55 Fax: +45 86 91 22 88 BBS: +45 86 91 11 00 E-mail: mariteam@inet.unic.dk	(1) Differential pressure (2) Acoustic	Data retrieval to computer. Data transmission via the public and mobile telephone network. Radio telemetry allows operation in both real time and offline.	Designed for operation in remote areas over extended periods on built-in alkaline batteries.

ocean-bottom pressure recorders (OBPR) and seismometers, and moored-type OBPRs such as DART systems. Whereas the availability of a continuous stream of data from the former scheme is possible because of the availability of electric power from a shore station, a fast-sampled data stream is available from the latter only on trigger—that is, only when an anomaly of ± 3 cm in sea-level measurement has been detected. This gap in the DART type system arises because of the necessity for electrical power consciousness imposed by the self-powered instrument package used at its bottom segment.

Measuring tsunamis using satellite altimetry is a rather recent development, and this scheme has shown its potential for detection and measurement of tsunamis on the high seas, free from coastal interactions. If the position of the satellite in the first few hours following tsunami generation is such that the tsunami signal can be captured before geometrical spreading on the spherical Earth and dispersive propagation significantly reduces the amplitude of the wave, then altimetry would have a real value. This would be a great advantage of satellite-based tsunami detection. However, the probability of occurrence of such a coincidence is small, and the chance of acquiring such valuable data is very remote. At present, the extent to which the satellite altimeter data can be used for tsunami analysis is debatable because the satellite altimeter time-series data approaches a "space series" more than a "time-series."

Chapter 20

Extracting Tsunami Signals From Sea-Level Records

A sea-level gauge records time-tagged sea-level elevation. In the event of a tsunami, the tsunami signal is superimposed on the tide. The sea-level record may also contain contributions from swells, local seiches, storm surges, and so on. Also, long surface gravity waves in the ocean modulate short gravity and gravity-capillary waves and alter the ocean surface roughness through air-sea interaction (Hara and Plant, 1994; Troitskaya, 1994), resulting in high-frequency sea-level oscillations. Furthermore, near-surface currents such as those associated with internal gravity waves can influence surface waves and thereby hydrodynamically modulate sea surface roughness (Hughes, 1978; Gotwols et al., 1988; Kropfli et al., 1999). Atmospheric disturbances such as winds and small-scale air pressure disturbances can also generate high-frequency oscillations in the sea-level measurements. High-frequency noise associated with infragravity waves generated by nonlinear interaction of wind waves and swell (Holman et al., 1978; Battjes, 1988; Rabinovich, 1993) caused serious difficulties in detecting and estimating the December 26, 2004, tsunami arrival times, periods, and amplitudes from the sea-level records of the North Pacific and North Atlantic oceans (Rabinovich et al., 2006).

Because of various practical issues, the measured sea-level records may have several problems, such as absence of some segments, occurrence of spikes and time shifts, and so on. Therefore, extraction of tsunami signal from sea-level records involves several steps. The first step is a preliminary analysis involving identification and rectification of obvious errors in the sea-level record. A method used to reduce the problem of high-frequency noise in the sea-level record is low-pass filtering the time series. A well accepted low-pass filtering method is the Kaiser-Bessel window (Emery and Thomson, 2003). To separate out the tsunami signal, the tides need to be removed from the measured (preferably low-pass filtered) sea-level record. The process of removing tides from the sea-level record is known as detiding. Detiding of the time-series is necessary to obtain accurate measurements of tsunami wave parameters and to maintain accurate historical tsunami databases. Detiding is done by calculating the tides based on the measured sea level and then subtracting the tides from the sea-level record. A tool widely used for detiding and recovery of tsunami

Tsunamis: Detection, Monitoring, and Early-Warning Technologies. DOI: 10.1016/B978-0-12-385053-9.10020-1

signals is a software package known as the tidal analysis software kit (TASK) (Bell et al., 2000). Tides are estimated using harmonic analysis. Tidal harmonic analysis is done by means of a least squares procedure using TIRA, which is a software developed at the Proudman Oceanographic Laboratory in the United Kingdom (Murray, 1964). TIRA allows analysis even in the presence of gaps. A few short gaps in the original data set, each smaller than 1 day, are thus interpolated by using predicted values. Usually, trend and mean are removed from the residual (i.e., detided sea-level record). The obvious presence of any low-frequency component in the detided signal can be removed by high-pass filtering the detided sea-level record. An optimal high-pass filter for tsunami signal recovery is the Kaiser-Bessel window (Emery and Thomson, 2001). To isolate the tsunami frequency band and simplify tsunami detection, experienced tsunami scientists follow the practice of high-pass filtering the residual sea-level time-series following application of a 4-hour Kaiser-Bessel window. The resulting filtered series is used for analyzing tsunami waves and constructing plots of tsunami records and estimating statistical characteristics of the tsunami waves. To examine the relative energy levels of the tsunami signal and the background signals, spectral analysis is usually performed on the time-series data. The background spectrum is computed for a long segment (several days) before the tsunami. Use of a sufficiently long segment results in high degrees of freedom (DoF) and, therefore, achieves an average background spectrum reflecting mainly the "mean" background conditions and the topographic resonant characteristics of the measurement site. The background spectrum derived from a reasonably long segment would enable an estimation of the topographic admittance function (see, for example, Rabinovich and Stephenson, 2004). On the contrary, selection of only a one- or two-day segment just before the tsunami would result in only a rough estimate of the background spectrum during the time of the tsunami event.

Time-frequency (wavelet) analysis of the residual sea-level data is used to investigate time-frequency variations of the sea-level oscillations and to specify tsunami arrivals, and to examine non-stationary properties of the tsunami waves. Results of wavelet analysis can also be used in those cases where it is difficult to identify tsunami wave arrival from the high-pass data alone (Rabinovich et al., 2006).

The results obtained from tsunami signal analysis for a given location are the so-called *statistical characteristics*, which include sign (i.e., crest/trough) of the first cycle of the tsunami wave stream at the given location, arrival time of the first wave after the occurrence of the tsunami at the source location, travel time of the first wave, height of the maximum wave, arrival time of the maximum wave, periods of the predominant tsunami spectral peaks, and periods of the predominant background spectral peaks. The period range of 30 to 60 minutes, with a peak value at 42–45 minutes was the most common for the December 2004 Sumatra tsunami observed all over the world oceans (Rabinovich and Thomson, 2007; Rabinovich et al., 2006). Apparently, these spectral peaks are related to the tsunami source characteristics. The spectral peaks with periods greater than 1 hour that have been seen at several sites in the Indian, Pacific, and Atlantic Oceans in the background spectrum are primarily related to the local topographic resonance at these sites.

In general, the ability to detect tsunami waves in a sea-level record depends strongly on the signal-to-noise ratio of the data. For the 2004 Indian Ocean tsunami, this ratio ranged from 40:1 to 20:1 in the Indian Ocean, so detection from the Indian Ocean records was straightforward. However, for both the North Pacific and North Atlantic records, the tsunami signal to background noise ratio ranged from 4:1 to 1:1 or even lower at some sites, making tsunami detection difficult. In fact, one of the main purposes of the preliminary analysis of the sea-level data (detiding, low-pass and high-pass filtering), is to diminish the background noise level and thereby improve this ratio. There are several additional factors that cause difficulties in tsunami detection (Rabinovich et al., 2006):

1. Nonlinear interaction of wind waves and swell, especially in stormy weather, generates infragravity (IG) waves. Typical periods of IG waves are from 0.5 to 5 minutes, although on stormy days these waves may have much longer periods (up to 35−40 minutes) (Kovalev et al., 1991). At exposed coastal sites, IG waves produce significant background noise at tsunami frequencies, creating serious problems in identifying weak tsunamis (Rabinovich and Stephenson, 2004).

2. Partially enclosed basins, such as bays, harbors, and fjords are usually well protected from wind waves and swell but have natural (eign) oscillations ("harbour oscillations" or "harbor seiches") that exist almost permanently (Honda et al., 1908; Nakano and Unoki, 1962; Wilson 1972). Incoming tsunami waves do not normally produce separate oscillations with different (additional) periods but simply amplify existing seiches. Quite often this amplification is not abrupt but gradual, making it difficult to define the exact arrival time and to separate seismically generated tsunami waves from common atmospherically generated seiches.

Atmospheric disturbances may generate longwave oscillations that have the same temporal and spatial scales as tsunami waves and sometimes even affect coasts in a similarly destructive way. Such meteorological tsunamis are relatively common both on the Pacific and the Atlantic coasts of North America.

Seismic source parameter estimation, upon which the decision to issue initial warning bulletins is based, can only be used to infer that conditions may exist for tsunami genesis. The tsunami warning centers rely on sea-level data to provide prima facie evidence for the existence or nonexistence of tsunami waves and to constrain tsunami wave height forecast models. The United States and other TWCs use near-real-time sea-level data carried by the WMO Global Telecommunications System (GTS) to monitor sea-level variations worldwide.

The processing of GTS sea-level messages is complicated because sea-level messages are structured in a rich variety of formats. In the aftermath of the

Sumatra disaster, the International Tsunami Information Center approached PTWC about developing a platform-independent, easy-to-use software package to give nascent warning centers the ability to process GTS sea-level messages and examine the resulting sea-level curves. The result was Tide Tool, which has steadily grown in sophistication to become the operational sea-level processing system at the PTWC (Weinstein et al., 2009). Tide Tool continuously decodes sea-level messages in real time and displays the time-series using the open-source platform-independent graphical toolkit scripting language Tcl/Tk. Mouse clickable functions include removal of the tide signal from the time-series, expansion of the time-series, and measurement of arrival time, period, amplitude of tsunamis, and other features. A station location map showing reverse tsunami travel time contours adds to the tool's usefulness as a decision-making tool for warning centers. Tide Tool consists of two main parts: the decoder that reads log files of GTS sea-level messages and dynamic map-based clients that allow the user to select a single station or groups of stations to display and analyze.

With Tide Tool, detiding is accomplished via the use of a set of tide harmonic coefficients routinely computed and updated at PTWC for many of the stations in PTWC's inventory (~400). Another method currently being explored at PTWC is using the decoded time-series files (previous 3—5 days' worth) to compute on-the-fly tide coefficients. This is useful in cases where the station is relatively new and a long-term stable set of tide coefficients is not available or cannot be easily obtained due to various nonastronomical effects.

Tide Tool is accompanied by a metadata set COMP_META that contains all of the information needed by Tide Tool to decode the sea-level messages and basic information, such as the geographical coordinates of the station. The COMP_META dataset can be viewed graphically with GoogleEarth using a kml file obtained from the Website: www.sealevelstations.org.

Chapter 21

Conclusions

A general introduction to tsunami waves, their generation and historical aspects, tsunami databases, tsunami dynamics, and their special features were discussed in this book. Tsunamis at islands and the influence of midocean ridges and continental shelves on their propagation, focusing, and amplification have been briefly touched upon. Attention has been drawn to the recent trend of tsunami generation and its widespread propagation in the Indian Ocean and beyond. A brief history of the past prominent tsunamis and their striking power; tsunami impacts on coastal and island habitats; and the protective role of coastal ecosystems have been given adequate attention for the benefit of the nonspecialist readers.

With these introductory aspects of general interest, the focus has been diverted primarily to the methods of earthquake detection and monitoring for early warnings of seismogenic tsunamis and the technology aspects highlighting instrumented methods for early tsunami warnings, together with the role of IOC-UNESCO in administering a global network for tsunami early warnings. This was followed by a fairly comprehensive exposure of early tsunami warning systems in various IOC-UNESCO member countries. The various techniques for detection of earthquakes, tsunamis, and inundations were examined. Additional topics briefly touched upon, which are relevant to tsunami warnings, include public announcement siren systems and numerical models. Special attention was given to the submarine cable-mounted systems.

Due importance was attached to the instrumented era that began some time prior to the Sumatra tsunami, especially because of the world's attention to the climate change and the role of sea-level monitoring in understanding the trend of climate variability. The reader was introduced to various methods used for tsunami detection from open ocean regions, islands, and coastal waters, and the significant progress made in this area over the years was addressed. Additionally, a brief description of some theoretically sound physical principles that have been confirmed through observations was provided because of their potential to possibly be implemented in the future for remote detection of tsunamis in the open ocean. The technological advancements made over the years in the field of sea-level measurements that brought to light the global extent of the December 2004 Sumatra tsunami were highlighted. It was emphasized that sea-level gauges incorporating stilling-wells with very small orifice-to-well diameter ratios and long, narrow pipes as water intake mechanisms (e.g., some types of

Tsunamis: Detection, Monitoring, and Early-Warning Technologies. DOI: 10.1016/B978-0-12-385053-9.10021-3

float-driven, bubbler, guided air-acoustic gauges) suffer from various inadequacies for measuring large-amplitude short-period sea-level oscillations such as tsunamis and storm surges. A fairly recently introduced downward-looking aerial microwave radar gauge appears to be a promising technology, with the advantages of relative ease in installation and insensitivity to ambient temperature changes, water quality, and water currents. However, its inherent height limit can be an impediment to tsunami and storm surge height measurements if its allowed maximum range is not adequately utilized during its installation. Also, while preparing the deployment site for radar gauges, adequate precautions must be taken to minimize or eliminate the possibility of contamination of sea-level measurements by surfactants such as floating debris, ice, and so on. Although pressure gauges are sensitive to large flows and suspended sediments, the use of subsurface pressure sensors with large enough measuring ranges as a backup to the microwave radar gauge is desirable in coastal sea-level measurements because of their ability to record any water level that exceeds the limited height of the radar sensor. Also, the pressure gauge measurements are least affected by surfactants. Further, the pressure sensor has advantages such as the ability to be sampled at a higher rate and linear response to large-amplitude short-period waves such as tsunamis and storm surges. Consequently, carefully installed subsurface pressure sensors can be regarded as the main "tsunami sensor" suitable for tsunami network applications. Thus, delivery of scientific inferences on tsunami and storm surge signals derived from subsurface pressure measurements in an "open" environment (i.e., without the use of stilling-wells or long and narrow water intake pipes) is expected to be more realistic than those inferred from conventional stilling-well and tube-mounted gauges, such as float-driven, bubbler, and perhaps even guided air-acoustic gauges.

Increased experience in tsunami analysis resulted in a deep-rooted realization and consensual opinion that real-time deep-sea and midocean sea-level measurements that are free of coastal contamination are vital for realistic tsunami forecasting and source identification. This realization, in turn, resulted in the great vision and concept of the use of buoy-mounted systems having two-way communication facilities via geostationary satellites. This was made possible, and deep-ocean buoy systems specifically designed for operational use have been practically implemented. Successful acoustic transmission of seafloor pressure measurements from 5000 m depth at 4800 bits/s with an error probability of 10^{-6} have been reported (Mackelburg et al., 1981). Commercially available deep-ocean quartz crystal pressure sensors with extremely high resolution and low noise levels, such as those produced by Hewlett-Packard and Paroscientific Inc., have enabled precise measurements of deep seafloor pressure. However, the inability to provide continuous electrical power supply precludes continuous long-term measurements with buoy-based tsunami warning systems (TWSs).

One of the limitations associated with buoy-based TWSs is reported to be a high level of vandalism. This problem, which is inherent in all types of buoy-based deep-sea TWSs, has been alleviated to a certain extent by the implementation of submarine cable-mounted seafloor observation systems. These systems also allow continuous electrical power supply, thereby enabling continuous long-term measurements with high temporal resolution. This technology has been effectively used by the Japanese and the Russians. Continuous observations of seafloor pressure variations at 2200 m depth, 110 km offshore, are routinely made by several research institutions and the meteorological agency of Japan since 1970s (Isozaki et al., 1980). In subsequent years, cable-mounted deployments were made at considerably larger depths and distances from the coast. The Sakhalin Complex Scientific Research Institute in Russia has operated a similar installation at 113 m depth, 8 km offshore of Shikotan Island, where a small tsunami was recorded in 1980 (Dykhan et al., 1983). However, the cable systems are prohibitively expensive. So far, the spatial extent of cable-mounted systems is limited to a few hundred km from the coast.

Whereas subsurface pressure sensor capsules and submarine cable-mounted seafloor observation systems are practical means for detecting sea-level changes, including those due to tsunamis and storm surges, the impracticability of deploying spatially dense arrays of such devices is a genuine concern for time-series measurements from the open ocean regions. However, this limitation is circumvented to a large extent by the use of satellite altimetric measurements. The advent of satellite altimetric methods for recording sea-level elevation has significantly expanded the capabilities of oceanography, particularly in terms of their applicability to the study of rare catastrophic events such as tsunamis and typhoons/tornados in the ocean. Currently, the application of satellite altimetry for high-precision radio measurements of sea-level elevation from Geosat, Topex/Poseidon, ERS-1, ERS-2, Jason, and Envisat satellites ensures a cardinal solution to the problem of determining sea-level variation from the open ocean, with precise correlation to the unified geodesic reference system. Clear detection of the December 26, 2004, Indian Ocean tsunami through the satellite altimetric method has demonstrated the feasibility of remotely monitoring such events from space. However, this technology suffers from a limitation that it can measure only along the satellite tracks and cannot provide a full picture of an event. Also, the altimeter data must be corrected for height variations due to mesoscale eddies and other error sources.

As emphasized before, the December 2004 tsunami has been a major driving force in accelerating the implementation of a tsunami forecast system within and beyond the Pacific Ocean. Before this tsunami episode, only a few nations, such as the United States, Japan, and Russia, had operational TWSs. However, at present, several other nations have either established or are in the process of establishing such systems.

In specialized applications such as monitoring of tsunamis and storm surges, sea-level data must be received in real time or near-real time from several locations so the received information can be used for efficient alert and warning purposes and will hasten the implementation of appropriate emergency measures. Whereas satellite-based reporting is one option, a simple and cost-effective method for real-time reporting of tsunamis from coastal regions and islands is cellular based GPRS technology.

From an operational point of view, an obvious advantage that accompanies real-time or near-real-time data reporting is the possibility of identifying instrumental malfunctions, if any, and initiating remedial measures more rapidly. This permits remote diagnostics and might also provide the ability to reprogram the system remotely.

Storm surges are of a problem in the Indian Ocean as hurricanes, as shown by Sidr, Nargis, and others. The various "tsunami" technologies, including telemetry and warning systems, can also be used for other marine hazards such as surges rather than tsunamis per se. Technologies for "multihazards" are becoming recognized as important aspects. In addition to hazard monitoring and management of coastal oceanogenic natural disasters, such as storm-surges and tsunamis, a network of real-time reporting sea-level communication systems has several routine operational benefits. These include safe navigation in shallow coastal waters, faster hydrographic surveys (automatic sounding correction), dredging operations, and port operations.

Although tsunami service has improved considerably after the December 2004 tsunami, it is still not completely free from its "regional" nature. Taking into account the limited capacities of a number of developing countries to provide early forecasts, and considering the scale of such catastrophes, it seems reasonable to develop a system that allows real-time observation of such catastrophes happening anywhere in the world. Many lives can be saved by providing real-time graphical displays of the time-series of such events through the Internet and broadcasting the relevant information via audiovisual media in those countries that are likely to be immediately affected by such catastrophes.

Recent developments in numerical modeling of tsunami and remote sensing technologies have enabled the identification and determination of detailed features of tsunami damage. Numerical models with high-resolution bathymetry/topography grids can predict the local hydrodynamic features of tsunamis, such as inundation depth and current velocity. Also, high-resolution satellite imagery provides detailed information of tsunami-affected areas. In recent years, a couple of nations have been expanding capabilities to comprehend the impact of major tsunami disasters by integrating several modeling (e.g., determination of tsunami inundation) and remote sensing technologies (e.g., assessment of posttsunami disasters using the information obtained from satellite imagery). Such an integrated approach (e.g., Koshimura et al., 2009) is expected to allow determining suitable relationships between hydrodynamic features of tsunami inundation flows and damage probabilities (the so-called fragility functions).

Rapid progress in the development of a new generation of tools and methods in information technologies has begun to accelerate the decision-making process by the emergency management personnel during the occurrence of a disastrous tsunami or storm surge event. Consequently, emergency management issues in terms of preparedness and response are constantly being improved.

References

Abe, K. (1973). Tsunami and mechanism of great earthquakes. *Phys. Earth Planet. Inter., 7,* 143—153.

Abe, K. (1979). Size of great earthquakes of 1873-1974 inferred from tsunami data. *J. Geophys. Res., 84,* 1561—1568.

Abe, K. (1981). Physical size of tsunamigenic earthquakes of the north-western Pacific. *Phys. Earth Planet. Inter., 27,* 194—205.

Abe, K. (1985). Quantification of major earthquake tsunamis of the Japan Sea. *Phys. Earth Planet. Inter., 38,* 214—223.

Abe, K., & Ishii, H. (1980). Propagation of tsunami on a linear slope between two flat regions, Part II reflection and transmission. *J. Phys. Earth, 28,* 543—552.

Ablain, M., Dorandeu, J., Le Traon, P.-Y., & Sladen, A. (2006). High resolution altimetry reveals new characteristics of the December 2004 Indian Ocean tsunami. *Geophys. Res. Lett., 33,* L21602, doi:10.1029/2006GL027533.

Allen, A. L., Bouchard, R. H., Burnett, W. H., Donoho, N. A., Gill, S. K., Kohler, C. A. et al. (2009). Operational tsunami detection for warnings and research: NOAA's network of offshore tsunameters and coastal tide gauges, Abstract; "*24th International Tsunami Symposium and Technical Workshop on Tsunami Measurements and Real-Time Detection*" held at Novosibirsk, Russia (14—16 July 2009).

Alongi, D. M. (2008). Mangrove forests: Resilience, protection from tsunamis, and responses to global climate change, Estuarine. *Coastal and Shelf Science, 76,* 1—13.

Anderson, K. D. (2000). Determination of water level and tides using interferometric observations of GPS signals. *J. Atmos. and Oceanic Technol., 17*(8), 1118—1127.

Andreev, A. K., Borodin, R. V., Kamaev, D. A., Chubarov, L. B., & Gusiakov, V. K. (2009). Automatic information and management system for the tsunami warning center, Abstract; "*24th International Tsunami Symposium and Technical Workshop on Tsunami Measurements and Real-Time Detection*" held at Novosibirsk, Russia (14—16 July 2009).

Androsov, A., Behrens, J., Harig, S., Wekerle, C., Schröter, J., & Danilov, S. (2009). Tsunami modelling with unstructured grids: Interaction between tides and tsunami waves, Abstract; "*24th International Tsunami Symposium and Technical Workshop on Tsunami Measurements and Real-Time Detection*" held at Novosibirsk, Russia (14—16 July 2009).

Anonymous. (1999). Tsunami test buoy reacts to California earthquake. *Sea Technol., 40*(12), 67—67.

Apel, J. R. (1982). Some recent scientific results from the Seasat altimeter. *Sea Technol., 23*(10), 21—27.

Araki, T., & Koshimura, Sh. (2009). Numerical modeling of free surface flow using a lattice Boltzmann method, Abstract; "*24th International Tsunami Symposium and Technical Workshop on Tsunami Measurements and Real-Time Detection*" held at Novosibirsk, Russia (14—16 July 2009).

Aswathanarayana, U. (2007). Integrated preparedness systems. In: T. S. Murthy, U. Aswathanarayana & N. Nirupama (Eds.), *The Indian Ocean Tsunami* (pp. 437—444). London, UK: Taylor and Francis, 2007.

Atwater, B. F., Musumi-Rokkaku, S., Satake, K., Tsuji, Y., Ueda, K., & Yamaguchi, D. K. (2005). The orphan tsunami of 1700: Japanese clues to a parent earthquake in North America. *U.S. Geol. Surv. Prob. Paper, 1707*, p. 144.

Ayers, R. A., & Cretzler, D. J. (1963). *A resistance wire water level measurement system, Marine Science Instrumentation, Vol. 2.* New York: Instrument Society of America, Plenum Press.

Baba, M. (2005). Occurrence of "Swell Waves" along the southwest coast of India from southern Indian Ocean storm. *J. Geol. Soc. India, 66*, 248–249.

Baba, T., Hirata, K., & Kaneda, Y. (2004). Tsunami magnitudes determined from ocean-bottom pressure gauge data around Japan. *Geophys. Res. Lett., 31*, L08303, doi:10.1029/2003GL019397.

Babajlov, V. V., Beisel, S. A., Chubarov, L. B, Eletsky, S. V., Fedotova, Z. I., Gusiakov, V. K., et al. (2009). Some aspects of the detailed numerical modelling of tsunami along the far east coast of the Russian Federation, Abstract; *"24th International Tsunami Symposium and Technical Workshop on Tsunami Measurements and Real-Time Detection"* held at Novosibirsk, Russia (14–16 July 2009).

Baker, D. J., Wearn, R. B., & Hill, W. (1973). Pressure and temperature measurements at the bottom of the Sargasso Sea. *Nature, 245*, 25–26.

Barberopoulou, A., Synolakis, C. E., Legg, M. R., & Uslu, B. (2009). Tsunami hazard of the state of California, Abstract; *"24th International Tsunami Symposium and Technical Workshop on Tsunami Measurements and Real-Time Detection"* held at Novosibirsk, Russia (14–16 July 2009).

Barrientos, S. E., & Ward, S. N. (2009). XIX Century Earthquakes in South America: What we can learn from Tsunamigrams. In: *Proceedings of the International Association of Seismology and Physics of the Earth's Interior (IASPEI) General Assembly 2009, Council for Geoscience* (abstract), January 10–16, 2009, Cape Town, South Africa: Cape Town International Convention Centre.

Battjes, J. A. (1988). Surf-zone dynamics. *Annu. Rev. Fluid Mech., 20*, 257–293.

Bell, C., Vassie, J. M., & Woodworth, P. L. (2000). *POL/PSMSL Tidal Analysis Software Kit 2000 (TASK-2000), Permanent Service for Mean Sea Level.* U.K: CCMS Proudman Oceanographic Laboratory.

Belusic, D., Grisogono, B., & Klaic, Z. B. (2007). Atmospheric origin of the devastating coupled air-sea event in the east Adriatic. *J. Geophys. Res., 112*, D17111, doi:10.1029/2006JD008204.

Belusic, D., & Strelec-Mahovic, N. (2009). Detecting and following atmospheric disturbances with a potential to generate meteotsunamis in the Adriatic. *Physics and Chemistry of the Earth, 34*, 918–927.

Benioff, H., Ewing, M., & Press, F. (1951). Sound waves in the atmosphere generated by a small earthquake. *Proc. US Natl. Acad. Sci., 37*, 600–603.

Beniston, M., Stephenson, D. B., Christensen, O. B., Ferro, C. A. T., Frei, C., Goyette, S., et al. (2007). Future extreme events in European climate: an exploration of regional climate model projections. *Climate Change, 81*, 71–95.

Ben-Menahem, A. (1961). Radiation of seismic surface waves from finite moving sources. *Bull. Seismol. Soc. Am., 51*, 401–435.

Ben-Menahem, A., & Rosenman, M. (1972). Amplitude patterns of tsunami wave from submarine earthquakes. *J. Geophys. Res., 77*(17), 3097–3128.

Berkman, S. C., & Symons, J. M. (1960). *The tsunami of May 22, 1960, as recorded at tide gauge stations*, U.S. Department of Commerce, Coast and Geodetic Survey, Washington, D.C., p. 79.

Bernard, E. N. (2005). National tsunami hazard mitigation program: a successful state-federal partnership. *Natural Hazard, 35*, 5–24.

Bernard, E. N. (2009). An educational tool for a new generation of tsunami scientists, Abstract; "*24th International Tsunami Symposium and Technical Workshop on Tsunami Measurements and Real-Time Detection*" held at Novosibirsk, Russia (14–16 July 2009).

Bernard, E. N. (2009). Tsunami science: Review of the past twenty years and discussion of future challenges, Abstract; "*24th International Tsunami Symposium and Technical Workshop on Tsunami Measurements and Real-Time Detection*" held at Novosibirsk, Russia (14–16 July 2009).

Bernard, E. N., Mofjeld, H. O., Titov, V. V., Synolakis, C. E., & Gonzalez, F. I. (2006). Tsunami: scientific frontiers, mitigation, forecasting and policy implications. *Philosophical Transactions of the Royal Society, A-364*, 1989–2007.

Bernard, E. N., & Robinson, A. R. (2009). Introduction: emergent findings and new directions in tsunami science. In: E. N. Bernard & A. R. Robinson (Eds.), *The Sea, Vol. 15* (pp. 1–22). Cambridge, MA: Harvard University Press.

Bhaskaran, P. K., Dube, S. K., Murty, T. S., Gangapadhyay, A., Chaudhuri, A., & Rao, A. D. (2005). *Tsunami Travel Time Atlas for the Indian Ocean*. Kharagpur, India: Indian Institute of Technology, p. 279.

Bhaskaran, P. K., Dube, S. K., Murty, T. S., Gangapadhyay, A., Chaudhuri, A., & Rao, A. D. (2007). Tsunami travel time atlas for the Indian Ocean. In: T. S. Murthy, U. Aswathanarayana & N. Nirupama (Eds.), *The Indian Ocean Tsunami* (pp. 273–291). London; UK: Taylor and Francis, 2007.

Bilichenko, S. V., Inchin, A. S., Kim, E. F., Pokhotelov, O. A., Puschaev, P. P., Stanev, G. A., et al. (1990). ULF response of ionosphere on the process of earthquake preparation. *Dokl. Akad. Nauk. SSSR, 311*, 1077–1080.

Bird, E. C. F., & Barson, M. M. (1977). Measurement of physiographic changes on mangrove-fringed estuarine coastlines. *Marine Research of Indonesia, 18*, 73–80.

Bjerrum, L. (1971). *Sub-aqueous slope failures in Norwegian fjords, Publ. 88*. Oslo: Norw. Geotech. Inst. p. 8.

Bornhold, B. D., Harper, J. R., McLaren, D., & Thomson, R. E. (2007). Destruction of the Precontact First Nations Village of Kwalate by a Rock Avalanche-generated tsunami. *Atmosphere-Ocean, 45*(2), 123–128, doi:10.3137/ao.450205.

Boss, E. F., & Gonzalez, F. I. (1995). Corrections to bottom pressure records for dynamic temperature response. *J. Atmos. Oceanic Technol., 12*, 915–922.

Botma, H. C., & Leenhouts J. B. (1993). A modern electric step gauge for accurate wave and tide measurement, In: *8th Hydrographic Symposium*.

Bower, J. R., Powers, T., & Whaley, H. L. (1992). Ultrasonic method and apparatus for determining water level in a closed vessel, United States Patent Number: 5,119,676, p. 8.

Braddock, R. D. (1980). Response of a conventional tide gauge to a tsunami. *Marine Geodesy, 4*(3), 223–236.

Brekhovskikh, L. M., & Goncharov, V. V. (1985). *Mechanics of Continua and Wave Dynamics, Springer Ser. Wave Phenom, Vol. 1*. New York: Springer-Verlag.

Bressan, L., Tinti, S., & Titov, V. (2009). Testing a real-time algorithm for the detection of tsunami signals on sea-level records, Abstract; "*24th International Tsunami Symposium and Technical Workshop on Tsunami Measurements and Real-Time Detection*" held at Novosibirsk, Russia (14–16 July 2009).

Brinkman, R. M., Massel, S. R., Ridd, P. V., & Furukawa, K. (1997). Surface wave attenuation in mangrove forests. In: *Proceedings of 13th Australasian Coastal and Ocean Engineering Conference, 2*, 941–949.

Bryant, E. (2001). *Tsunami: The underrated Hazard*. Cambridge: Cambridge University Press, p. 320.

Busse, D. W. (1978). *Digital quartz pressure transducers for air data applications*, V-7-1 to V-7-15.

Busse, D. W. (1987). Quartz transducers for precision under pressure. *Mech. Eng., 109*(5), 1–5.

Butler, D. (1989). Tide Gauge Telemetry Keeps Dredge Rigs on Schedule in St. Lawrence River Project, *Endeco Currents, 1*(1), 1–2.

Callahan, P. S., & Daffer, W. H. (1994). Search for earthquake effects in TOPEX/POSEIDON data (abstract), Eos, *Trans. AGU, 75*(44), Fall Mtg. Supp., 357.

Candela, J., Mazzola, S., Sammari, C., Limeburner, R., Lozano, C. J., Patti, B., et al. (1999). The "Mad Sea" phenomenon in the Strait of Sicily. *J. Phys. Oceanogr., 29*, 2210–2231.

Candella, R. N. (2009). Meteorologically induced strong seiches observed at Arraial do Cabo, RJ, Brazil, *Physics and Chemistry of the Earth, 34*, 989–997, Elsevier.

Candella, R. N. (2005). Sea level at Arraial do Cabo, RJ, Brazil. http://www.pmel.noaa.gov/tsunami/sumatra20041226.html

Candella, R. N., Rabinovich, A. B., & Thomson, R. E. (2008). The 2004 Sumatra tsunami as recorded on the Atlantic coast of South America. *Adv. Geosci., 14*, 117–128.

Cerni, R. H., & Foster, L. E. (1970). Variable capacitance transducers. In: *Instrumentation for Engineering Measurement*. John Wiley & Sons, Inc., 91–93.

Chadha, R. K. (2007). Tsunamigenic sources in the Indian Ocean: Factors and impact on the Indian landmass. In: T. S. Murthy, U. Aswathanarayana & N. Nirupama (Eds.), *The Indian Ocean Tsunami* (pp. 33–48). London, UK: Taylor and Francis, 2007.

Chadha, R. K., Latha, G., Yeh, H., Peterson, C., & Katada, T. (2005). The tsunami of the great Sumatra earthquake of M.9.0 on 26 December 2004 — Impact on the east coast of India. *Curr. Sci., 88*(8), 1297–1300.

Chakrabarti, S. K., Saha, M., Khan, R., Mandal, S., Acharyya, K., & Saha, R. (2005). Possible detection of ionospheric disturbances during the Sumatra-Andaman Islands of earthquakes of December 2004. *Indian J. Radio and Space Physics, 34*, 314–318.

Chapman, V. J. (1976). Mangrove Vegetation, Cramer, Vaduz, p. 447.

Charalampakis, M., Daskalaki, E., Fokaefs, A., & Orfanogiannaki, K. (2009). The Decision Matrix for Early Tsunami Warning in the Mediterranean Sea: is revision needed after the 2008 strong earthquake activity in Greece?, In: *Proceedings of the International Association of Seismology and Physics of the Earth's Interior (IASPEI) General Assembly 2009, Council for Geoscience* (abstract), January 10–16, 2009, Cape Town, South Africa: Cape Town International Convention Centre.

Cherniawsky, J. Y., Titov, V. V., Wang, K., & Thomson, R. E. (2007). Numerical simulation of tsunami waves and currents for southern Vancouver Island from a Cascadia megathrust earthquake. *Pure Appl. Geophys., 164*, 465–492.

Chiswell, S. M. (1991). Dynamic response of CTD pressure transducers to temperature. *J. Atmos. Oceanic Technol., 8*, 659–668.

Choi, B. H., Pelinovsky, E., Kim, K. O., & Lee, J. S. (2003). Simulation of the trans-oceanic tsunami propagation due to the 1883 Krakatau volcanic eruption. *Natural Hazards and Earth System Sciences, 3*, 321–332.

Choudhury, S. (2005). Development of remote sensing based geothermic techniques in earthquake studies, *Ph.D Thesis*, Roorkee, India: Indian Institute of Technology.

Clague, J. J. (2001). Tsunamis. In: G. R. Brooks (Ed.), *A Synthesis of Geological Hazards in Canada. Geol. Surv. Canada, Bull, 548* (pp. 27–42).

Clague, J. J., Munro, A., & Murty, T. S. (2003). Tsunami hazard and risk in Canada. *Natural Hazards, 28*(2–3), 407–434.

Clague, J. J., & Bobrowsky, P. T. (1999). The geological signature of great earthquakes off Canada's west coast. *Geoscience, 26*(1), 1−15.

Clague, J. J., Bobrowski, P. T., & Hamilton, T. S. (1994). A sand sheet deposited by the 1964 Alaska tsunami at Port Alberni British Columbia, Estuarine, Coast. *Shelf Sci., 38*, 413−421.

Clark, J. R. (1996). *Coastal Zone Management Handbook*. USA: Lewis Publication, p. 694.

Cohen, J. E., & Tilman, D. (1996). Biosphere and biodiversity: the lessons so far. *Science, 74*, 1150−1151.

Cohen, J. E., & Belcher, S. E. (1999). Turbulent shear flow over fast-moving waves. *J. Fluid Mech., 386*, 345−371.

Collar, P. G., & Cartwright, D. E. (1972). Open sea tidal measurements near the edge of the Northwest European Continental Shelf. *Deep-Sea Res., 19*, 673−689.

Cotanza, R., d'Arge, R., de Groot, R., Farber, S., Grasso, M., & Hannoh, B. (1997). The value of the world's ecosystem services and natural capital. *Nature, 387*, 253−260.

Cox, C., Deaton, T., & Webb, S. (1984). A deep-sea differential pressure gauge. *J. Atmos. Oceanic Technol., 1*, 237−246.

Cross, R. H. (1968). Tide gage frequency response. *J. Waterways Harbor Div. Proc. Am. Soc. Civil Eng., 94*, 317−330.

Curtis, G. D. (1986). Design and development of a coastal tsunami gauge. *Sci. of Tsunami Hazards, 4*, 173−182.

Dahdouh-Guebas, F., Jayatissa, L. P., Nitto, D., Bosire, J. O., Seen, D. L., & Koedam, N. (2005). How effective were mangroves as a defence against the recent tsunami? *Current Biology, 15*, R443−R447.

Dahl-Jensen, T., Larsen, L. M., Pedersen, S. A. S., Pedersen, J., Jepsen, H. F., Pedersen, G. K., et al. (2004). Landslide and tsunami of 21 November 2000 in Paatut, West Greenland. *Natural Hazards, 31*, 277−287.

Daily, G. C. (1997). *Societal dependence on natural ecosystems, Nature's Services*. Washington DC: Island Press.

Danielsen, F., Sorensen, M. K., Olwig, M. F., Selvam, V., Parish, F., Burgess, N. D., et al. (2005). The Asian tsunami: A protective role for coastal vegetation. *Science, 310*, 643−643.

Dao, M. H., & Tkalich, P. (2007). Tsunami propagation modeling — a sensitivity study. *Nat. Hazards Earth Syst. Sci., 7*, 741−754.

Darienzo, M., Aya, A., Crawford, G. L., Gibbs, D., Whitmore, P. M., Wilde, T., et al. (2005). Local tsunami warning in the Pacific coastal United States. *Natural Hazards, 35*, 111−119.

Darienzo, M. E., & Peterson, C. D. (1990). Episodic tectonic subsidence of late Holocene salt marshes, northern Oregon, central Cascadia margin. *Tectonics, 9*, 1−22.

Dasgupta, S. (2005). *Satellite geothermic techniques in earthquake studies, M. Tech Dissertation*, Department of Earth Sciences, Indian Institute of Technology, Roorkee, India.

Dawson, A. G. (1994). A geomorphological effect of tsunami run-up and backwash. *Geomorphology, 10*, 83−94.

Defant, A. (1961). *Physical Oceanography, Vol. 2*. Oxford, U.K: Pergamon Press.

de Jong, M. P. C., & Battjes, J. A. (2004). Low-frequency sea waves generated by atmospheric convection cells. *J. Geophys. Res., 109*, C01011, doi: 10.1029/2003JC001931.

De Lange, W. P., & Healy, T. R. (2001). Tsunami hazards for the Auckland region and Hauraki Gulf, New Zealand. *Natural Hazards, 24*, 267, doi: 10.1023/A:1012051113852.

De Lange, W. P., Magill C. R., Nairn I. A., & Hodgson, K. A. (2002). Tsunami generated by pyroclastic flows entering Lake Tarawera, EOS Trans. AGU, 83(22), Western Pacific Meet. Suppl., OS51C-10: WP54.

Desa, E. S., Joseph, A., Rodrigues, D., Chodankar, V. N., & Tengali, S. (1999). An improved hydraulic coupling device for use with in-water pressure based systems, Indian Patent No. 215625.

Didenkulova, I. I. (2005). Tsunami in Russian lakes and rivers, Izv. AIN RF, series: Applied Mathematics and Mechanics, 14, p. 82.

Didenkulova, I. I. (2006). *Modelling of long wave run-up over a flat slope and analysis of real-events, Candidate Dissertation in Mathematics and Physics.* Nizhniy Novgorod: State Technical University, p. 199.

Didenkulova, I. I., & Pelinovsky, E. N. (2002). The 1597 tsunami in the River Volga∗, *Proceedings of the International Workshop on Local Tsunami Warning and Mitigation*, pp. 17–22, Janus-K, Moscow. September 10–15, 2002. ∗Translated by O.I. Yakovenko, edited by A.B. Rabinovich and W. Rapatz

Didenkulova, I. I., & Pelinovsky, E. N. (2006). Phenomena similar to tsunami in Russian internal basins. In: *Russian Journal of Earth Sciences, Vol. 8*. ES6002, doi:10.2205/2006ES000211, 2006.

Didenkulova, I. I., & Pelinovsky, E. N. (2009). Abnormal amplification of long waves in the coastal zone, Abstract; "*24th International Tsunami Symposium and Technical Workshop on Tsunami Measurements and Real-Time Detection*" held at Novosibirsk, Russia (14–16 July 2009).

Didenkulova, I. I., Pelinovsky, E. N., & Soomere, T. (2009). Can the waves generated by fast ferries be a physical model of tsunami? Abstract; "*24th International Tsunami Symposium and Technical Workshop on Tsunami Measurements and Real-Time Detection*" held at Novosibirsk, Russia (14–16 July 2009).

Dietrich, G., Kalle, K., Kraus, W., & Siedler, G. (1980). Measurement of water level variations, In: General Oceanography: An Introduction, 128–131.

Dilmen, D. I., Yalciner, A. C., Zaytsev, A., Chernov, A., Ozer, C., Insel, I., et al. (2009). Development of GIS based inundation maps; application to Fethiye town, Turkey, Abstract; "*24th International Tsunami Symposium and Technical Workshop on Tsunami Measurements and Real-Time Detection*" held at Novosibirsk, Russia (14–16 July 2009).

Dobson, F. W. (1980). Air pressure measurement techniques. In: F. Dobson, L. Hasse & R. Davis (Eds.), *Air-Sea Interaction; Instruments and Methods* (pp. 231–253). Plenum Press.

DOD (2005). *Preliminary assessment of impact of tsunami in selected coastal areas of India*, Report, p. 42; Compiled by Department of Ocean Development (DOD), ICMAM Project Directorate, Chennai, India.

Doelling, N. (1980). *A new underwater communication system*, The MIT Marine Industry Collegium, Opportunity Brief #19, Marine Industry Advisory Services, MIT Sea Grant Program, Cambridge, Massachusetts, Report No. 80-6, p. 17.

Dominey-Howes, D., & Goff, J. (2009). Hanging on the line — on the need to assess the risk to global submarine telecommunications infrastructure — an example of the Hawaiian "bottle-neck" and Australia. *Nat. Hazards Earth Syst. Sci., 9*, 605–607.

Donn, W. L., & Ewing, M. (1956). Stokes' edge waves in Lake Michigan. *Science, 124*, 1238–1242.

Donn, W. L., & Posmentier, E. S. (1964). Ground-coupled air-waves from the Great Alaskan Earthquake. *J. Geophys. Res., 69*, 5357–5361.

Douglas, C. K. M. (1929). The line-squall and Channel wave of July 20th, 1929. *Meteorological Magazine, 64*, 187–189.

Dragani, W. C. (2007). Numerical experiments on the generation of long ocean waves in coastal waters of the Buenos Aires province, Argentina. *Continental Shelf Research, 27*, 699–712.

Dragani, W. C., D'Onofrio, E. E., Grismeyer, W., & Fiore, M. E. (2006). Tide gauge observations of the Indian Ocean tsunami, December 26, 2004, in Buenos Aires coastal waters, Argentina *Cont. Shelf Res.*, 26, 1543−1550.

Drago, A. (2008). Numerical modelling of coastal seiches in Malta, *Physics and Chemistry of the Earth, 33,* 260−275.

Dudley, W. C., & Lee, M. (1998). *Tsunami!* Honolulu: University of Hawaii Press, pp. 5, 302−303, 321−322.

Dunbar, P. (2009). Integrated tsunami data for better hazard assessments. *EOS, 90*(22), 189−190.

Dunbar, P., & Stroker, K. (2009). Statistics of global historical tsunamis, Abstract; *"24th International Tsunami Symposium and Technical Workshop on Tsunami Measurements and Real-Time Detection"* held at Novosibirsk, Russia (14−16 July 2009).

Dunbar, P., & Stroker, K. (2009). Historical Significant Earthquakes, Volcanic Eruptions, and Tsunamis in Africa. In: *Proceedings of the International Association of Seismology and Physics of the Earth's Interior (IASPEI) General Assembly 2009, Council for Geoscience* (abstract), January 10−16, 2009, Cape Town, South Africa: Cape Town International Convention Centre.

Dupuy, P. Y., (1992). The SHOM ultrasonic tide gauge, Workshop Report No. 81, Intergovernmental Oceanographic Commission, pp. 8−12.

Dykhan, B. D., Jaque, V. M., Kulikov, E. A., Lappo, S. S., Mitrofanov, V. N., Poplavsky, A. A., et al. (1983). Registration of tsunamis in the open ocean. *Marine Geodesy, 6*(3−4), 303−309.

Dzienovski, A., Bloch, S., & Landisman, M. (1969). *Bull. Seism. Soc., Am., 59,* 427.

Efimov, (V. V.), Kulikov, E. A., Rabinovich, A. B., & Fain, I. V. (1985). Waves in boundary regions of the Oceans (Gidrometeoizdat, Leningrad) [in Russian].

Emery, W. J., & Thomson, R. E. (2001). *Data Analysis Methods in Physical Oceanography* (2nd ed.). New York: Elsevier, p. 638.

Emery, W. J., & Thomson, R. E. (2003). *Data analysis methods in physical oceanography* (2nd ed.). New York: Elsevier, p. 638.

Eving, M., & Press, F. (1955). Tide gauge disturbances from the great eruption of Krakatau. *Trans. Amer. Geophys. Un., 36*(1), 53−60.

Ewel, K. C., Twilley, R. R., & Ong, J. E. (1998). Different kinds of mangrove forest provide different kind of goods and services. *Global Ecology and Biogeography Letters, 7,* 83−94.

Eyries, M. (1968). Maregraphes de grandes profondeurs. *Cahiers Oceanogr, 20,* 355−368.

Farre, R. (2004). Experience with SRD tide gauges and reasoning behind change to radar tide gauges. In: S. Holgate & T. Aarup (Eds.), *IOC Workshop Report No. 193,* Intergovernmental Oceanographic Commission of UNESCO, 9−12.

Filloux, J. H. (1969). Bourdon tube deep sea tide gauges, Tsunami in the Pacific Ocean. In: W. M. Sams (Ed.), (pp. 223−238). East-West Center Press.

Filloux, J. H. (1970). Deep-sea tide gauge with optical readout of bourdon tube rotation. *Nature, 226,* 936−938.

Filloux, J. H. (1971). Deep-sea tide observations from the Northeastern Pacific. *Deep-Sea Res, 18,* 275−284.

Filloux, J. H. (1980). Pressure fluctuations on the open ocean floor over a broad frequency range: New program and early results. *J. Phys. Oceanogr., 10,* 1959−1971.

Filloux, J. H. (1982). Tsunami recorded on the open ocean floor. *Geophys. Res. Lett., 9*(1), 25−28.

Filloux, J. H. (1983). Pressure fluctuations on the open-ocean floor off the Gulf of California: Tides, earthquakes, tsunamis. *J. Phys. Oceanogr., 13*(5), 783−796.

Fine, I. V., Thomson, R. E., & Rabinovich, A. B. (2009). Meteorological tsunamis on the Pacific coast of North America, Abstract; "*24th International Tsunami Symposium and Technical Workshop on Tsunami Measurements and Real-Time Detection*" held at Novosibirsk, Russia (14—16 July 2009).

Fine, I. V., Rabinovich, A. B., Bornhold, B. D., Thomson, R. E., & Kulikov, E. A. (2005). The Grand Banks landslide-generated tsunami of November 18, 1929: Preliminary analysis and numerical modeling, *Marine Geology, 215*, 45—57.

Fokaefs, A., Novikova, T., & Papadopoulos, G. A. (2009). Tsunami risk mapping for the city of Rhodes, Greece, Abstract; "*24th International Tsunami Symposium and Technical Workshop on Tsunami Measurements and Real-Time Detection*" held at Novosibirsk, Russia (14—16 July 2009).

Franca, C. A. S., & de Mesquita, A. R. (2007). The December 26th 2004 tsunami recorded along the southeastern coast of Brazil. *Nat. Hazards, 40*, 209—222.

Freeman, N. G., Hamblin, P. F., & Murty, T. S. (1974). Helmholtz resonance in harbours of the great lakes, Proceedings of 17th Conference on Great Lakes Research International Association. *Great Lakes Res. Proc., 15*, 399—411.

Fu, L.-L., & Cazenave, A. (2001). *Satellite Altimetry and Earth Sciences: A Handbook of Techniques and Applications.* San Diego: Academic Press, 95—99, 453—456.

Fujii, Y., & Satake, K. (2007). Tsunami sources of November 2006 and January 2007 great Kuril earthquakes. *Bull. Seism. Soc. Amer.*

Fujiwara, N., Momma, H., Kawaguchi, K., Iwase, R., & Kinoshita, H. (1998). Comprehensive deep seafloor monitoring system in JAMSTEC, In: *Proc. Underwater Technology '98*, 383—388.

Furukawa, K., Wolanski, E., & Mueller, H. (1997). Currents and sediment transport in mangrove forests, Estuarine. *Coastal and Shelf Science, 44*, 301—310.

Garcies, M., Gomis, D., & Monserrat, S. (1996). Pressure-forced seiches of large amplitude in inlets of the Balearic Islands, Part II: Observational study. *J. Geophys. Res., 101*, 6453—6467.

Garrett, C. J. R. (1970). A theory of the Krakatoa tide gauge disturbances. *Tellus, 22*(1), 43—52.

Gatsysky, A. (2001). Nizhegorodsky Letopisets, *Nizhegorodskaya Yarmarka*, Nizhniy Novgorod, p. 716.

Geist, E. L., Titov, V. V., & Synolakis, C. E. (2006). Wave of change. *Scientific American*, 56—63.

George, P.-C. (2003). Near- and far-field effects of tsunamis generated by the paroxysmal eruptions, explosions, caldera collapses and massive slope failures of the Krakatau volcano in Indonesia on August 26-27, 1883. *Sci. Tsunami Hazards, 21*(4), 191—222.

Gieles, A. C. M., & Somers, G. H. J. (1973). Miniature pressure transducers with a silicon diaphragm. *Philips Tech. Rev., 33*(1), 4—20.

Gisler, G. R., Weaver, R. P. & Gittings, M. L. (2009). Calculations of asteroid impacts into deep and shallow water, Abstract; "*24th International Tsunami Symposium and Technical Workshop on Tsunami Measurements and Real-Time Detection*" held at Novosibirsk, Russia (14—16 July 2009).

Glazman, R. E. (1981). Radio interferential measurements of sea level oscillations with large tidal amplitudes. *IEEE J. Oceanic Eng., OE-7*, 73—76.

Glazman, R. E. (1982). An experimental implementation of interferometric techniques for sea level variation measurements and reflection coefficient phase determination. *IEEE J. Oceanic Eng., OE-7*(4), 155—160.

Goda, Y. (1995). *Random Seas and Design of Maritime Structures.* Tokyo, Japan: University of Tokyo Press.

Godin, O. A. (2003). Influence of long gravity waves on wind velocity in the near-water layer and feasibility of early tsunami detection. *Dokl. Earth Sci., 391*, 841—844.

Godin, O. A. (2004). Air-sea interaction and feasibility of tsunami detection in the open ocean. *J. Geophys. Res., 109,* C05002, doi:10.1029/2003JC002030.

Godin, O. A. (2005). Wind over fast waves and feasibility of early tsunami detection from space, In: A. Litvak (Ed.), *Frontiers of Nonlinear Physics, Inst. Appl. Phys.,* Nizhny Novgorod, 210—215.

Godin, O. A., & Irisov, V. G. (2003). A perturbation model radiometric manifestations of ocean currents. *Radio Sci., 38,* 8070, doi: 10.1029/2002RS002642.

Godin, O. A., Irisov, V. G., Leben, R. R., Hamlington, B. D., & Wick, G. A. (2009). Variations in sea surface roughness induced by the 2004 Sumatra-Andaman tsunami. *Nat. Hazards Earth Syst. Sci., 9,* 1135—1147.

Gomez, B. P., & Maldonado, J. D. L. (2004). Experience with sonar research and development (SRD) acoustic gauges in Spain. In: S. Holgate & T. Aarup (Eds.), *IOC Workshop Report No. 193,* Intergovernmental Oceanographic Commission of UNESCO, 4—8.

Gomis, D., Monserrat, S., & Tintore, J. (1993). Pressure-forced seiches of large amplitude in inlets of the Balearic Islands. *J. Geophys. Res., 98,* 14437—14445.

Gonzalez, F. I., Satake, K., Boss, E. F., & Mofjeld, H. (1995). Edge wave and non-trapped modes of the 25 April 1992 Cape Mendocino tsunami. *Pure Appl. Geophys., 144,* 409—426.

Gonzalez, F. I., & Kulikov, Ye. A. (1993). Tsunami dispersion observed in the deep ocean, In: S. Tinti (Ed.), *Tsunamis in the World,* Norwell, Mass: Kluwer Acad, pp. 7—16.

Gonzalez, F. I., Mader, C. L., Ebel, M. C., & Bernard, E. N. (1991). The 1987-88 Alaskan Bight tsunamis: Deep ocean data and model comparisons. *Natural Hazards, 4,* 119—140.

Gonzalez, J. I., Fareras, S. F., & Ochoa, J. (2001). Seismic and meteorological tsunami contributions in the Manzanillo and Cabo San Lukas seiches. *Marine Geodesy, 24,* 219—227.

Goring, D. G. (2005). Rissaga (long-wave events) on New Zealand's eastern seaboard: a hazard for navigation, In: *Proceedings 17th Australasian Coastal Ocean Eng. Conference,* 20—23 September 2005, Adelaide, Australia, pp. 137—141.

Goring, D. G. (2009). Meteotsunami resulting from the propagation of synoptic-scale weather systems, *Physics and Chemistry of the Earth, 34,* 1009—1015, Elsevier.

Gorny, V. I., Salman, A. G., Tronin, A. A., & Shilin, B. B. (1988). The earth outgoing IR radiation of the earth as an indicator of seismic activity. *Proc. Acad. Sci., USSR, 301,* 67—69.

Gossard, E. E., & Hooke, B. H. (1975). *Waves in Atmosphere.* New York: Elsevier, p. 456.

Gossard, E. E., & Munk, W. H. (1954). On gravity waves in the atmosphere. *J. Meteorol, 11,* 259—269.

Gotwols, B. L., Sterner, R. E., II, & Thompson, D. R. (1988). Measurement and interpretation of surface roughness changes induced by internal waves during the Joint Canada-US Ocean Wave Investigation Project. *J. Geophys. Res., 93,* 12,265—12,281.

Gower, J., & González, F. (2006). U.S. Warning system detected the Sumatra tsunami. *Eos Trans. AGU, 87*(10), 105—108.

Gower, J. (2005). Jason 1 detects the 26 December 2004 tsunami. *Eos, 86*(4), 37—38.

Gower, J. (2007). The 26 December 2004 tsunami measured by satellite altimetry. *Int. J. Remote Sens., 28*(13—14), 2897—2913, doi:10.1080/01431160601094484.

Greenslade, D. J. M., & Jarrott, K. (2009). The Indian Ocean tsunameter network, Abstract; "*24th International Tsunami Symposium and Technical Workshop on Tsunami Measurements and Real-Time Detection*" held at Novosibirsk, Russia (14—16 July 2009).

Greenslade, D. J. M., Simanjuntak, M. A., & Allen, S. C. R. (2009). Modeling and forecasting for the Australian tsunami warning system, Abstract; "*24th International Tsunami Symposium and Technical Workshop on Tsunami Measurements and Real-Time Detection*" held at Novosibirsk, Russia (14—16 July 2009).

Greenspan, H. P. (1956). The generation of edge waves by moving pressure disturbances. *J. Fluid Mech.*, *1*, 574—592.

Gufeld, I. L., Gusev, G., & Pokhotetov, O. (1994). Electromagnetic phenomena related to earthquake prediction. In: M. Hayakawa & Y. Fujinawa (Eds.), (pp. 381). Tokyo, Japan: TERRAPUB.

Gusiakov, V. K. (2006). *Historical tsunami database for the Atlantic*, Version 2.0, CD- ROM, Tsunami Laboratory, ICMMG RAS, Novosibirsk, Russia.

Gusiakov, V. K. (2009). Historical data in application to the tsunami hazard and risk assessment, Abstract; "*24th International Tsunami Symposium and Technical Workshop on Tsunami Measurements and Real-Time Detection*" held at Novosibirsk, Russia (14—16 July 2009).

Gusiakov, V. K. (2009). Tsunami impact on the African continent: historical overview. In: *Proceedings of the International Association of Seismology and Physics of the Earth's Interior (IASPEI) General Assembly 2009, Council for Geoscience* (abstract), January 10—16, 2009, Cape Town, South Africa: Cape Town International Convention Centre.

Gusiakov, V., Dallas, A. H., Edward, B., & Bruce, M. W. (2009). *Mega Tsunami in the Indian Ocean as the Evidence of Recent Oceanic Bolide Impacts*, Chevron Dune Formation and Rapid Climate Change. In: *Proceedings of the International Association of Seismology and Physics of the Earth's Interior (IASPEI) General Assembly 2009, Council for Geoscience* (abstract), January 10—16, 2009, Cape Town, South Africa: Cape Town International Convention Centre.

Guth, J. E. (1981). *Photo-tide level recorder*, United States Patent No. 4,268,839 dated May 19, 1981, p. 3.

Gwal, A. K., Sarkar, S., Bhattacharya, S., & Parrot, M. (2007). Seismo-electromagnetic precursors registered by DEMETER satellite. In: T. S. Murthy, U. Aswathanarayana & N. Nirupama (Eds.), *The Indian Ocean Tsunami* (pp. 235—246). London; UK: Taylor and Francis, 2007.

Gwilliam, T. J. P. & Collar, P. G. (1974). A strain gauge pressure sensor for measuring tides on the continental shelf, *Institute of Ocean Sciences Report No. 14*, p. 27.

Hamilton, T. S., & Wigen, S. O. (1987). The foreslope hills of the Fraser River delta: Implication of tsunamis in Georgia Strait. Sci. Tsunami Hazards, 5, 15—33.

Hammack, J. L., & Segur, H. (1974). The Korteweg-de Vries equations and water waves, Part 2. Comparison with experiments. *J. Fluid Mech., 65*, 289—314.

Hamzah, L., Harada, K., & Imamura, F. (1999). Experimental and numerical study on the effect of mangroves to reduce tsunami. *Tohoku Journal of Natural Disaster Science, 35*, 127—132.

Hanson, D. R., & Dioneff, J. (1973). A diode-quad bridge circuit for use with capacitance transducers. *Review of Scientific Instruments, 44*(10), 1468—1471.

Hara, T., & Plant, W. J. (1994). Hydrodynamic modulation of short wind-wave spectra by long waves and its measurement using microwave backscatter. *J. Geophys. Res., 99*, 9767—9784.

Harada, K., & Imamura, F. (2005). Effects of coastal forest on tsunami hazard mitigation — a preliminary investigation. *Advances in Natural and Technological Hazards Research, 23*, 279—292.

Harada, K., Imamura, F., & Hiraishi, T. (2002). Experimental study on the effect in reducing tsunami by the coastal permeable structures. *Final Proceedings of the International Offshore and Polar Engineering Conference USA*, 652—658.

Harata, M., Sato, T., Hirabayashi, S., & Goto, K. (2010). Numerical simulation of dispersion of volcanic CO_2 seeped from seafloor by using multi-scale ocean model, *OCEANS' 10 — IEEE Sydney Technical Conference*; 24—28 May 2010; 978-1-4244-5222-4/10, 2010 IEEE, p. 5.

Hartnady, C. J. H., Brundrit, G., Hunter, I., Luger, S., Saunders, I., & Wonnacott, R. (2009). The Cape West Coast Tsunami of 20-21 August 2008. In: *Proceedings of the International*

Association of Seismology and Physics of the Earth's Interior (IASPEI) General Assembly 2009, Council for Geoscience (abstract), January 10–16, 2009, Cape Town, South Africa: Cape Town International Convention Centre.

Hashimoto, T., Koshimura, Sh., & Kobayashi, E. (2009). Analysis of ship drifting motion by tsunami, Abstract; "*24th International Tsunami Symposium and Technical Workshop on Tsunami Measurements and Real-Time Detection*" held at Novosibirsk, Russia (14–16 July 2009).

Haslett, S. K., Mellor, H. E., & Bryant, E. A. (2009). *Meteo-tsunami hazard associated with summer thunderstorms in the United Kingdom*, Physics and Chemistry of the Earth, 34, 1016–1022, Elsevier.

Hayakawa, M., Molchanov, O. V., Shima, N., Shvets, A. V. & Yamamoto, N. (2003). *Seismo Electromagnetics: Lithosphere Atmosphere-Ionosphere Couplings*, In: M. Hayakawa & O. A. Molchanov (Eds.), (p. 223) (TERRAPUB, Tokyo, Japan).

Hayes, S. P. (1979). Variability of current and bottom pressure across the continental shelf in the northeast Gulf of Alaska. *J. Phys. Oceanogr., 9*, 88–103.

Hebenstreit, G. T., & Murty, T. S. (1989). Tsunami amplitudes from local earthquakes in the Pacific Northwest Region of North America, Part 1: The outer coast. *Marine Geodesy., 13*, 101–146.

Heezen, B. C., & Ewing, M. (1952). Turbidity currents and submarine slumps and the 1929 Grand Banks earthquake. *Am. J. Sci., 250*, 849–873.

Heezen, B. C., Ericson, D. B., & Ewing, M. (1954). Further evidence for turbidity current following the 1929 Grand Banks earthquake. *Deep Sea Res., 1*, 193–202.

Hemphil-Haley, E. (1995). Diatom evidence for earthquake-induced subsidence and tsunami 300 yr ago in southern coastal Washington. *Geol. Soc. Am. Bull, 107*, 367–378.

Hibiya, T., & Kajiura, K. (1982). Origin of 'Abiki' phenomenon (kind of seiches) in Nagasaki Bay. *J. Oceanogr., Soc. Japan, 38*, 172–182.

Hicks, S. D., Goodheart, A. J., & Iseley, C. W. (1965). Observations of the tides on the Atlantic continental shelf. *J. Geophys. Res., 70*, 827–1830.

Hillebrandt-Andrade, C. G. V. (2009). The Puerto Rico seismic network earthquake and tsunami information and warning system. In: *Proceedings of the International Association of Seismology and Physics of the Earth's Interior (IASPEI) General Assembly 2009, Council for Geoscience* (abstract), January 10–16, 2009, Cape Town, South Africa: Cape Town International Convention Centre.

Hillebrandt-Andrade, C. G. V., Velez, L., Huerfano, V., & Mercado, A. (2008). *Final Technical Report: Hazard mitigation grant program GAR-PR-1552-PR08*, Acquisition and Installation of Puerto Rico Tsunami Ready Tide Gauge Network, pp. 1–34.

Hino, R., Tanioka, Y., Kanazawa, T., Sakai, S., Nishino, M., & Suyehiro, K. (2001). Micro-tsunami from a local interplate earthquake detected by cabled offshore tsunami observation in northeastern Japan. *Geophys. Res. Lett., 28*(18), 3533–3536.

Hiraishi, T., & Harada, K. (2003). Greenbelt tsunami prevention in South Pacific region, Report of the Port and Airport Research Institute, 42, 1–23, Available from http://eqtap.edm.bosai.go.jp/useful_outputs/report/hiraishi/data/papers/greenbelt.pdf.

Hirata, K. (2009). Recent off-shore tsunami observations in Japan: 30 years of experience, Abstract; "*24th International Tsunami Symposium and Technical Workshop on Tsunami Measurements and Real-Time Detection*" held at Novosibirsk, Russia (14–16 July 2009).

Hirata, K., & Baba, T. (2006). Transient thermal response in ocean bottom pressure measurement, *Geophysical Res. Lett., 33*, L10606, doi:10.1029/2006GL026084.

Hirata, K., Takahashi, H., Geist, E., Satake, K., Tanioka, Y., Sugioka, H., et al. (2003). Source depth dependence of micro-tsunamis recorded with ocean-bottom pressure gauges: the

January 28, 2000 Mw 6.8 earthquake off Nemuro Peninsula, Japan, *Earth and Planetary Science Letters*, 208, 305—318

Hirata, K., Takahashi, H., Geist, E., Satake, K., Tanioka, Y., Sugioka, H., et al. (2003). Source depth dependence of micro-tsunamis recorded with ocean-bottom pressure gauges: the January 28, 2000 Mw 6.8 earthquake off Nemuro Peninsula, Japan, *Earth and Planetary Science Letters*, 209, 305—318.

Hirata, K., Aoyagi, M., Mikada, H., Kawaguchi, K., Kaiho, Y., Iwase, R., et al. (2002). Real-time geophysical measurements on the deep seafloor using submarine cable in the southern Kurile subduction zone. *IEEE J. Oceanic Eng.*, 27(2), 170—181.

Hiroi (1979). A formula for evaluating breaking wave pressure intensity in the case of breaking waves, *J. Coll. Eng.*, Tokyo Imperial University, Tokyo, 11—21.

Hodzic (1979/1980). Occurrences of exceptional sea-level oscillations in the Vela Luka Bay (in Croatian), *Priroda*, 68, 52—53.

Holgate, S. J., Foden, P., & Pugh, J. (2007). Tsunami monitoring system: Implementing global real time telemetry. *Sea Technol.*, 48, 37—40.

Holgate, S. J., Woodworth, P. L., Foden, P. R., & Pugh, J. (2008). A study of delays in making Tide gauge data available to tsunami warning centers. *J. Atmos. Oceanic Technol.*, 25, 475—481.

Holman, R. A., Huntley, D. A., & Bowen, A. J. (1978). Infragravity waves in storm conditions. *Proc. 16th Coastal Eng. Conf.*, Hamburg, pp. 268—284.

Honda, K., Terada, T., Yoshida, Y., & Isitani, D. (1908). An investigation on the secondary undulations of oceanic tides. *J. College Sci., Imper Univ.*, Tokyo, p. 108.

Hughes, B. A. (1978). The effect of internal waves on surface wind waves: 2, Theoretical analysis. *J. Geophys. Res.*, 83, 455—465.

Huguenin, J. E., & Ansuini, F. J. (1975). The advantages and limitations of using copper material in marine aquaculture. *Proc. IEEE Oceans'*, 75, 444—453.

Hutchinson, I., Guilbault, J. P., Clauge, J. J., & Bobrowski, P. T. (2000). Tsunamis and the tectonic deformation at the northern Cascadia margin: A 3000 year record from Deserted Lake, Vancouver Island, British Columbia, Canada. *Holocene, 10*, 249—439.

ICG-NEAMTWS (2008). Tsunami Early Warning and Mitigation System in the North Eastern Atlantic, the Mediterranean and Connected Seas, NEAMTWS, Implementation Plan, IOC Technical Series No. 73, UNESCO, p. 46.

Igarashi, Y., Kong, L., & Yamamoto, M. (2009). Compilation of historical tsunami mareograms, Abstract; "*24th International Tsunami Symposium and Technical Workshop on Tsunami Measurements and Real-Time Detection*" held at Novosibirsk, Russia (14—16 July 2009).

Igarashi, Y., Yamamoto, M. & Kong, L. (2009). Anatomy of tsunamis: 1960 Chile earthquake and tsunami, Abstract; "*24th International Tsunami Symposium and Technical Workshop on Tsunami Measurements and Real-Time Detection*" held at Novosibirsk, Russia (14—16 July 2009).

Imamura, F., & Imteaz, M. M. A. (1995). Long waves in two layers: governing equations and numerical model. *Sci. Tsunami Hazards, 13*(1), 3—24.

Inoue, S., Wijeyewickrema, A. C., Matsumoto, H., Miura, H., Gunaratne, P. P., Madurapperuma, M., et al. (2007). Field survey of tsunami effects in Sri Lanka due to the Sumatra-Andaman earthquake of December 26, 2004. *Pure Appl. Geophys., 164*, 395—412.

Irish, J. D., & Snodgrass, F. E. (1972). Quartz crystals as multi-purpose oceanographic sensors, I. Pressure. *Deep-Sea Res., 19*, 165—169.

Isozaki, I., Den, N., Iinuma, T., Matumoto, H., Takahashi, M., & Tsukakoshi, T. (1980). Deep sea observation and its application to pelagic tide analysis. *Papers Meteorology and Geophysic, 31*(2), 87—96.

Ivelskaya, T. N. (2009). Monitoring of marine hazards in the Kholmsk and the Korsakov ports, Sakhalin island, Abstract; "*24th International Tsunami Symposium and Technical Workshop on Tsunami Measurements and Real-Time Detection*" held at Novosibirsk, Russia (14–16 July 2009).

Iwabuchi, Y., Sugino, H., Ebisawa, K., & Imamura, F. (2009). The development of tsunami trace database in which reliability is taken into account based on the trace data on the coast of Japan, Abstract; "*24th International Tsunami Symposium and Technical Workshop on Tsunami Measurements and Real-Time Detection*" held at Novosibirsk, Russia (14–16 July 2009).

Iyer, C. S. P. (2007). Ecological impact of Indian Ocean tsunami. In: T. S. Murthy, U. Aswathanarayana & N. Nirupama (Eds.), *The Indian Ocean Tsunami* (pp. 339–350). London; UK: Taylor and Francis, 2007.

Jansa, A. (1986). Marine response to mesoscale meteorological disturbances: The June 21, 1984 event in Ciutadella (Menorca), *Rev. Meteorol., 7*, 5–29.

Jansa, A., Monserrat, S., & Gomis, D. (2007). The Rissaga of 15 June 2006 in Ciutadella (Menorca), a meteorological tsunami. *Advances in Geosciences, 12*, 1–4.

Jaque, V. M., & Soloviev, S. L. (1971). Distant registration of small waves of tsunami type on the shelf of the Kuril Islands (in Russian), Doklady Akademii Nauk SSR, *198*(4).

Joseph, A. (1999a). *Modern techniques of sea level measurement, Encyclopedia of Microcomputers*. New York: Marcel Dekker, Inc., *23*, 319–344.

Joseph, A., & Desa, E. S. (1984). A microprocessor based tide measuring system. *J. Phys. E: Sci. Instrum.*, 1135–1138.

Joseph, A., & Prabhudesai, R. G. (2005). Need of a disaster alert system for India through a network of real time monitoring of sea level and other meteorological events. *Curr. Sci., 89*(5), 864–869.

Joseph, A., Desa, E., Desa, E. S., Smith, D., Peshwe, V. B., Vijaykumar, et al. (1999c). Evaluation of pressure transducers under turbid natural waters. *J. Atmos. Oceanic Technol., 16*(8), 1150–1155.

Joseph, A., Desa, E., Kumar, V., Desa, E. S., Prabhudesai, R. G., & Prabhudesai, S. (2004). In S. Holgate & T. Aarup (Eds.), *Pressure gauge experiments in India, Workshop Report No. 193, Intergovernmental Oceanographic Commission of UNESCO* (pp. 22–37).

Joseph, A., Desa, E. S., Prabhudesai, R. G., Kumar, V., Desa, E., & Peshwe, V. B. (1999b). Development of a sea level recorder for measurements at harbours and jetties. *Proceedings of the International Conference in trends in Industrial Measurements and Automation (TEMA-99)*, Chennai, pp. 205–214.

Joseph, A., Odametey, J. T., Nkebi, E. K., Pereira, A., Prabhudesai, R. G., Mehra, P., et al. (2006). The 26 December 2004 Sumatra tsunami recorded on the coast of West Africa. *African J. Marine Sci., 28*(3&4), 705–712.

Joseph, A., Desa, J. A. E., Foden, P., Taylor, K., McKeown, J., & Desa, E. (2000). Evaluation and performance enhancement of a pressure transducer under flows, waves, and a combination of flows and waves. *J. Atmos. Oceanic Technol., 17*(3), 357–365.

Joseph, A., Peshwe, V. B., Vijaykumar, E., Desa, S., & Desa, E. (1997). Effects of water trapping and temperature gradient in a NGWLMS gauge deployed in Zuari estuary, Goa, *Proc. SYNPOL-97*, Cochin University of Science & Technology, pp. 73–82.

Joseph, A., Vijaykumar, Desa, E. S., Desa, E., & Peshwe, V. B. (2002). Over-estimation of sea level measurements arising from water density anomalies within tide-wells — A case study at Zuari estuary, Goa. *J. Coastal Res., 18*(2), 362–371.

Kajiura, K. (1963). The leading wave of a tsunami. *Bull. Earthquake Res. Inst.*, Univ. of Tokyo, *41*, 535–571.

Kajiura, K. (1970). Tsunami source, energy and the directivity of wave radiation. *Bull. Earthquake Res. Inst.*, Univ. of Tokyo, 48, 835–869.

Kajiura, K. (1972). The directivity of energy radiation of the tsunami generated in the vicinity of a continental shelf. *J. Oceanogr. Soc. Japan, 28*(6), 32–49.

Kamigaichi, O. (2009). Tsunami forecasting and warning, Encyclopedia of Complexity and Systems Science. R. A. Meyers (Ed.), 9592–9617.

Kanamori, H. (1972). Mechanism of tsunami earthquakes. *Phys. Earth Planet. Inter., 6*, 346–359.

Kanamori, H., & Anderson, D. L. (1975). Theoretical basis of some empirical relations in seismology. *Bull. Seismol. Soc. Am., 65*, 1073–1095.

Kanazawa, T. (2000). Real time seismic observation in oceanic area. *Earthquake Journal, 29*, 34–44, (in Japanese).

Kanazawa, T., & Hasegawa, A. (1997). Ocean-bottom observatory for earthquakes and tsunami off Sanriku, North-eastern Japan using submarine cable, In: *Proc. Int. Workshop on Scientific Use of Submarine Cables*, 208–209.

Kânoğlu, U., Titov, V. V., Aydın, B., & Synolakis, C. E. (2009). Propagation of finite strip sources over a flat bottom, Abstract; "*24th International Tsunami Symposium and Technical Workshop on Tsunami Measurements and Real-Time Detection*" held at Novosibirsk, Russia (14–16 July 2009).

Karlsrud, K., & Edgers, L. (1980). Some aspects of submarine slope stability. In: *Marine Slides and Other Mass Movements* (pp. 61–81). New York: Plenum.

Karrer, H. E., & Leach, J. (1969). A quartz resonator pressure transducer. *IEEE Trans. Ind. Electron. Control Instrum., 16*, 44–50.

Kasahara, J., Chave, A., & Mikada, H. (2003). Exploring the use of submarine cables and related technologies. *Eos Trans. AGU, 84*(50), doi:10.1029/2003EO500006.

Kathiresan, K. (2003). How do mangrove forests induce sedimentation? *Revista de Biologia Tropical, 51*, 355–360.

Kathiresan, K., & Bingham, B. L. (2001). Biology of mangroves and mangrove ecosystems. *Advances in Marine Biology, 40*, 81–251.

Kathiresan, K., & Rajendran, N. (2005). Coastal mangrove forests mitigated tsunami, Estuarine. *Coastal and Shelf Science, 65*, 601–606.

Kawaguchi, K., Hirata, K., Mikada, H., Kaiho, Y., & Iwase, R. (2000). An expandable deep seafloor monitoring system for earthquake and tsunami observation network, In: *Proc. Oceans '00*, 2000, CD-ROM.

Kawaguchi, K., Hirata, K., Nishida, T., Obana, S., & Mikada, H. (2002). A new approach for mobile and expendable real-time deep seafloor observation — Adaptable observation system. *IEEE J. Oceanic Eng., 27*(2), 182–192.

Kaystrenko, V. M., Khramushin, V. N., Zolotukhin, D. E. (2009). Tsunami hazard zoning for the southern Kuril islands, Abstract; "*24th International Tsunami Symposium and Technical Workshop on Tsunami Measurements and Real-Time Detection*" held at Novosibirsk, Russia (14–16 July 2009).

Kharif, C., & Pelinovsky, E. (2003). Physical mechanisms of the rogue wave phenomenon. *European Journal of Mechanics: B, Fluids, 22*, 603–634.

Kharif, Ch., & Pelinovsky, E. (2005). Asteroid impact tsunamis. *Comptes Rendus Physique, 6*, 361.

Kobayashi, E. S. Koshimura, Yoneda, S. & Wakabayashi, N. (2008). Guidelines and considerations for ship evacuation from a tsunami attack. *Proc.of the 9th Annual General Assembly International Association of Maritime Universities*, San Francisco, CA USA, October 19–22, pp. 385–398.

Komarov, A., Korobkin, A. A., & Sturova, I. V. (2009). Interaction between a tsunami wave and a very large elastic platform floating in shallow water of variable depth, Abstract; "*24th International Tsunami Symposium and Technical Workshop on Tsunami Measurements and Real-Time Detection*" held at Novosibirsk, Russia (14−16 July 2009).

Korolev, Yu. (2009a). An approximate method of tsunami early warning, Abstract; "*24th International Tsunami Symposium and Technical Workshop on Tsunami Measurements and Real-Time Detection*" held at Novosibirsk, Russia (14−16 July 2009).

Korolev, Yu. (2009b). Application of sea-level data (DART system) for short-term tsunami forecast near Russian and US coasts during 2006, 2007 and 2009 events, Abstract; "*24th International Tsunami Symposium and Technical Workshop on Tsunami Measurements and Real-Time Detection*" held at Novosibirsk, Russia (14−16 July 2009).

Koshimura, S., Yanagisawa, H., & Miyagi, T. (2009). Mangrove's fragility against tsunami, inferred from high-resolution satellite imagery and numerical modeling, Abstract; "*24th International Tsunami Symposium and Technical Workshop on Tsunami Measurements and Real-Time Detection*" held at Novosibirsk, Russia (14−16 July 2009).

Kovalev, P. D., Rabinovich, A. B., & Shevchenko, G. V. (1991). Investigation of long waves in the tsunami frequency band on the southwestern shelf of Kamchatka. *Nat. Hazards, 4*, 141−159.

Kowalik, Z., Horillo, J., Knight, W., & Logan, T. (2008). The Kuril Islands tsunamis of November 2006: Part I: Impact at Crescent city by distant scattering. *J. Geophys. Res., 113*, C01020.

Kowalik, Z., Knight, W., & Whitmore, P. (2005a). Numerical modeling of the global tsunami: Indonesian tsunami of 26 December 2004. *Sci. Tsunami Hazards, 23*(1), 40−56.

Kowalik, Z., Knight, W., Logan, T., & Whitmore, P. (2005b). The tsunami of 26 December 2004: Numerical modeling and energy considerations. In: G. A. Papadopoulos & K. Satake (Eds.), *Proceedings of the International Tsunami Symposium, Chania, Greece, 27−29 June 2005* (pp. 140−150).

Kowalik, Z., Knight, W., Logan, T., & Whitmore, P. (2007a). Numerical modeling of the Indian Ocean tsunami. In: T. S. Murthy, U. Aswathanarayana & N. Nirupama (Eds.), *The Indian Ocean Tsunami* (pp. 97−122). London; UK: Taylor and Francis, 2007.

Kowalik, Z., Knight, W., Logan, T., & Whitmore, P. (2007b). The tsunami of 26 December 2004: Numerical modeling and energy considerations. *Pure Appl. Geophys, 164*, 379−393.

Krishnakumar (1997). The coral reef ecosystem of the Andaman and Nicobar islands, Problems and prospects and the World Wide Fund for Nature — India initiatives for its conservation. In: M. S. Vineeta Hoon, (Ed.), *Proceedings of the Regional Workshop on the Conser. Sustain. Manag. Coral Reef Ecosystem* , Swaminathan Research Foundation and BOBP of FAO/UN, 29−47.

Kropfli, R. A., Ostrovski, L. A., Stanton, T. P., Skirta, E. A., Keane, A. N., & Irisov, V. (1999). Relationships between strong internal waves in the coastal zone and their radar and radiometric signatures. *J. Geophys. Res., 104*, 3133−3148.

Kubo, M., Ik-Soon, Cho, Sakakibara, S., Kobayashi, E., & Koshimura, S. (2005). The influence of tsunamis on moored ships and ports. *International Journal of Navigation and Port Research, 29*(4), 319−325.

Kudryavtsev, V. N., Mastenbroek, C., & Makin, V. K. (1997). Modulation of wind ripples by long surface waves via the air flow: a feedback mechanism. *Bound. Lay. Meteorol., 83*, 99−116.

Kulikov, E. A. & Fine, I. V. (2009). Tsunami risk estimation in the Tatar Strait; Abstract; "*24th International Tsunami Symposium and Technical Workshop on Tsunami Measurements and Real-Time Detection*" held at Novosibirsk, Russia (14−16 July 2009).

Kulikov, E. A., Medvedev, P. P., & Lappo, S. S. (2005). Satellite recording of the Indian Ocean tsunami on December 26, 2004. *Doklady Earth Sciences, 401 A*(3), 444−448.

Kulikov, E. A., Rabinovich, A. B., Thomson, R. E., & Bornhold, B. B. (1996). The landslide tsunami of November 3, 1994, Skagway Harbor, Alaska. *J. Geophys. Res., 101*, 6609−6615.

Kumar, K. A., Achyuthan, H., & Shankar, N. (2007). Paleo-tsunami and storm surge deposits. In: T. S. Murthy, U. Aswathanarayana & N. Nirupama (Eds.), *The Indian Ocean Tsunami* (pp. 49−55). London; UK: Taylor and Francis, 2007.

Kurian, N. P., Nirupama, N., Baba, M., & Thomas, K. V. (2009). Coastal flooding due to synoptic scale, meso-scale and remote forcing. *Natural Hazards, 48*, 259−273.

Kusters, J. A. (1976). Transient thermal compensation for quartz resonators. *IEEE Trans. Son. Ultrason, SU-23*, 273−276.

Lander, J. (1995). Nonseismic tsunami event in Skagway, Alaska. *Tsunami Newsl.* 8−9.

Lander, J. F., & Lockridge, P. A. (1989). United States Tsunamis, 1690−1988, p. 26, *Natl., Geophys. Data Cent., Natl. Oceanic and Atmos. Admin.*, Boulder, Colo.

Landisman, M., Dziewonski, A., & Sato, Y. (1969). Recent improvements in the analysis of surface wave observations. *Geophys. J. Roy. Astr. Soc., 17*, 369−403.

Larkina, V. I., Migulin, V. V., Molchanov, O. A., Kharkov, I. P., Inchin, A. S., & Sshevtsova, V. B. (1989). Some statistical results on very low frequency radiowave emissions in the upper ionosphere over earthquake zones. *Phys. Earth Planet. Inter., 57*, 100−109.

Latief, H., & Hadi, S. (2007). The role of forests and trees in protecting coastal areas against tsunamis. In: S. Braatz, S. Fortuna, J. Broadhead & R. Leslie (Eds.), *Coastal protection in the aftermath of the Indian Ocean Tsunami: What role for Forests and trees? Proceedings of the Regional Technical Workshop, Khao Lak, Thailand, 28−31 August 2006* (pp. 5−35). Bangkok: FAO. http://www.fao.org/forestry/site/coastalprotection/en/.

Laverov, N. P., Lappo, S. S., Lobkovskii, L. I., & Kulikov, E. A. (2006a). *Strongest underwater earthquakes and catastrophic tsunamis: Analysis, modeling, forecast.* In: Fundamental Investigations of Oceans and Seas (Nauka, Moscow), 1, 191−209 [in Russian].

Laverov, N. P., Lappo, S. S., Lobkovskii, L. I., et al., (2006b). "The central Kuril gap": Structure and seismic potential. *Dokl. Earth Sci., 409*, 787−790 [*Dokl. Akad. Nauk, 408*, 1−4(2006)].

Lay, T., Kanamori, H., Ammon, C. J., Nettles, M., Ward, S. N., et al. (2005). The great Sumatra-Andaman Earthquake of 26 December 2004. *Science, 308*, 1127−1133.

LeBlond, P. H., & Mysak, L. A. (1978). *Waves in the Ocean*, Elsevier, Amsterdam, p. 602.

Lee, J. J., & Raichlen, F. (1972). Oscillations in harbours with connected basins, J. Waterway, harbours and coastal engineering division. *Proceedings of the ASCE, 98*, 311−332.

Lennon, G. W. (1970). Sea level instrumentation, its limitations and the optimization of the performance of conventional gauges in Great Britain. *Int. Hydrogr. Rev., 48*(2), 129−147.

Lennon, G. W., & Mitchell, W. Mcl. (1992). The stilling well — A help or hindrance?, *Workshop Report No. 81*, Intergovernmental Oceanographic Commission, 52−64.

Levin, B. V., Kaistrenko, V. M., Rybin, A. V. et al. (2008). Manifestations of the tsunami on November 15, 2006 on the central Kuril Islands and results of the run-up heights modeling. *Dokl. Earth Sci., 419*, 335−338 [*Dokl. Akad. Nauk, 419*, 118−122 (2008)].

Lindzen, R. S., & Tung, K.-K. (1976). Banded convective activity and ducted gravity waves. *Monthly Weather Review, 104*, 1602−1617.

Ling, S. C., & Pao H. P. (1994). *Precision water level and sediment monitoring systems.* In: First International Symposium on Hydraulic Measurements, 1−33.

Liu, P. L., Synolakis, C. E., & Yeh, H. H. (1991). Report on the international workshop on long-wave run-up. *J. Fluid Mech., 229*, 675−688.

Lobkovsky, L. I. (2005). Catastrophic earthquake and tsunami of December 26, 2004 in the northern part of the Sunda Island arc: Geodynamical analysis and similarity with the central Kuril islands. *Vestnik RAEN, 2*, 53−61.

Lobkovsky, L. I., Rabinovich, A. B., Kulikov, E. A., Ivashchenko, A. I., Fine, I. V., Thomson, R. E., et al. (2009). The Kuril earthquakes and tsunamis of November 15, 2006, and January 13, 2007: Observations, analysis, and numerical modeling. *Oceanology, 49*(2), 166−181.

Lobkovsky, L. I., & Kulikov, E. (2006a). Analysis of hypothetical strong earthquake and tsunami in the central Kuril arc, *Geophys. Res. Abstracts*, 8, 04138: SRef-ID: 1607−7962/gra/EGU06-A-04138.

Lobkovsky, L. I., Mazova, R. Kh., Kataeva, Yu. L., & Baranov, B. V. (2006b). Generation and propagation of catastrophic tsunamis in the Sea of Okhotsk Basin: Possible scenarios, *Dokl. Earth Sci., 410*, 1156−1159 [*Dokl. Akad. Nauk, 410*, 528−531].

Lobkovsky, L. I., Kulikov, E. A., Rabinovich, A. B., et al. (2006c). Earthquakes and tsunamis (November 15, 2006 and January 13, 2007) in the central Kuril islands region: A justified prediction, *Dokl. Earth Sci., 419*, 320−324 [*Dokl. Akad. Nauk, 418*, 829−833(2006)].

Lockridge, P. A. (1989). Tsunami: trouble for mariners. *Sea Technol., 30*(4), 53−57.

Lockridge, P. A., Whiteside, L. S., & Lander, J. F. (2002). Tsunamis and tsunami-like waves of the eastern United States. *Sci Tsunami Hazards, 20*(3), 120−157.

Longva, O., Janbu, N., Blikra, L. H., & Boe, R. (2003). The 1996 Finneid fjord slide; seafloor failure and slide dynamics. In: J. Locat & J. Mienert (Eds.), *Submarine Mass Movements and their Consequences* (pp. 531−538). Dordrecht, The Netherlands: Kluwer Academic Publishers.

Loomis, H. G. (1966). Spectral analysis of tsunami records from stations in the Hawaiian islands. *Bull. Seis. Soc. Amer., 56*, 697−713.

Loomis, H. G. (1983). The nonlinear response of a tide gauge to a tsunami. *Proc. 1983 Tsunami Symp.*, 177−185.

Low, K. C. (2009). The usage of ICT in Malaysian meteorological department's multi-hazard early warning system, *Fifth Technical Conference on Management of Meteorological and Hydrological Services in Regional Association V*, held at Petaling Jaya, Malaysia during 20−24 April 2009.

Lowe, D. J., & de Lange, W. P. (2000). Volcano-meteorological tsunamis, the c. AD 200 Taupo eruption (New Zealand) and the possibility of a global tsunami. *The Holocene, 10*, 401−407.

Mackelburg, G. R., Watson, S. J., & Gordon, A. (1981). Benthic 4800 bits/s acoustic telemetry, Proc. OCEANS 81 IEEE-MTS Conf. (Boston), Sept. 1981, p. 72.

Madsen, O. S., & Mei, C. C. (1969). The transformation of a solitary wave over an uneven bottom. *J. Fluid Mech., 39*, 781−791.

Maramai, A., Graziani, L., & Tinti, S. (2007). Investigation on tsunami effects in the central Adriatic Sea during the last century — a contribution. *Natural Hazards and Earth System Sciences, 7*, 15−19.

Marchuk, An. G. & Titov, V. V. (1990). Source configuration and the process of tsunami waves forming. In: *Proc. Fourth Pacific Congress on Marine Science and Technology*, PACON-90, Tokyo, Japan; July 16−20, 1990, pp. 156−161.

Marchuk, An. G. (2009). Directivity of tsunami generated by subduction zone sources, Abstract; "*24th International Tsunami Symposium and Technical Workshop on Tsunami Measurements and Real-Time Detection*" held at Novosibirsk, Russia (14−16 July 2009).

Marcos, M., Monserrat, S., Medina, R., Orfila, A., & Olabarrieta, M. (2009). External forcing of meteorological tsunamis at the coast of the Balearic Islands. *Physics and Chemistry of the Earth, 34*, 938−947.

Masaitis, V. L. (2002). The middle Devonian Kaluga impact crater (Russia): new interpretation of marine setting. *Deep-Sea Research II, 49,* 1157, doi:10.1016/S0967-0645(01)00142-4.

Mascarenhas, A., & Jaykumar, S. (2007). Protective role of coastal ecosystems in the context of the tsunami in Tamil Nadu coast, India. In: T. S. Murthy, U. Aswathanarayana & N. Nirupama (Eds.), *The Indian Ocean Tsunami* (pp. 423–436). London; UK: Taylor and Francis, 2007.

Massel, S. R. (1999). Tides and waves in mangrove forests. In: *Fluid Mechanics for Marine Ecologists* (pp. 418–425). Springer-Verlag.

Massel, S. R., Furukawa, K., & Brinkman, R. M. (1999). Surface wave propagation in mangrove forests. *Fluid Dynamics Research, 24,* 219–249.

Matias, L., Carrilho, F., Baptista, M. A., Annunziatto, A., Omira, R., & Luis, J. (2009). Towards a Regional Tsunami Warning System in the Gulf of Cadiz. In: *Proceedings of the International Association of Seismology and Physics of the Earth's Interior (IASPEI) General Assembly 2009, Council for Geoscience* (abstract), January 10–16, 2009, Cape Town, South Africa: Cape Town International Convention Centre.

Mazda, Y., Wolanski, E., King, B., Sase, A., Ohtsuka, D., & Magi, M. (1997). Drag force due to vegetation in mangrove swamps. *Mangroves and Salt Marshes, 1,* 193–199.

Mazda, Y., Wolanski, E., & Ridd, P. V. (2007). *The role of physical processes in mangrove environments, Manual for the preservation and utilization of mangrove ecosystems.* Tokyo: Terrapub, 598.

Mazda, Y., Magi, M., Kogo, M., & Hong, P. N. (1997). Mangrove on coastal protection from waves in the Tong King Delta, Vietnam. *Mangroves and Salt Marshes, 1,* 127–135.

Mazda, Y., Magi, M., Ikeda, Y., Kurokawa, T., & Asano, T. (2006). Wave reduction in a mangrove forest dominated by Sonneratia sp. *Wetlands Ecology and Management, 14,* 365–378.

McCreery, C. S. (2005). Impact of the national tsunami hazard mitigation program on operations of the Richard H. Hagemeyer Pacific tsunami warning center. *Natural Hazards, 35,* 73–88.

Melo, E., & Rocha, C. (2005). Sumatra tsunami detected in Southern Brazil, Maritime Hydraulics Laboratory, Federal University of Santa Catarina, Brazil, Unpublished Report. http://www.pmel.noaa.gov/tsunami/sumatra20041226.html

Merrifield, M. A., Firing, Y. L., Aarup, T., Agricole, W., Brundrit, G., Chang-Seng, D., et al. (2005). Tide gauge observations of the Indian Ocean tsunami, December 26, 2004. *Geophys. Res. Lett., 32,* L09603, doi:10.1029/2005GL022610.

Metzner, M., Gade, M., Hennings, I., & Rabinovich, A. B. (2000). The observation of seiches in the Baltic Sea using a multi data set of water levels. *Journal of Marine Systems, 24,* 67–84.

Miles, J., & Munk, W. (1961). Harbour paradox. *Journal of Waterways Harbours, ASCE, 87,* 111–130.

Miller, D. J. (1960). The Alaska earthquake on July 10, 1958: Giant wave in Lituya Bay. *Bull. Seismol. Soc. Am., 50,* 253–266.

Miller, G. R. (1964). *Tsunamis and tides,* PhD Thesis, Univ. of Calif. San Diego, p. 120.

Miller, G. R. (1972). Relative of tsunamis. *Hawaii Inst. Geophys. HIG-72-8,* p. 7.

Miller, G. R., Munk, W. H., & Snodgrass, F. E. (1962). Long-period waves over California's continental borderland, II, Tsunamis. *J. Marine Res., 20*(1), 31–41.

Mitsubayashi, H., & Honda, T. (1974). The high-frequency spectrum of wind-generated waves. *J. Oceanogr. Soc. Jpn., 30,* 185–198.

Mofjeld, H. O. (2009). Tsunami measurements. In: Bernard & Robinson (Eds.), *The Sea, Vol. 15.* Tsunamis, Harvard University Press.

Mofjeld, H. O., Gonzalez, F. I., & Newman, J. C. (1999b). Tsunami prediction in U.S. coastal regions. In: C. N. K. Mooers (Ed.), *Coastal Ocean Prediction, Coastal and Estuarine Studies 56* (pp. 353–375). Washington: AGU.

Mofjeld, H. O., Gonzalez, F. I., Bernard, E. N., & Newman, J. C. (2000). Forecasting the heights of later waves in Pacific-wide tsunamis. *Natural Hazards, 22,* 71–89.

Mofjeld, H. O., Titov, V. V., Gonzalez, F. I., & Newman, J. C. (1999a). Tsunami wave scattering in the north Pacific, *IUGG-99 Abstracts,* July 26–30, B. 1326.

Mohan, P. M. (December 2005). Tsunami and tsunamigenic sediments in Andaman and Nicobar islands. S. P. Raj (Ed.). *National Commemorative Conference on Tsunami, Madurai,* 26–29 December 2005. Proceedings. (Natl. Commemorative Conf. on Tsunami; Madurai; India; 26–29 Dec 2006). Madurai, India: Madurai Kamaraj University, 2006, 88–91.

Molchanov, O. A. (1993). Wave and plasma phenomena inside the ionosphere and the magnetosphere associated with earthquakes. In: W. R. Stone (Ed.), *Review of Radio Science 1990–1992* (pp. 591–600). New York: Oxford University Press.

Molchanov, O. A., & Hayakawa, M. (1994). Generation of ULF seismogenic electromagnetic emission: A natural consequence of microfracturing process. In: M. Hayakawa & Y. Fujinawa (Eds.), *Electromagnetic Phenomena Related to Earthquake Prediction.* Tokyo: Terra Scientific.

Molchanov, O. V., & Hayakawa, M. (1998). Subionospheric VLF signal perturbations possibly related to earthquakes. *J. Geophys. Res., 103*(A8), 17489–17504.

Molchanov, O. A., Mazhaeva, O. A., Golyavin, A. N., & Hayakawa, M. (1993). Observation by Intercosmos-24 satellite of ELF-VLF electromagnetic emissions associated with earthquakes. *Ann. Geophys., 11,* 431–440.

Molchanov, O. V., Hayakawa, M., & Rafalsky, V. (1995). Penetration characteristics of electromagnetic emissions from an underground seismic source into the atmosphere, ionosphere, and magnetosphere. *J. Geophys. Res., 106,* 1691–1712.

Momma, H., Kawaguchi, K., & Iwase, R. (1999). A new approach for long-term seafloor monitoring and data recovery, In: *Proc. 9th Inl. Offshore and Polar Eng. Conf.,* 603–610.

Momma, H., Fujiwara, N., Iwase, R., Kawaguchi, K., Suzuki, S., & Kinoshita, H. (1997). Monitoring system for submarine earthquakes and deep sea environment, In: *Proc. MTS/IEEE OCEANS'97, 2,* 1453–1459.

Momma, H., Fujiwara, N., & Suzuki, S. (1998). Deep-sea monitoring system for submarine earthquakes, environment. *Sea Technol., 39*(6), 72–76.

Monserrat, S., Ibberson, A., & Thorpe, A. J. (1991a). Atmospheric gravity waves and the "Rissaga" phenomenon. *Quart. J. Roy. Meteorol. Soc., 117,* 553–570.

Monserrat, S., Ramis, C., & Thorpe, A. J. (1991b). Large-amplitude pressure oscillations in the Western Mediterranean. *Geophys. Res. Lett., 18,* 183–186.

Monserrat, S., Rabinovich, A. B., & Casas, B. (1998). On the reconstruction of the transfer function for atmospherically generated seiches. *Geophysical Research Letters, 25*(12), 2197–2200.

Monserrat, S., & Thorpe, A. J. (1992). Gravity-wave observations using an array of microbarographs in the Balearic Islands. *Quarterly Journal of the Royal Meteorological Society, 118,* 259–282.

Monserrat, S., & Thorpe, A. J. (1996). Use of ducting theory in an observed case of gravity waves. *J. Atmos. Sci., 53,* 1724–1736.

Monserrat, S., Vilibić, I., & Rabinovich, A. B. (2006). Meteotsunamis: atmospherically induced destructive ocean waves in the tsunami frequency band. *Nat. Hazards Earth Syst. Sci., 6,* 1035–1051.

Munk, W. H. (1947). Increase in period of waves traveling over large distances: with applications to tsunamis, swell, and seismic surface waves. *Trans. Am. Geophys. Union, 28,* 198–217.

Murray, M. T. (1964). A general method for the analysis of hourly heights of the tide. *International Hydrographic Review, 41*(2), 91–101.

Murty, T. S. (1977). Seismic Sea Waves — Tsunamis. *Bull. Fish. Res. Board Canada, 198,* 337.

Murty, T. S. (1979). Submarine slide-generated water waves in Kitimat Inlet, British Columbia. *J. Geophys. Res., 84*(C12), 7777–7779.

Murty, T. S. (1992). Tsunami threat to the British Columbia coast. In: *Geotechnique and Natural Hazards* (pp. 81–89). Vancouver, B.C: BiTech. Publ.

Murty, T.S. & Kurian, N. P. (2006). A possible explanation for the flooding several times in 2005 on the coast of Kerala and Tamil Nadu, *J. Geol. Soc. India, 67*, 535–536.

Murty, T. S., & Loomis, H. G. (1980). A new objective tsunami magnitude scale. *Mar. Geod., 4*, 267–282.

Murty, T. S., Nirupama, N., Nistor, I., & Rao, A. D. (2005a). *Leakage of the Indian Ocean tsunami energy into the Atlantic and Pacific Ocean*, CSEG Recorder. December, 2005, 33–36.

Murty, T. S., Nirupama, N., Nistor, I., & Hamdi, S. (2005b). Far field characteristics of the tsunami of 26 December 2004. *ISET J. Earthq. Technol., 42*(4), 213–217.

Murty, T. S., Nirupama, N., Nistor, I., & Hamdi, S. (2005c). Far field characteristics of the tsunami of 26 December 2004. *ISET J. Earthq. Technol., 42*(4), 213–217.

Murty, T. S., Nirupama, N., & Rao, A. D. (2005). Why the earthquakes of 26th December 2004 and the 27th March 2005 differed so drastically in their tsunamigenic potential. *Newslett. Voice Pacific, 21*(2), 2–4.

Murty, T. S., Rao, A. D., Nirupama, N., & Nistor, I. (2006). Numerical modeling concepts for the tsunami warning systems. *Curr. Sci., 90*(8), 1073–1081.

Murty, T. S., Nirupama, N., Rao, A. D., & Nistor, I. (2007a). Possible detection in the ionosphere of the signals from earthquake and tsunamis. In: T. S. Murthy, U. Aswathanarayana & N. Nirupama (Eds.), *The Indian Ocean Tsunami* (pp. 227–234). London; UK: Taylor and Francis, 2007.

Murty, T. S., Nirupama, N., Rao, A. D., & Nistor, I. (2007b). Methodologies for tsunami detection, In: T. S. Murthy (Eds.), *The Indian Ocean Tsunami*, Aswathanarayana.

Nagai, T., Ogawa, H., Terada, Y., Kato, T., & Kudaka, M. (2004). GPS buoy application to offshore wave, tsunami, and tide observation, ASCE. In: *Proc. Of the 29th International Conference on Coastal Engineering (ICCE'04)*, 1, 1093–1105.

Nagai, T., Satomi, S., Terada, Y., Kato, T., Nukada, K., & Kudaka, M. (2005). GPS buoy and seabed installed wave gauge application to offshore tsunami observation. In: *Proc. of the 15th International Offshore and Polar Engineering Conference*, 3, 292–299, 2005.

Nagai, T., Shimizu, K., Sasaki, M., & Murakami, A. (2008). Improvement of the Japanese NOWPHAS Network by Introducing Advanced GPS Buoys. In: *Proceedings of the Eighteenth (2008) International Offshore and Polar Engineering Conference*, Vancouver, BC, Canada, July 6–11, 2008, pp. 558–564.

Nagarajan, B., Suresh, I., Sundar, D., Sharma, R., Lal, A. K., Neetu, S., et al. (2006). The great tsunami of 26 December 2004: A description based on tide-gauge data from the Indian subcontinent and surrounding areas. *Earth Planets Space, 58*, 211–215.

Nakano, M., & Unoki, S. (1962). On the seiches (secondary undulations of tides) along the coast of Japan. *Records Oceanogr. Works Japan Spec., 6*, 169–214.

Namegaya, Y., Tanioka, Y., & Satake, K. (2009). Configuration of tsunami source in waveform inversion, Abstract; "*24th International Tsunami Symposium and Technical Workshop on Tsunami Measurements and Real-Time Detection*" held at Novosibirsk, Russia (14–16 July 2009).

Narayana, A. C., & Tatavarti, R. (2005). High wave activity on the Kerala coast. *J. Geol. Soc. India, 66*, 249–250.

Nayak, S., & Kumar, T. S. (2008). The first tsunami early warning centre in the Indian Ocean, In: Risk Wise, Tudor Rose Publishers, p. 1536.

Nekrasov, A. V. (1970). Transformation of tsunami on the continental shelf. In: W. M. Adams (Ed.), *Tsunami in the Pacific Ocean* (pp. 337–350). Honolulu: East-West Center Press.

Newman, A. V., & Okal, E. A. (1998). Teleseismic estimates of radiated seismic energy: The E/M_0 discriminant for tsunami earthquakes. *J. Geophys. Res, 103*, 26885–26898.

Nicolsky, D. J., Suleimani, E. N., West, D. A., & Hansen, R. A. (2009). Tsunami modeling and inundation mapping in Alaska: validation and verification of a numerical model, Abstract; *"24th International Tsunami Symposium and Technical Workshop on Tsunami Measurements and Real-Time Detection"* held at Novosibirsk, Russia (14–16 July 2009).

Nikonov, A. A. (2009). Tsunami and tsunami-like water excitations in large inland basins of Russia, Abstract; *"24th International Tsunami Symposium and Technical Workshop on Tsunami Measurements and Real-Time Detection"* held at Novosibirsk, Russia (14–16 July 2009).

Nirupama, N., Murty, T. S., Rao, A. D., & Nistor, I. (2005). Numerical tsunami models for the Indian Ocean countries and states. *Indian Ocean Survey, 2*(1), 1–14.

Nirupama, N., Murty, T. S., Nistor, I., & Rao, A. D. (2006). The energetics of the tsunami of 26 December 2004 in the Indian Ocean: a brief review. *Mar. Geod., 29*(1), 39–48.

Nirupama, N., Murty, T. S., Nistor, I., & Rao, A. D. (2007a). A review of classical concepts on phase and amplitude dispersion: Application to tsunamis. In: T. S. Murthy, U. Aswathanarayana & N. Nirupama (Eds.), *The Indian Ocean Tsunami* (pp. 91–95). London; UK: Taylor and Francis, 2007.

Nirupama, N., Murty, T. S., Nistor, I., & Rao, A. D. (2007b). The Cauchy-Poisson problem: Application to tsunami generation and propagation. In: T. S. Murthy, U. Aswathanarayana & N. Nirupama (Eds.), *The Indian Ocean Tsunami* (pp. 175–184). London; UK: Taylor and Francis.

Nirupama, N., Murty, T. S., Nistor, I., & Rao, A. D. (2007c). Helmholtz mode and K-S-P waves: application to tsunamis. In: T. S. Murthy, U. Aswathanarayana & N. Nirupama (Eds.), *The Indian Ocean Tsunami* (pp. 151–157). London; UK: Taylor and Francis, 2007.

Nirupama, N., Murty, T. S., Nistor, I., & Rao, A. D. (2007d). Possible amplification of tsunami through coupling with internal waves. In: T. S. Murthy, U. Aswathanarayana & N. Nirupama (Eds.), *The Indian Ocean Tsunami* (pp. 91–95). London; UK: Taylor and Francis, 2007.

Nirupama, N., Murty, T. S., Rao, A. D., & Nistor, I. (2007e). Normal modes and tsunami coastal effects. In: T. S. Murthy, U. Aswathanarayana & N. Nirupama (Eds.), *The Indian Ocean Tsunami* (pp. 143–150). London; UK: Taylor and Francis, 2007.

Nirupama, N., Murty, T. S., Rao, A. D., & Nistor, I. (2007f). A partial explanation of the initial withdrawal of the ocean during a tsunami. In: T. S. Murthy, U. Aswathanarayana & N. Nirupama (Eds.), *The Indian Ocean Tsunami* (pp. 73–80). London; UK: Taylor and Francis, 2007.

Nishimura, Y., & Nakamura, Y. (2009). Spatial distribution and lithofacies of pumiceous-sand tsunami deposits, Abstract; *"24th International Tsunami Symposium and Technical Workshop on Tsunami Measurements and Real-Time Detection"* held at Novosibirsk, Russia (14–16 July 2009).

Nomitsu, T. (1935). *A theory of tsunamis and seiches produced by wind and barometric gradient, Memoirs of the College of Science.* Kyoto Imperial University, *A 18*(4), 201–214.

Nordstrom, K. F. (2000). *Beaches and Dunes of Developed Coasts.* Cambridge, UK: Cambridge University Press, p. 338.

Nosov, M. A., & Kolesov, S. V. (2009). Optimal initial conditions for simulation of seismotectonic tsunamis, Abstract; *"24th International Tsunami Symposium and Technical Workshop on Tsunami Measurements and Real-Time Detection"* held at Novosibirsk, Russia (14–16 July 2009).

Nowroozi, A. A., Sutton, G. H., & Auld, B. (1966). Oceanic tides recorded on the seafloor. *Ann. Geophys., 22*, 512—517.

Noye, B. J. (1974). Tide-well systems I: Some non-linear effects of the conventional tide-well. *J. Marine Res., 32*(2), 129—153.

Okada, Y. (1985). Surface deformation due to shear and tensile faults in a half-space. *Bull. Seismol. Soc. Am., 75*, 1135—1154.

Okada, M. (1995). Tsunami observation by ocean bottom pressure gauge. In: Y. Tsuchiya & N. Shuto (Eds.), *Tsunami: Progress in prediction, disaster prevention and warning* (pp. 287—303). New York: Kluwer Academic, Dordrecht; Springer.

Okal, E. A. (1982). Mode-wave equivalence and other asymptotic problems in tsunami theory. *Phys. Earth Planet Inter., 30*, 1—11.

Okal, E. A. (1988). Seismic parameters controlling far-field tsunami amplitudes: A review. *Natural Hazards, 1*, 67—96.

Okal, E. A., & Hartnady, C. J. H. (2009). Tsunami hazard in the South Atlantic and SW Indian Ocean from large normal- and thrust-faulting earthquakes along the South Sandwich arc. In: *Proceedings of the International Association of Seismology and Physics of the Earth's Interior (IASPEI) General Assembly 2009, Council for Geoscience* (abstract), January 10—16, 2009, Cape Town, South Africa: Cape Town International Convention Centre.

Okal, E. A., Piatanesi, A., & Heinrich, P. (1999). Tsunami detection by satellite altimetry. *J. Geophys. Res., 104*(B1), 599—616.

Okal, E. A., & Talandier, J. (1989). Mm: A variable-period magnitude. *J. Geophys. Res., 94*(B4), 4169—4193.

Okal, E. A., Plafker, G., Synolakis, C. E., & Borrero, J. C. (2002). Near-field survey of the 1946 Aleutian tsunami on Unimak and Sanak islands. *Bull. Seismol. Soc. Am., 93*, 1226—1234.

Okamoto, T. (1994). Location of shallow subduction-zone earthquakes inferred from teleseismic body waveforms. *Bull. Seismol. Soc. Am., 84*, 264—268.

Ouzounov, D., & Freund, F. (2004). Mid-infrared emission prior to strong earthquakes analyzed by remote sensing data. *Adv. Space Res., 33*, 268—273.

Palmer, H. R. (1831). Description of graphical register of tides and winds. *Philos. Trans. Roy. Soc. London, 121*, 209—213.

Papadopoulos, G. A. (1993). On some exceptional seismic (?) sea-waves in the Greek archipelago. *Science of Tsunami Hazards, 11*, 25—34.

Papadopoulos, G. A. (2009). NATNEG: Building up the national tsunami network of Greece — Present status, Abstract; *"24th International Tsunami Symposium and Technical Workshop on Tsunami Measurements and Real-Time Detection"* held at Novosibirsk, Russia (14—16 July 2009).

Papadopoulos, G. A., Daskalaki, E., & Fokaefs, A. (2009). Tsunami intensity: A valuable parameter of multiple usefulness, Abstract; *"24th International Tsunami Symposium and Technical Workshop on Tsunami Measurements and Real-Time Detection"* held at Novosibirsk, Russia (14—16 July 2009).

Papadopoulos, G. A., & Imamura, F. (2001). A proposal for a new tsunami intensity scale, *ITS 2001 Proceedings, Session* 5, No. 5-1, 569—577.

Paros, J. M., (1976). *Digital pressure transducers*, Measurements & Data, *10*(2), p. 6.

Paros, J. M. (1984). *Digital pressure transducer*, U.S. Patent No. 4,455,874.

Parrot, M., & Lefeuvre, F. (1985). Correlation between GEOS VLF emissions and earthquakes. *Ann. Geophys, 3*, 737—748.

Pathak, A. G., & Ramadass, G. A. (2002). *Device for measuring liquid level preferably measuring tide level in Sea*, United States Patent Number: 6,360,599 B1, p. 16.

Pattiaratchi, C. B., & Wijeratne, E. M. S. (2009). Tide gauge observations of 2004-2007 Indian Ocean tsunamis from Sri Lanka and Western Australia. *Pure appl. Geophys., 166*, 233—258.

Pearce, F. (1996). Living sea walls keep floods at bay. *New Scientist, 150*, 7.

Peharda, M., Zupan, I., Bavcevic, L., Frankic, A., & Klanjscek, T. (2007). Growth and condition index of mussel *Mytilus galloprovincialis* in experimental integrated aquaculture. *Aquaculture Research, 38*, 1714—1720.

Pelinovsky, E., Choi, B. H., Stromkov, A., Didenkulova, I., & Kim, H. S. (2005). Analysis of tide-gauge records of the 1883 Krakatau tsunami. In: K. Satake (Ed.), *Tsunamis: Case studies and recent developments* (pp. 57—77). Dordrecht: Springer.

Peshwe, V. B., Diwan, S. G., Joseph, A., & Desa, E. (1980). Wave and tide gauge. *Indian J. Marine Sci., 9*, 73—76.

Pilkey, O. H., Bush, D. M., & Neal, W. J. (2000). Storms and the coast. In: R. Pielke (Ed.), *Storms, Vol. 1* (pp. 427—448). New York: Routledge Hazards and Disaster Series.

Platzman, G. W. (1972). Two-dimensional free oscillations in natural basins. *J. Phys. Oceanogr, 2*(2), 117—138.

Porter, D. L., & Shih, H. H. (1996). Investigations of temperature effects on NOAA's next generation water level measurement system. *J. Atmos. Oceanic Technol., 13*, 714—725.

Powers, J. G., & Reed, R. J. (1993). Numerical simulations of the large amplitude mesoscale gravity-wave event of 15 December 1987 in the central United States. *Monthly Weather Review, 121*, 2285—2308.

Prabhudesai, R. G., Joseph, A., Agarvadekar, Y., Dabholkar, N., Mehra, P., Gouveia, A., et al. (2006). Development and implementation of cellular-based real-time reporting and Internet accessible coastal sea level gauge — A vital tool for monitoring storm surge and tsunami. *Curr. Sci., 90*(10), 1413—1418.

Prabhudesai, R. G., Joseph, A., Agarwadekar, Y., Mehra, P., Kumar, K. V., & Ryan, L. (2010). Integrated Coastal Observation Network (ICON) for real-time monitoring of sea-level, sea-state, and surface-meteorological data, OCEANS' 10 — *IEEE* Seattle Technical Conference; September 2010; IEEE, p. 9.

Prabhudesai, R. G., Joseph, A., Mehra, P., Agarvadekar, Y., Tengali, S., & Kumar, V. (2008). Cellular-based and Internet-enabled real-time reporting of the tsunami at Goa and Kavaratti Island due to M_w 8.4 earthquake in Sumatra on 12 September 2007. *Curr. Sci., 94*(9), 1151—1157.

Prior, D. B., Bornhold, B. D., Coleman, J. M., & Bryant, W. R. (1982). Morphology of a submarine slide, Kitimat Arm, British Columbia, Geol., 10, 588—592.

Proudman, J. (1929). The effects on the sea of changes in atmospheric pressure. *Geophys. Suppl. Mon. Notices R. Astr. Soc., 2*(4), 197—209.

Pugh, D. T. (1972). The physics of pneumatic tide gauges. *Int. Oceanogr. Rev., 49*(2), 71—97.

Pugh, D. T. (1987). *Tides, surges and mean sea-level: A handbook for engineers and scientists* (Wiley, Chichester), p. 472.

Pulinets, S., & Boyarchuk, K. (2004). *Ionospheric Precursors of Earthquakes* Springer: Heidelberg, Germany.

Purcell, G. H., Jr., Young, L. E., Wolf, S. K., Meehan, T. K., Duncan, C. B., Fisher, S. S., Spiess, F. N., et al. (1990). Accurate GPS Measurement of the Location and Orientation of a Floating Platform. *Marine Geodesy, 14*(3—4), 255—264.

Quartel, S., Kroon, A., Augustinus, P. G. E. F., Santen, P. V., & Tri, N. H. (2007). Wave attenuation in coastal mangroves in the Red River Delta, Vietnam. *Journal of Asian Earth Sciences, 29*, 576—584.

Rabinovich, A. B. (1993). Long ocean gravity waves: trapping, resonance, and leaking (in Russian), Gidrometeoizdat, St. Petersburg, p. 325.

Rabinovich, A. B. (1997). Spectral analysis of tsunami waves: Separation of source and topography effects. *J. Geophys. Res., 102*(C6), 12,663–12,676.

Rabinovich, A. B. (2005). Web compilation of tsunami amplitudes and arrival times, www-sci.pac. dfo-mpo.gc.ca/osap/projects/tsunami/tsunamiasia_e.htm.

Rabinovich, A. B. (2006b). Scientific and methodological basis and application technology of hydrophysical stations with bottom-mounted recorders. Unpublished report, P.P. Shirshov Institute of Oceanology, RAS, Moscow, p. 87.

Rabinovich, A. B. (2009). Seiches and harbour oscillations. In: Y. C. Kim (Ed.), *Handbook of Coastal and Ocean Engineering*, Singapore: World Scientific.

Rabinovich, A. B. (2009d). Coastal and deep-ocean tsunami measurements: Experience based on recent trans-oceanic events, Abstract; *"24th International Tsunami Symposium and Technical Workshop on Tsunami Measurements and Real-Time Detection"* held at Novosibirsk, Russia (14–16 July 2009).

Rabinovich, A. B., & Candella, R. (2009c). Analysis of the 2004 Sumatra tsunami records for the coast of South Africa. In: *Proceedings of the International Association of Seismology and Physics of the Earth's Interior (IASPEI) General Assembly 2009, Council for Geoscience* (abstract), January 10–16, 2009, Cape Town, South Africa: Cape Town International Convention Centre.

Rabinovich, A. B., & Monseratt, S. (1996). Meteorological tsunamis near the Balearic and Kuril Islands: Descriptive and statistical analysis. *Natural Hazards, 13*(1), 55–90.

Rabinovich, A. B., & Monserrat, S. (1998). Generation of meteorological tsunamis (large amplitude seiches) near the Balearic and Kuril Islands, Natural Hazards, *18*(1), 27–55.

Rabinovich, A. B., Thomson, R. E., Kulikov, E. A., Bornhold, B. D., & Fine, I. V. (1999). The landslide-generated tsunami of November 3, 1994 in Skagway Harbor, Alaska: A case study, *Geophysical Research Letters, 26*(19), 3009–3012.

Rabinovich, A. B., & Stephenson, F. E. (2004). Longwave measurements for the coast of British Columbia and improvements to the tsunami warning capability. *Natural Hazards, Vol. 32,* 313–343.

Rabinovich, A. B., & Thomson, R. E. (2007). The 26 December 2004 Sumatra tsunami: Analysis of tide gauge data from the World Ocean — Part 1. Indian Ocean and South Africa. *Pure and Applied Geophysics, 164,* 2–3, 261–308.

Rabinovich, A. B., Thomson, R. E., & Stephenson, F. E. (2006a). The Sumatra tsunami of 26 December 2004 as observed in the North Pacific and North Atlantic Oceans. *Surveys in Geophysics, 27,* 647–677. doi:10.1007/s10712-006-9000-9.

Rabinovich, A. B., Vilibic, I., & Tinti, S. (2009e). Meteorological tsunamis: Atmospherically induced destructive ocean waves in the tsunami frequency band. *Physics and Chemistry of the Earth, 34,* 891–893, Elsevier.

Rabinovich, A. B., Lobkovsky, L. I., Fine, I. V., Thomson, R. E., Ivelskaya, T. N., & Kulikov, E. A. (2008). Near-source observations and modeling of the Kuril Islands tsunamis of 15 November 2006 and 13 January 2007. *Adv. Geosci., 14,* 105–116.

Rabinovich, A. B., Stroker, K., Thomson, R. E., & Davis, E. (2009b). High resolution deep-ocean observations in the northeast pacific reveal important properties of the 2004 Sumatra tsunami, Abstract; *"24th International Tsunami Symposium and Technical Workshop on Tsunami Measurements and Real-Time Detection"* held at Novosibirsk, Russia (14–16 July 2009).

Rabinovich, A. B., Thomson, R. E. & Candella, R. (2009a). Energy decay of the 2004 Sumatra tsunami in the Indian and Atlantic oceans, Abstract; *"24th International Tsunami Symposium and Technical Workshop on Tsunami Measurements and Real-Time Detection"* held at Novosibirsk, Russia (14–16 July 2009).

Rae, J. B. (1976). The design of instrumentation for the measurement of tides offshore, Hydrographic Society, In: *Proceedings of Tidal Symposium*, 111–133, 1976.

Raichlen, F., Lepelletier, T. G., & Tam, C. K. (1983). *The excitation of harbors by tsunamis*, In: K. Iida & T. Iwasaki (Eds.), Tsunamis – Their Science and Engineering, Terra Sci., Tokyo, 359–385.

Ramsden, J. D., & Raichlen, F. (1990). Forces on a vertical wall due to incident bores. *J. Waterway, Port, Coastal and Ocean Engng Div. ASCE, 116*, 252–266.

Rangan, C. S., Sarma, G. R., & Mani, V. S.V. (1983). Pressure, In: Instrumentation; Devices and Systems, p.529

Rao, D. V. S., Ingole, B., Tang, D., Satyanarayana, B., & Zhao, H. (2007). Tsunamis and marine life. In: T. S. Murthy, U. Aswathanarayana & N. Nirupama (Eds.), *The Indian Ocean Tsunami* (pp. 373–391). London; UK: Taylor and Francis, 2007.

Rao, P. C. (2005). Earthquakes — Vulnerability, awareness and preparedness, National Commemorative Conference on Tsunami, Madurai, 26–29 December 2005. In: S. P. Raj (Ed.), *(Natl. Commemorative Conf. on Tsunami; Madurai; India; 26–29 Dec 2006)*. Madurai, India: Madurai Kamaraj University, 2006, 10–20.

Rapatz, W. J., & Murty, T. S. (1987). Tsunami warning system for the Pacific coast of Canada. *Marine Geodesy, 11*, 213–220.

Rastogi, B. K. (2007). A historical account of the earthquakes and tsunamis in the Indian Ocean. In: T. S. Murthy, U. Aswathanarayana & N. Nirupama (Eds.), *The Indian Ocean Tsunami* (pp. 3–18). London; UK: Taylor and Francis, 2007.

Ritsema, J., Ward, S. N., & Gonzalez, F. I. (1995). Inversion of deep-ocean tsunami records for 1987 to 1988 Gulf of Alaska earthquake parameters. *Bull. Seismol. Soc. Am., 85*, 747–754.

Roberts, D. L. (2009). *Historical and Palaeo-record of Tsunami in Southern Africa-A Review, IASPEI (International Association of Seismology and Physics of the Earth's Interior) Symposium*, Cape Town, South Africa, 10–16 January 2009 (abstract).

Roger, J., Baptista, M. A., & Hébert, H. (2009). Comparison of the different proposed sources for the 1755 Lisbon tsunami: Modeling in the West Indies, Abstract; "*24th International Tsunami Symposium and Technical Workshop on Tsunami Measurements and Real-Time Detection*" held at Novosibirsk, Russia (14–16 July 2009).

Romano, M., Liong, S. Y., Vu, M. T., Zemskyy, P., Doan, C. D., Dao, M. H., et al., (2009). Artificial neural network for tsunami forecasting, *Journal of Asian Earth Sciences* (in press).

Rowan, L. (2004). Tsunami and its shadow. *Science, 304*, 1569.

Saatcioglu, M., Ghobarah, A., & Nistor, I. (2007). Performance of structures affected by the 2004 Sumatra tsunami in Thailand and Indonesia. In: T. S. Murthy, U. Aswathanarayana & N. Nirupama (Eds.), *The Indian Ocean Tsunami* (pp. 297–322). London; UK: Taylor and Francis, 2006.

SAC (2005). Assessment of damages to coastal ecosystems due to the recent tsunami — Summary Report; Prepared by Indian Space Application Centre (SAC), Ahmedabad, p. 36.

Sadhuram, Y., Murthy, T. V. R., & Rao, B. P. (2007). Hydrophysical manifestations of the Indian Ocean tsunami. In: T. S. Murthy, U. Aswathanarayana & N. Nirupama (Eds.), *The Indian Ocean Tsunami* (pp. 365–372). London; UK: Taylor and Francis, 2007.

Sahal, A., Roger, J., Allgeyer, S., Hébert, H., Schindelé, F., & Lavigne, F. (2009). The effects of the 2003 Zemmouri tsunami: database of observations in French harbors and influence of resonance effects, Abstract; "*24th International Tsunami Symposium and Technical Workshop on Tsunami Measurements and Real-Time Detection*" held at Novosibirsk, Russia (14–16 July 2009).

Sakakibara, S., Kubo, M., Kobayashi, E., & Koshimura, S. (2005). Dynamic behaviour of moored ship motions induced by initial attack of large scaled tsunami. Proceedings of OMAE2005, *24th International Conference on Offshore Mechanics and Arctic Engineering (OMAE2005)*, Halkidiki, Greece, June 12—16.

Sallenger Jr., A. H., List, J. H., Gelfenbaum, G., Stumpf, R. P., & Hansen, M. (1995). Large Wave at Dayton Beach, Florida, explained as a squall-line surge. *J. Coastal Res., 11*(4), 1383—1388.

Saraf, A. K., & Choudhury, S. (2003). Earthquakes and thermal anomalies. *Geospatial Today, 2*(2), 18—20.

Saraf, A. K., & Choudhury, S. (2004). Thermal remote sensing technique in the study of pre-earthquake thermal anomalies. *J. Indian Geophys. Union, 9*(3), 197—207.

Saraf, A. K., & Choudhury, S. (2005). NOAA-AVHRR detects thermal anomaly associated with 26 January 2001 Bhuj Earthquake, Gujarat, India. *Int. J. Remote Sens., 26*(6), 1065—1073.

Saraf, A. K., Choudhury, S., & Dasgupta, S. (2007). Satellite detection of pre-earthquake thermal anomaly and sea water turbidity associated with the great Sumatra earthquake. In: T. S. Murthy, U. Aswathanarayana, & N. Nirupama (Eds.), *The Indian Ocean Tsunami* (pp. 215—225). London, UK: Taylor and Francis, 2007.

Sarang, K. (2005). Tsunami impact assessment of coral reefs in the Andaman and Nicobar. *CARDIO News, Interim Report* 1—6.

Satake, K. (1995). Linear and nonlinear computations of the 1992 Nicaragua earthquake tsunami. *Pure Appl. Geophys., 144*, 455—470.

Satake, K. (1987). Inversion of tsunami waveforms for the estimation of a fault heterogeneity: method and numerical experiments. *J. Phys. Earth, 35*, 241—254.

Satake, K. (2002). Tsunamis. In: W. H. K. Lee et al., (Eds.), *International Handbook of Earthquake and Engineering Seismology.* pp. 437—451. *Int. Assoc. of Seismol. and Phys. Of the Earth's Inter.*, Boulder, Colo.

Satake, K., Hasegawa, Y., Nishimae, Y., & Igarashi, Y. (2009). Recent tsunamis that affected the Japanese coasts and evaluation of JMA's tsunami warning. In: *Proceedings of the International Association of Seismology and Physics of the Earth's Interior (IASPEI) General Assembly 2009, Council for Geoscience* (abstract), January 10—16, 2009, Cape Town, South Africa: Cape Town International Convention Centre.

Satake, K., Okada, M., & Abe, K. (1988). Tide gauge response to tsunamis: Measurements at 40 tide gauge stations in Japan. *J. Marine Res., 46*, 557—571.

Satake, K., Okal, E. A., & Borrero, J. C. (2007). Tsunami and its hazard in the Indian and Pacific Oceans: Introduction. *Pure Appl. Geophys., 164*, 249—259.

Saxena, N. K., & Zielinski, A. (1981). Deep ocean system to measure tsunami wave height. *Marine Geodesy, 5*(1), 55—62.

Schaad, T. (2000). Eleven-year test of barometer long-term stability. www.paroscientific.com/11yeartest.htm

Schulte, P., Alegret, L., Arenillas, I., Arz, J. A., Barton, P. J., Bown, P. R., et al. (2010). The Chicxulub asteroid impact and mass extinction at the Cretaceous-Paleogene boundary. *Science, 327*, 1214—1218.

SDMRI (2005). Rapid environmental impact assessment after tsunami in the inter-tidal, sub-tidal, and coastal areas including water bodies and lakes along Tamil Nadu coast, Final Report, p. 30.

Seeling, W. N. (1977). Stilling-well design for accurate water level measurement, Technical Paper No. 77—2, U.S. Army Corps of Engineers, *Coastal Engineering Research Centre*, 1—21.

Sepic, J., Vilibic, I., & Belusic, D. (2009). The source of the 2007 Ist meteotsunami (Adriatic Sea). *J. Geophys. Res.- Oceans, 114,* doi: 10.1029/2008JC005092.

Serebryakova, O. N., Bilichenko, S. V., Chmyrev, V. M., Parrot, M., Rauch, J. L., Lefeuvre, F., et al. (1992). Electromagnetic ELF radiation from earthquake region as observed by low-altitude satellites. *Geophys. Res. Lett., 19*, 91–94.

Shaw, M. S. B. (2005). Tsunami: The nemesis and neurosis — an analysis with special focus on Andaman archipelago, *National Commemorative Conference on Tsunami, Madurai*, 26–29 December 2005. In: Raj, S. P. (Ed.), *Proceedings. Natl. Commemorative Conf. on Tsunami; Madurai; India; 26-29 Dec 2006.* Madurai, India: Madurai Kamaraj University, 2006, 111–115.

Shevchenko, G. V., Chernov, A. G., Levin, B. W., Kovalev, P. D., Kovalev, D. P., Kurkin, A. A., et al. (2009). Tsunami-range long-wave measurements in the area of the south Kuril Islands, Abstract; "*24th International Tsunami Symposium and Technical Workshop on Tsunami Measurements and Real-Time Detection*" held at Novosibirsk, Russia (14–16 July 2009).

Shih, H. H., & Porter, D. L. (1981). Error models for stilling-well — Float type tide gauges. *Proceedings of IEEE Oceans'81*, 1118–1123.

Shih, H. H., & Baer, L. (1991). Some errors in tide measurement caused by dynamic environment. In: B. B. Parker (Ed.), *Tidal Hydrodynamics* (pp. 641–671). New York: John Wiley and Sons.

Shimazaki, K., Kim, H., Ishibe, T., Tsuji, Y., Satake, K., Imai, K., et al. (2009). Recurrence of Kanto earthquakes revealed from tsunami deposits in Miura peninsula, Japan, Abstract; "*24th International Tsunami Symposium and Technical Workshop on Tsunami Measurements and Real-Time Detection*" held at Novosibirsk, Russia (14–16 July 2009).

Shirman, B. (2003). Comparison of the OTT radar gauge to the float gauge installed at the Haifa port (Israel), *Workshop on New Technical Developments in Sea and Land Level Observing Systems*, 14–16 October 2003, IOC, UNESCO.

Shuvalov, V., Dypvik, H., & Tsikalas, F. (2002). Numerical simulations of the Mjolnir marine impact crater. *J. Geophys. Res., 107*(E7), 5047, doi:10.1029/2001JE001698.

Sidor, T. (1983). Simple and accurate strain gauge signal-to-frequency converter. *J. Phys. E: Sci. Instrum., 16*, 253–255.

Sieberg, A. (1927). Geologische, physikalische und angewandte Erdbebenkunde, Verlag von Gustav Fischer, Jena.

Simonenko, V. A., Skorkin, N. A., Minaev, I. V., Abramov, A.V., & Abramov, E. A. (2009). Mathematical modeling of asteroid falling into the ocean, Abstract; "*24th International Tsunami Symposium and Technical Workshop on Tsunami Measurements and Real-Time Detection*" held at Novosibirsk, Russia (14–16 July 2009).

Sinclair, D. (1994). Liquid level gauging, *United States Patent Number.* 5,357,801; p. 4.

Smith, W. H. F., Scharroo, R., Titov, V. V., Arcas, D., & Arbic, B. K. (2005). Satellite altimeters measure tsunami. *Oceanography, 18*, 11–13.

Snodgrass, F. E. (1968). Deep-sea instrument capsule. *Science, 162*, 78–87.

Solandt, J. L., Goodwin, L., Beger, M., & Harborne, A. R. (2001). Sedimentation and related habitat characteristics in the vicinity of Danjugan Island, Negros Occidental, *Danjugan Island Survey, Summary Report*, p. 5.

Soloviev, S. L. (1978). Tsunamis, In: "Technics of the Youth", No. 8, 38–43 (in Russian)

Spencer, R., & Foden, P. (1996). Data from the deep ocean via releasable data capsules. *Sea Technol., 37*(2), 10–14.

Spencer, R., Foden, P. R., & Vassie, J. M. (1994). Development of a multi-year deep sea bottom pressure recorder. *Electron Eng. Oceanogr., EEO-394*, 175–180.

Sridhar, P. N., Surendran, A., Jain, S., & Narayan, B. V. (2007). Tsunami impact on coastal habitats of India. In: T. S. Murthy, U. Aswathanarayana & N. Nirupama (Eds.), *The Indian Ocean Tsunami* (pp. 393–403). London, UK: Taylor and Francis, 2007.

Stein, S., & Okal, E. A. (2005). Speed and size of the Sumatra earthquake. *Nature, 434*, 581–582.

Stephenson, F., Rabinovich, A. B., Solovieva, O. N., Kulikov, E. A. & Yakovenko, O. I. (2007). Catalogue of tsunamis, British Columbia, Canada: 1700—2007, p. 133

Stroker, K. J., Bouchard, R. H., Eble, M. C., Rabinovich, A. B. (2009). High-resolution deep-ocean tsunami data retrieved from DART® systems in the North Pacific, Abstract; *"24th International Tsunami Symposium and Technical Workshop on Tsunami Measurements and Real-Time Detection"* held at Novosibirsk, Russia (14—16 July 2009).

Symons, G. J. (1888). *The eruption of Krakatoa and subsequent phenomena.* London: Triibner and Co., p. 494.

Synolakis, C. E. (1987). The run-up of solitary waves. *J. Fluid Mech., 185*, 523—545.

Synolakis, C. E. (1991). Green's law and the evolution of solitary waves. *Phys. Fluids, A3*, 490—491.

Synolakis, C. E., Bernard, E. N., Titov, V. V., Kanoglu, U., & Gonzalez, F. I. (2007). Standards, criteria, and procedures for NOAA evaluation of tsunami numerical models, *NOAA Tech. Memo., ERL PMEL-135*, p. 55.

Synolakis, C. E., & Kânoğlu, U. (2009). Tsunami modelling: development of benchmark models, Proceedings of the *24th International Tsunami Symposium, July 14—16*, 2009, Dom Uchenykh (House of Scientists), 23, Morskoy Prospect, Novosibirsk, Russia.

Szmant, A. M. (2002). Nutrient enrichment on coral reefs: Is it a major cause of coral reef decline? *Estuaries, 25*(4b), 743—766.

Tadepalli, S., & Synolakis, C. E. (1994). The run-up of N-waves on sloping beaches. *Proc. Roy. Soc. Lond., 445*, 99—112.

Taira, K., Kawaguchi, H., Shikama, N., & Teramoto, T. (1977). Analysis of sea level variations in Otsuchi Bay, *Otsuchi Marine Research Center Report, 3*, 50—58.

Taira, K., Teramoto, T., & Kitagawa, S. (1985). Measurements of ocean bottom pressure with a quartz sensor. *Journal of the Oceanographic Society of Japan, 41*, 181—192.

Takahashi, M. (1981). Telemetry bottom pressure observation system at a depth of 2,200 meter. *J. Phys. Earth, 29*, 77—88.

Takahashi, M. (1983). Thermal response of the bottom pressure sensor off the coast of Tokai District, central Honshu and its application to oceanographic analysis. *Pap. Meteorol. Geophys., 33*, 245—255.

Takahashi, R., & Hatori, T. (1961). A summary report on the Chilean tsunami of May 1960. In: *Report on the Chilean tsunami, Comm* (pp. 23—34). Tokyo: Field Investigations on Chilean Tsunami.

Tanaka, N., Sasaki, Y., Mowjood, M. I. M., Jinadasa, K. B. S. N., & Homchuen, S. (2007). Coastal vegetation structures and their functions in tsunami protection: experience of the recent Indian Ocean tsunami, Landscape Ecology and Engineering, 3, 33—45.

Tanioka, Y., Hasegawa, Y., & Kuwayama, T. (2008). Tsunami waveform analyses of the 2006 underthrust and 2007 outer-rise Kuril earthquakes. *Adv. Geosci., 14*, 129—134.

Tanioka, Y., & Ioki, K. (2009). Tsunami analyses of great interplate earthquakes along the central Kurile subduction zone, Abstract; *"24th International Tsunami Symposium and Technical Workshop on Tsunami Measurements and Real-Time Detection"* held at Novosibirsk, Russia (14—16 July 2009).

Tatehata, H. (1997). In G. Hebenstreit (Ed.), *Perspectives on Tsunami Hazard Reduction* (pp. 175—188). New York: Springer.

Terdre, N. (1995). Saab sensor improves accuracy of wave behavior measurements, Offshore Magazine, W.S. Ocean Systems, U.K. p. 1.

Thomson, R. E., Rabinovich, A. B., & Krassovski, M. V. (2007). Double jeopardy: Concurrent arrival of the 2004 Sumatra tsunami and storm-generated waves on the Atlantic coast of the United States and Canada. *Geophys. Res. Lett, 34*, L15607, doi: 10.1029/2007GL030685.

Tinti, S., Armigliato, A., Gallazzi, S., Manucci, A., Pagnoni, G., Tonini, R., et al. (2009a), Scenarios of tsunami impact in the town of Alexandria, Egypt. In: *Proceedings of the International Association of Seismology and Physics of the Earth's Interior (IASPEI) General Assembly 2009, Council for Geoscience* (abstract), January 10—16, 2009, Cape Town, South Africa: Cape Town International Convention Centre.

Tinti, S., Armigliato, A., Pagnoni, G., & Tonini, R. (2009b). On tsunami attacks on the southern Italian coast: From hazard to vulnerability and risk, Abstract; "*24th International Tsunami Symposium and Technical Workshop on Tsunami Measurements and Real-Time Detection*" held at Novosibirsk, Russia (14—16 July 2009).

Tinti, S., Maramai, A., & Graziani, L. (2004). The new catalogue of the Italian tsunamis. *Natural Hazards, 33*, 439—465.

Tintore, J., Gomis, D., Alonso, S., & Wang, D. P. (1988). A theoretical study of large sea level oscillations in the western Mediterranean. *J. Geophy. Res., C9*, 2804—2830.

Titov, V. V. (2009a). Progress in tsunami forecasting, Abstract; "*24th International Tsunami Symposium and Technical Workshop on Tsunami Measurements and Real-Time Detection*" held at Novosibirsk, Russia (14-16 July 2009).

Titov, V. V. (2009b). Dart data for estimating tsunami magnitude during real-time tsunami forecast, Abstract; "*24th International Tsunami Symposium and Technical Workshop on Tsunami Measurements and Real-Time Detection*" held at Novosibirsk, Russia (14—16 July 2009).

Titov, V. V., & Synolakis, C. E. (1997). Extreme inundation flows during the Hokkaido—Nansei—Oki tsunami. *Geophys. Res. Lett., 24*(11), 1315—1318.

Titov, V. V, & Gonzalez, F. I. (1997a), Implementation and testing of the Method of Splitting Tsunami (MOST) model, *Technical Report NOAA Tech. Memo*, ERL PMEL-112 (PB98-122773), NOAA/Pacific Marine Environmental Laboratory, Seattle, WA.

Titov, V. V., & Gonzalez, F. I. (1997b). Implementation and testing of the Method of Splitting Tsunami (MOST) model, *NOAA Tech. Memo., ERL PMEL*-112, p. 11

Titov, V., Rabinovich, A. B., Mofjeld, H. O., Thomson, R. E., & Gonzalez, F. I. (2005a). The global reach of the 26 December 2004 Sumatra tsunami. *Science, 309*, 2045—2048.

Titov, V. V., Gonzalez, F. I., Bernard, E. N., Eble, M. C., Mofjeld, H. O., Newman, J. C., et al. (2005b). Real-time tsunami forecasting: Challenges and solutions. *Natural Hazards, 35*, 41—58.

Titov, V. V., Mofjeld, H. O., Gonzalez, F. I., & Newman, J. C. (1999). Offshore forecasting of Alaska-Aleutian Subduction Zone tsunamis in Hawaii, *NOAA Tech. Memo, ERL PMEL-114*, NOAA/Pacific Marine Environmental Laboratory, Seattle, WA.

Titov, V. V., Mofjeld, H. O., Gonzalez, F. I., & Newman, J. C. (2001). Offshore forecasting of Alaskan tsunamis in Hawaii. In: G. T. Hebenstreit (Ed.), *Tsunami Research at the End of a Critical Decade* (pp. 75—90). Amsterdam: Kluwer Acad.

Tkalich, P., Ha, D. M., & Soon, C. E. (2007). Tsunami propagation modeling and forecasting for early warning system. *Journal of Earthquake and Tsunami, 1*(1), 87—98.

Troitskaya, Y. I. (1994). Modulation of the growth rate of short surface capillary-gravity wind waves by a long wave. *J. Fluid Mech., 273*, 169—187.

Troitskaya, Y. I., & Ermakov, S. A. (2005). Recording of the December 26, 2004 tsunami in the open ocean based on variations in radar scattering section. *Dokl. Earth Sci., 405A*, 1384—1387.

Tronin, A. A. (2000). Thermal IR satellite sensor data application for earthquake research in China. *Int. J. Remote Sens, 21*(16), 3169—3177.

Tsuboi, S., Abe, K., Takano, K., & Yamananka, Y. (1995). Rapid determination of Mw from broadband P waveforms. *Bull. Seismol. Soc. Am., 85*, 606–613.

Tsuji, Y. (2009). Secondary Tsunamis Induced by Submarine Slope Slumping Triggered by Earthquakes in Tropical Countries. In: *Proceedings of the International Association of Seismology and Physics of the Earth's Interior (IASPEI) General Assembly 2009, Council for Geoscience* (abstract), January 10–16, 2009, Cape Town, South Africa: Cape Town International Convention Centre.

Tsuji, Y., Tachibana, T. (2009). Possibility of generation of tsunamis caused by burst of methane hydrate in sea bed induced by shaking of earthquakes, Abstract; *"24th International Tsunami Symposium and Technical Workshop on Tsunami Measurements and Real-Time Detection"* held at Novosibirsk, Russia (14–16 July 2009).

Tsuru, T., Park, J.-O., Takahashi, N., Kodaira, S., Kido, Y., Kaneda, Y., et al. (2000). Tectonic features of the Japan Trench convergent margin off Sanriku, northeastern Japan, revealed by multichannel seismic reflection data. *J. Geophys. Res., 105*, 16403–16413.

UNEP-WCMC. (2006). *In the front line: Shoreline protection and other ecosystem services from mangroves and coral reefs*. Cambridge: UNEP-WCMC, p. 33.

Valiela, I., & Cole, M. L. (2002). Comparative evidence that salt marshes and mangroves may protect seagrass meadows from land-derived nitrogen loads. *Ecosystems, 5*, 92–102.

Van Dorn, W. G. (1965). Tsunamis. In: V. T. Chow (Ed.), *Advances in Hydroscience, 2* (pp. 319–366). New York: Academic Press.

Van Dorn, W. G. (1968). Tsunamis. *Contemporary Physics, 9*(2), 145–164.

Van Dorn, W. G. (1984). Some tsunami characteristics deducible from tide records. *J. Phys. Oceanogr., 14*, 353–363.

Van Dorn, W. G. (1987). Tide gauge response to tsunamis, Part II: Other Oceans and Smaller Seas. *J. Phys. Oceanogr., 17*, 1507–1516.

Varner, J. & Dunbar, P. (2009). NGDC historical hazards database: standalone GIS software based on UDIG, Abstract; *"24th International Tsunami Symposium and Technical Workshop on Tsunami Measurements and Real-Time Detection"* held at Novosibirsk, Russia (14–16 July 2009).

Vassie, J. M., Woodworth, P. L., Smith, D. E. & Spencer, R. (1992). Comparison of NGWLMS, bubbler and float gauges at Holyhead, *Joint IAPSO-IOC Workshop on Sea Level Measurements and Quality Control*, 40–51.

Vijaykumar, K., Joseph, A., Prabhudesai, R. G., Prabhudesai, S., Nagvekar, S., & Damodaran, V. (2005). Performance evaluation of Honeywell silicon piezoresistive pressure transducers for oceanographic and limnological measurements. *J. Atmos. and Oceanic Technol., 22*(12), 1933–1939.

Vilibic, I. (2008). Numerical simulations of the Proudman resonance. *Continental Shelf Res., 28*, 574–581.

Vilibic, I., & Mihanovic, H. (2005b). Resonance in Ploce Harbor (Adriatic Sea). *Acta Adriatica, 42*(2), 125–136.

Vilibic, I., & Sepic, J. (2009). Destructive meteotsunamis along the eastern Adriatic coast: Overview, *Physics and Chemistry of the Earth, 34*, 904–917, Elsevier.

Vilibic, I., Domijan, N., & Cupic, S. (2005a). Wind versus air-pressure seiche triggering in the Middle Adriatic coastal waters. *J. Mar. Syst., 57*, 189–200.

Vilibic, I., Domijan, N., Orlic, M., Leder, N., & Pasaric, M. (2004). Resonant coupling of a traveling air-pressure disturbance with the east Adriatic coastal waters. *J. Geophys. Res., 109*, C10001, doi: 10.1029/2004JC002279.

Voronina, T. A. (2009). The inverse problem of reconstructing a tsunami source with numerical simulation, Abstract; "*24th International Tsunami Symposium and Technical Workshop on Tsunami Measurements and Real-Time Detection*" held at Novosibirsk, Russia (14–16 July 2009).

Vucetic, T., Vilibic, I., Tinti, S., & Maramai, A. (2009). The great Adriatic flood of 21 June 1978 revisited: An overview of the reports. *Physics and Chemistry of the Earth, 34*, 894–903, Elsevier.

Walker, D. A. (1996). Observations of tsunami "shadows": A new technique for assessing tsunami wave heights? *Science of Tsunami Hazards, 14*, 3–11.

Walker, D. A., & Cessaro, R. K. (2002). Locally generated tsunamis in Hawai: A low cost, real time warning system with world wide applications. *Sci. Tsu. Haz, 20*(4), 177–186.

Walters, B. B. (2003). People and mangroves in the Philippines: fifty years of coastal environmental change. *Environ. Conserv., 30*, 293–303.

Walters, B. B. (2004). Local management of mangrove forests in the Philippines: successful conservation or efficient resource exploitation? *Hum. Ecol, 32*, 177–195.

Wang, X., Li, K., Yu, Z., & Wu, J. (1987). Statistical characteristics of seiches in Longkou Harbour. *J. Phys. Oceanogr., 17*, 1063–1065.

Ward, S. N. (1980). Relationships of tsunami generation and an earthquake source. *J. Phys. Earth, 28*, 441–474.

Ward, G. A., Smith, T. J., III, Whelan, K. R. T., & Doyle, T. W. (2006). Regional processes in mangrove ecosystems: spatial scaling relationships, biomass, and turnover rates following catastrophic disturbance. *Hydrobiologia, 569*, 517–527.

Wearn, Jr., R. B., & Paros, J. M. (1988). Measurements of Dead Weight Tester Performance Using High Resolution Quartz Crystal Pressure Transducers, *Presented at 34th International Instrumentation Symposium, Instrument Society of America*, p. 8.

Wearn, R. B., & Larson, N. G. (1982). Measurements of the sensitivity and drift of digiquartz pressure transducers. *Deep-Sea Res., 29 A*, 111–134.

Webb, S. C. (1998). Broadband seismology and noise under the ocean. *Rev. Geophys., 36*, 105–142.

Webb, S. C., Zhang, X., & Crawford, W. C. (1991). Infragravity waves in the deep ocean. *J. Geophys. Res., 96*, 2723–2736.

Wei, Y., Bernard, E. N., Tang, L., Weiss, R., Titov, V. V., Moore, C., et al. (2008). Real-time experimental forecast of the Peruvian tsunami of August 2007 for U.S. coastlines. *Geophys. Res. Lett., 35,* L04609, doi: 10.1029/2007GL032250.

Wei, Y., Cheung, K. F., Curtis, G. D., & McCreery, C. S. (2003). Inverse algorithm for tsunami forecasts. *J. Waterw. Port Coast. Ocean Eng., 129*(3), 60–69.

Weichert, D., Horner, R. B., & Evans, S. G. (1994). Seismic signatures of landslides: The 1990 Brenda Mine collapse and the 1965 Hope rockslides. *Bull. Seismol. Soc. Am., 84*, 1523–1532.

Weinstein, S. A. (2009). Coastal sea level stations vs. DARTs: competing technology or complimentary technology? Abstract; "*24th International Tsunami Symposium and Technical Workshop on Tsunami Measurements and Real-Time Detection*" held at Novosibirsk, Russia (14–16 July 2009).

Weinstein, S. A., Kong, L. S. L. & Wang, D. (2009). Tide tool: Software to analyze GTS sea-level data, Abstract; "24th *International Tsunami Symposium and Technical Workshop on Tsunami Measurements and Real-Time Detection*" *held at Novosibirsk, Russia* (14–16 July 2009).

Whitmore, P. M, Knight, W. K. & Weinstein, S. A. (2009). Recent United States tsunami warning system improvements, Abstract; "*24th International Tsunami Symposium and Technical Workshop on Tsunami Measurements and Real-Time Detection*" held at Novosibirsk, Russia (14–16 July 2009).

Whitmore, P. M., & Sokolowski, T. J. (1996). Predicting tsunami amplitudes along the North American coast from tsunamis generated in the northwest Pacific during tsunami warnings. *Sci. Tsunami Haz., 14*, 147–166.

Wigen, S. O. (1960). *Tsunami of May 22*, 1960, West coast of Canada. *Unpublished Report*, Canadian Hydrographic Service, Sidney, BC.

Wilson, B. W. (1964). *Generation and disperson characteristics of tsunami, studies on oceanography, Ocean Research Institute*. University of Tokyo, 413–444.

Wolanski, E. (1995). Transport of sediment in mangrove swamps. *Hydrobiologia, 295*, 31–42.

Wolanski, E., Mazda, Y., & Ridd, P. (1992). Mangrove hydrodynamics. In: A. I. Robertson & D. M. Alongi (Eds.), *Coastal and Estuarine Studies: Tropical Mangrove Ecosystem* (pp. 43–62). Washington DC, USA: American Geophysical Union.

Woodroffe, C. (1992). Mangrove sediments and geomorphology. In: A. I. Robertson & D. M. Alongi (Eds.), *Coastal and Estuarine Studies: Tropical Mangrove Ecosystem* (pp. 7–41). Washington DC, USA: American Geophysical Union.

Woodworth, P. L., & Smith, D. (2003). A one year comparison of radar and bubbler tide gauges at Liverpool. *International Hydrographic Review, 4*(3), 2–9.

Woodworth, P. L., Aman, A., & Aarup, T. (2007). Sea level monitoring in Africa. *Afr. J. Mar. Sci, 29*(3), 321–330.

Woodworth, P. L., Blackman, D. L., Foden, P., Holgate, S., Horsburgh, K., Knight, P. J., et al. (2005). Evidence for the Indonesian tsunami in British tidal records. *Weather, 60*(9), 263–267.

Woodworth, P. L., Holgate, S., Foden, P., & Pugh, J. (2007). Africa: Tsunami alerts and rising sea levels. *Planet Earth*, 24–25.

Wrathall, P. N. (1991). Mixing it up: Electromagnetic communications via multiple media. *Sea Technology, 32*(5), 10–11.

Wunsch, C., & Dahlen, J. (1974). A moored temperature and pressure recorder. *Deep-Sea Res., 21*, 145–154.

Wunsch, C., & Stammer, D. (1997). Atmospheric loading and the oceanic 'Inverted Barometer' effect. *Rev. Geophys., 35*, 79–107.

Yamashita, T., & Sato, R. (1974). Generation of tsunami by a fault model. *J. Phys. Earth., 22*, 415–440.

Yanuma, T., & Tsuji, Y. (1998). Observation of edge waves trapped on the continental shelf In the vicinity of makurazaki harbour. *Kyushu, Japan, J. Oceanogr., 54*, 9–18.

Yilmaz, M. (2004). Calculation of temperature compensated pressure from Digiquartz frequency output transducers, Technical Note, Paroscientific, *Inc. Precision Pressure Instrumentation*, Doc. No. G8038 Rev. NC 10 February 2004, p. 3.

Zabusky, N. J., & Galvin, C. J. (1971). Shallow water waves, the Korteweg-de Vries equations and solitons. *J. Fluid Mech., 47*, 811–824.

Zahibo, N., Pelinovsky, E., Yalciner, A., Zaytsev, A., Talipova, T., Nikolkina, I., et al. (2009). The 1755 Lisbon tsunami propagation in the atlantics and its effect in the Caribbean Sea, Abstract; *"24th International Tsunami Symposium and Technical Workshop on Tsunami Measurements and Real-Time Detection"* held at Novosibirsk, Russia (14–16 July 2009).

Zaichenko, M. Y., Kulikov, E. A., Levin, B. V., & Medvedev, P. P. (2005). On the possibility of registration of tsunami waves in the open ocean with the use of a satellite altimeter. *Oceanology, 45*, 194–201.

Zelt, J. A., & Raichlen, F. (1991). Overland flow from solitary waves. *J. Waterway, Port, Coastal and Ocean Engng Div. ASCE, 117*, 247–263.

Zielinski, A., & Saxena, N. (1983). Rationale for measurement of midocean tsunami signature. *Marine Geodesy, 6*(3-4), 331–337.

1. Anonymous (2001). Tsunami: The Great Waves. www.nws.noaa.gov/om/tsunami.htm.
2. Anonymous (2001). Physics of Tsunamis. <http://wcatwc.gov/physics.htm>.
3. Anonymous (1993). Tsunami Devastates Japanese Coastal Region. Eos Transactions, *American Geophysical Union, 74*(37), 417−432.
4. Anonymous (2001). What Cause Tsunamis?, www.nws.noaa.gov/om/tsunami2.htm, p. 1−2.
5. Ayers, R. A., & Cretzler, D. J. (1963). A Resistance Wire Water Level Measurement System. *Marine Science Instrumentation, Vol. 2*. New York: Instrument Society of America, Plemun Press.
6. Baker, D. J. (1969). On the History of the High Seas Tide Gauge. *Woods Hole Technical Memorandum* 5−69.
7. Beaumariage, D. C., & Scherer, W. D. (1987). New Technology Enhances Water Level Measurement. *Sea Technology, 28*(5), 29−32.
8. Bruce, O. (1989). Tsunami! *Canadian Geographic, 109*(1), 46−53.
9. Butler, D. (1989). Tide Gauge Telemetry Keeps Dredge Rigs on Schedule in St. Lawrence River Project. *Endeco Currents, 1*(1), 1−2.
10. Cheney, R. E., & Marsh, J. G. (1981). Oceanographic Evaluation of Geoid Surfaces in the Western North Atlantic. In: J. Gower (Ed.), *Oceanography From Space* (pp. 855−864). New York: Plenum Press.
11. Cheney, R. E., Marsh, J. G., & Beckley, B. D. (1983). Global Mesoscale Variability from Repeat Tracks of Seasat Altimeter Data. *J. Geophys. Res., 88*, 4343.
12. Cheney, R. E., Douglas, B., Agreen, R., Miller, L., Milbert, D., & Porter, D. (1986). The GEOSAT Altimeter Mission: A Milestone in Satellite Oceanography. *Eos, 67*(48), 1354−1355.
13. Cheney, R. E., Douglas, B. C., & Miller, L. (1989). Evaluation of GEOSAT Altimeter Data with Application to Tropical Pacific Sea Level Variability. *J. Geophys. Res., 94*, 4737−4748.
14. Cheney, R. E., Emery, W. J., Haines, B. J., & Wentz, F. (1991). Recent Improvements in Geosat Altimeter Data. *Eos, 72*(51), 577−580.
15. Cheney, B., Miller, L., Agreen, R., Doyle, N., & Lillibridge, J. (1994). TOPEX/POSEIDEN: The 2−cm Solution. *J. Geophys. Res., 99*, 24555−24564.
16. Chiswell, S. M., Wimbush, M., & Lukas, R. (1988). Comparison of Dynamic Height Measurements from an Inverted Echo Sounder and an Island Tide Gauge in the Central Pacific. *J. Geophys. Res., 93*, 2277−2283.
17. Chiswell, S. M., & Lukas, R. (1989). The Low−Frequency Drift of Paroscientific Pressure Transducers. *J. Atmos. Oceanic Technol., 6*, 389−395.
18. Clark, S. K., & Wise, K. D. (1979). Pressure Sensitivity in Anisotropically Etched Thin−Diaphragm Pressure Sensors. *Transactions on Electron Devices, 26*(12), 1887−1895.
19. Cutting, E., Born, G. H., & Frautnik, J. C. (1978). Orbit Analysis for SEASAT−A. *The Journal of the Astronautical Sciences, XXVI*, 315−342.
20. Dean, M., III, & Douglas, R. D. (1962). *Semiconductor and Conventional Strain Gauges*. New York: Academic Press.
21. Desa, E., Peshwe, V. B., Joseph, A., Mehra, P., Naik, G. P., Kumar, V., et al. (2001). A Compact Self−Recording Pressure Based Sea Level Gauge Suitable for Deployments at

Harbor and Offshore Environments. In: *Proceedings SYMPOL—'2001* (pp. 1—15). Cochin University of Science & Technology.

22. Desa, E., Naik, G. P., Joseph, A., Desa, E. S., Mehra, P., Kumar, V., et al. (2003). Pressure Housing for In—Water Pressure Based Systems. U.S. Patent No. 6,568,266 B1 *dated May 27, 2003*.

23. Delcroix, T., Boulanger, J. P., Masia, F., & Menkes, C. (1994). GEOSAT—Derived Sea Level and Surface Current Anomalies in the Equatorial Pacific During the 1986—1989 El Nino and La Nina. *J. Geophys. Res., 99*, 25093—25107.

24. Diamante, J. M., Douglas, B. C., Porter, D. L., & Masterson, R. P. (1982). Tidal and Geodetic Observations for the Seasat Altimeter Calibration Experiment. *J. Geophys. Res., 87*(C5), 3199—3206.

25. Diamante, J. M., Pyle, T. E., Carter, W. E., & Scherer, W. D. (1987). Global Change and the Measurement of Absolute Sea Level. In: *Progress in Oceanography*. Pergamon Press.

26. Doodson, A. T. (1927). The Analysis of Tidal Observations. *Philosophical Transaction of the Royal Society, London, A 227*, 223—279.

27. Doodson, A. T., & Warburg. (1941). *Admiralty Manual of Tides* (p. 270). London: HMSO.

28. Douglas, B. C., & Cheney, R. E. (1990). Geosat: Beginning a New Era In Satellite Ocean-ography. *J. Geophys. Res., 95*(C3), 2833—2835.

29. EerNisse, E. P. (1980). Miniature Quartz Resonator Force Transducer. U.S. Patent No. 4,215,570.

30. EerNisse, E. P., & Paros, J. M. (1983). Resonator Force Transducer. U.S. Patent No. 4,372,173.

31. EerNisse, E. P., & Wiggens, R. B. (1986). Resonator Temperature Transducer. U.S. Patent No. 4,592,663.

32. Evans, J. J., & Pugh, D. T. (1982). Analysing Clipped Sea-Level Records for Harmonic Tidal Constituents. *Int. Hydrogr. Rev, 59*, 115—122.

33. Fine, I., Cherniawsky, J. Y., Rabinovich, A. B., & Stephenson, F. (2009). Numerical Modeling and Observations of Tsunami Waves in Alberni Inlet and Barkley Sound, British Columbia. *Pure Appl. Geophys., 165*, 1019—2044, doi:10.1007/s00024—008—0414—9.

34. Fine, I. V., Rabinovich, A. B., & Thomson, R. E. (2005). The dual source region for the 2004 Sumatra tsunami. *Geophys. Res. Lett., 32*, L16602, doi:10.102g/2005GL023521.

35. Fine, I. V., Rabinovich, A. B., Thomson, R. E., & Kulikov, E. A. (2003). Numerical Modeling of Tsunami Generation by Submarine and Subaerial Landslides. In: C. Ahmet et al. (Eds.), *NATO Science Series, Underwater Ground Failures On Tsunami Generation, Modeling, Risk and Mitigation* (pp. 69—88). Kluwer.

36. Gieles, A. C. M. (1969). Subminiature Silicon Pressure Transducer. *IEEE Int. Solid—State Circuits Conf. Digest Tech., Papers* 108—109.

37. Gower, J. F. (2005). Jason 1 detects December 26, 2004 tsunami. *Eos, 86*(4), 37—38.

38. Gower, J. F. (2007). The 26 December 2004 tsunami measured by satellite altimetry. *International Journal of Remote Sensing, 28*(13), 2897—2913, doi:10.1080/01431160601094484.

39. Groves, G. W. (1955). Numeric Filters for Discrimination Against Tidal Periodicities. *Transaction of American Geophysical Union, 36*, 1073—1084.

40. Hogle, L. (1988). Investigation of the Potential Application of GPS for Precision Approaches. *Navigation, 35*(3), 317—334.

41. Hoffman, C. (1997). Checking On 'Seaquake,' *Sea Technol., 38*(8), 74.

42. Hsiao, T. T. (1986). Test and Evaluation Software for a Prototype Network of Water Level Measurement Stations. In: *Proceedings. IEEE Oceans' 86*, 358—363.

43. Hwang, L. S., & Divoky, D. (1971). Tsunami. *Underwater J.*, 207—219.

44. Ippen, A. T., & Goda, Y. (1963). *Wave Induced Oscillations in Harbors: The Solution for a Rectangular Harbor Connected to the Open Sea*. Cambridge, Mass: TR.59, Parsons Laboratory, MIT.

45. Irish, J. D., & Snodgrass, F. E. (1971). Quartz Crystals as Multi-Purpose Oceanographic Sensors — I. Pressure *Deep-Sea Res., 19*, 165–169.

46. Jacob, K. H. (1984). Estimates of Long-Term Probabilities for Future Great Earthquakes in the Aleutians. *Geophys. Res. Lett., 11*, 295–298.

47. Joseph, A., Desa, J. A. E. Desa, E. Foden, P. & Taylor, K. (1995). Flow Effects on a Pressure Transducer. In: *Proceedings of the Symposium on Ocean Electronics (Sympol—'95)*, pp. 11–16.

48. Joseph, A., Desa, J. A. E., Desa, E., McKeown, J., & Peshwe, V. B. (1996). Wave Effects on a Pressure Sensor. In: *Proceedings of the International Conference in Ocean Engineering (ICOE '96)*, pp. 568–572.

49. Joseph, A., Peshwe, V. B. Vijaykumar, Elgar Desa & Ehrlich Desa (1997). Effects of Water Trapping and Temperature Gradient in the Sounding Tube of An NGWLMS Gauge Deployed in the Zuari Estuary, Goa. In: *Proceedings of the National Symposium on Ocean Electronics (SYMPOL'97)*, Cochin, pp.73–82.

50. Keller, J. B. (1961). Tsunamis — Water Waves Produced by Earthquakes. In: *Proc. Pacific Sci. Cong., 10th Honolulu IUGG Monograph, 24*, 154–166.

51. Kulikov, E. A., Rabinovich, A. B., & Thomson, R. E. (2005). Estimation of Tsunami Risk for the Coasts of Peru and Northern Chile. *Natural Hazards., 35*, 185–209.

52. Kulikov, E. A., Rabinovich, A. B., & Thomson, R. E. (2005). On Long–Term Tsunami Forecasting. *Oceanology., 45*(4), 488–499, (Translated from Okeanologiya. 45 (4):518–530).

53. Kroebel, W. (1996). Precision 600–Bar Marine Pressure Sensor. *Sea Technol., 37*(4), 39–42.

54. Liu, P. L. F., Monserrat, S., Marcos, M., & Rabinovich, A. B. (2003). Coupling between two inlets: Observation and modeling. *J. Geophys. Res., 108*(C3), 3069, doi:10.1029/2002JC001478. (Correction).

55. Lockridge, P. A. (1989). Tsunami: Trouble for Mariners. *Sea Technol., 30*(4), 53–57.

56. Maudie, W. (1994). The Night the Sea Smashed Lord's Cove. *Canadian Geographic, 114*(6), 70–73.

57. Miche, M., Miller, G. R., Munk, W. H., & Snodgrass, F. E. (1962). Long-Period Waves over California's Continental Borderland. Part II: Tsunamis. *J. Mar. Res., 20*, 31–41.

58. Miller, G. R. (1972). Relative of tsunamis. *Hawaii Inst. Geophys. HIG–72–8, 7.*

59. Miller, L., Cheney, R. E., & Douglas, B. C. (1988). Geosat Altimeter Observations of Kelvin Waves and the 1986–87 El Nino *Science, 239*, 52–54.

60. Miller, L., & Cheney, R. E. (1990). Large-scale Meridional Transport in the Tropical Pacific Ocean During the 1986–87 El Nino. *J. Geophys. Res., 95*, 17,905.

61. Mitchum, G. T. (1994). Comparison of TOPEX Sea Surface Heights and Tide Gauge Sea Levels. *J. Geophys. Res., 99*(C12), 24, 541–24 553.

62. Mofjeld, H. O., & Wimbush, M. (1977). Bottom Pressure Observations in the Gulf of Mexico and Caribbean Sea. *Deep-Sea Res., 24*, 987–1004.

63. Munk, W. H. (1961). Some Comments Regarding Diffusion and Absorption of Tsunamis. 10th Honolulu IUGG Monograph, *Proc. Pacific. Sci. Confr.*, (24), 53–72.

64. Noye, B. J. (1974). Tide-well systems II: The frequency response of a linear tide–well system. *J. Marine Res., 32*(2), 155–181.

65. Parker, B. B., Cheney, R. E., & Carter, W. E. (1992). NOAA Global Sea Level Program. *Sea Technol., 33*(6), 55–62.

66. Paros, J. M. (1973). Precision Digital Pressure Transducer. *ISA Transactions, 12*(2), 173–179.

67. Paros, J. M. (1983). Mounting System for Applying Forces to Load—Sensitive Resonators. U.S. Patent No. 4,384,495.

68. Paros, J. M. (1983). Isolating and Temperature Compensating System For Resonators. U.S. Patent No. 4,406,966.

69. Paros, J. M., & Busse, D. W. (1982). Longitudinal Isolation System for Flexurally Vibrating Force Transducers. U.S. Patent No. 4,321,500.

70. Paros, J. M., Wearn, R. B. & Tonn, J. F. (1987). Mounting and Isolating System for Tuning Fork Temperature Sensor. U.S. Patent No. 4,706,259.

71. Pigeon, N. B., & Denner, W. W. (1968). A Resistance Tide Gauge. *Marine Science Instrumentation, Vol. 4*, (pp. 694—698). Instrument Society of America, Plenum Press.

72. Ponte, R. M., & Lyard, F. (2002). Effects of Unresolved High—Frequency Signals in Altimeter Records Inferred from Tide Gauge Data. *J. Atmos. and Oceanic Technol., 19*(4), 534—539.

73. Powell, R. J. (1986). Relative Vertical Positioning Using Ground—Level Transponders with the ERS—1 Altimeter. *IEEE Trans. Geosci. Remote Sensing, 24*(3), 421—425.

74. Powell, R. J. (1992). Measurement of Mid-ocean Surface Levels to ± 3 cm with Respect to Mid-Continent Reference Points Using Transponders with the ERS-1 and TOPEX Altimeters—A Developing Technique. *Workshop Report No. 81, Intergovernmental Oceanographic Commission*, pp. 111—119.

75. Pugh, D. T. (1987). *Tides, Surges and Mean Sea-Level: A Handbook for Engineers and Scientists*, (pp. 1—472). New York: John Wiley and Sons.

76. Purcell, G. H., Jr., Young, L. E., Wolf, S. K., Meehan, T. K., Duncan, C. B., Fisher, S. S., et al. (1990). Accurate GPS Measurement of the Location and Orientation of a Floating Platform. *Marine Geodesy, 14*(3—4), 255—264.

77. Rabinovich, A. B., Stephenson, F. E., & Thomson, R. E. (2006). The California Tsunami of 15 June 2005 along the Coast of North America. *Atmosphere-Ocean, 44*(4), 415—427.

78. Rabinovich, A. B., Thomson, R. E., Titov, V. V., Stephenson, F. E., & Rogers, G. C. (2008). Locally Generated Tsunamis Recorded on the Coast of British Columbia. *Atmosphere-Ocean, 46*(3), 343—360, doi:10.3137/ao.460304.

79. Schwiderski, E. W. (1991). High—Precision Modeling of Mean Sea Level, Ocean Tides, and Dynamic Ocean Variations with Geosat Altimeter Signals. In: B. B. Parker (Ed.), *Tidal Hydrodynamics* (pp. 593—616). New York: John Wiley and Sons.

80. Sennot, J. W., & Pietraszewski, D. (1987). Experimental Measurement and Characterisation of Ionospheric and Multipath Errors in Differential GPS. *Navigation, 34*(2), 160—173.

81. Shepard, F. P., MacDonald, G. A., & Cox, D. C. (1950). The Tsunami of April 1, 1946. *Bull. Scripps Inst. Oceanogr., 5*(6), 391—528.

82. Shih, H. H., Deitemyer, D. H. & Allen, M. W. (1984). Field Evaluation of a New Tide Gauge Stilling Well Design. *Proceedings of 3rd International Offshore Mechanics and Arctic Engineering Symposium*, New Orleans, 98—105.

83. Smith, C. S. (1954). Piezoresistance Effect in Germanium and Silicon. *Phys. Rev., 94*, 42—45.

84. Snodgrass, F., Brown, W., & Munk, W. (1975). MODE: IGPP Measurements of Bottom Pressure and Temperature. *J. Phys. Oceanogr., 5*, 63—74.

85. Tai, C. K., White, W. B., & Pazan, S. E. (1989). Geosat Crossover Analysis in the Tropical Pacific, 2. Verification Analysis of Altimetric Sea Level Maps with Expendable Bathythermograph and Island Sea Level Data. *J. Geophys. Res., 94*, 897.

86. Tapley, B. D., Born, G. H., Hager, H. H., Lorell, J., Parke, M. E., Diamante, J. M., et al. (1979). Seasat Altimeter Calibration: Initial Results. *Science, 204*(4400), 1410—1412.

87. Tufte, O. N., Chapman, P. W., & Long, D. (1962). Silicon Diffused-Element Piezoresistive Diaphragms. *J. Appl. Phys., 33*, 3322—3327.

88. Tufte, O. N., & Stelzer, E. L. (1963). Piezoresistive Properties of Silicon Diffused Layers. *J. Appl. Phys, 34,* 313–318.

89. Van Dorn, W. G. (1966). Tsunami. In: R. W. Fairbridge (Ed.), *The Encyclopedia of Oceanography, Encyclopedia of Earth Science Series, Vol. 1* (pp. 941–942). Van Nostrand Reinhold Company.

90. Van Dorn, W. G. (1982). Tsunami. *McGraw-Hill Encyclopedia of Science & Technology, 14,* 140–142.

91. Vitousek, M., & Miller, G. (1970). An Instrumentation System for Measuring Tsunamis in the Deep Ocean, *Tsunamis in the Pacific Ocean* (pp. 239–252). East–West Center Press.

92. Watts, D. R., & Kontoyiannis, H. (1990). Deep-Ocean Bottom Pressure Measurement: Drift Removal and Performance. *J. Atmos. and Oceanic Technol., 7,* 296–306.

93. Wearn, R. B., Jr., & Baker, D. J., Jr. (1980). Bottom Pressure Measurements Across the Antarctic Circumpolar Current and Their Relation to the Wind. *Deep-Sea Res., 27,* 875–888.

94. Wearn, R. B., Jr., & Paros, J. M. (1988). Measurements of Dead Weight Tester Performance Using High Resolution Quartz Crystal Pressure Transducers. *Presented at 34th International Instrumentation Symposium,* (p. 8), Instrument Society of America, Plenum Press.

95. Wimbush, M. (1977). An Inexpensive Sea-Floor Precision Pressure Recorder. *Deep-Sea Res., 24,* 493–497.

96. Wunsch, C., & Dahlen, J. (1974). A Moored Temperature and Pressure Recorder. *Deep-Sea Res., 21,* 145–154.

97. Ych, H., Imamura, F., Synolakis, C., Tsuji, Y., Liu, P., & Shi, S. (1993). The Flores Island Tsunami. *Eos Transactions, American Geophysical Union, 74*(33), 371–373.

98. Zetler, B. D., Schuldt, M. D., Whipple, R. W., & Hicks, S. D. (1965). Harmonic Analysis of Tides from Data Randomly Spaced in Time. *J. Geophys. Res., 70,* 2805–2811.

Index

Printed and bound by CPI Group (UK) Ltd, Croydon, CR0 4YY

13/05/2025

01869571-0001